How to Solve
GENERAL CHEMISTRY PROBLEMS

How to Solve
GENERAL CHEMISTRY PROBLEMS

fifth edition

C. H. Sorum

University of Wisconsin

R. S. Boikess

Douglass College,
Rutgers University

PRENTICE-HALL, INC.

Englewood Cliffs, New Jersey

Library of Congress Cataloging in Publication Data

SORUM, CLARENCE HARVEY, (date)
How to solve general chemistry problems.

Includes index.
1. Chemistry—Problems, exercises, etc.
I. Boikess, R. S., joint author.   II. Title.
QD42.S62 1976   540'.76   75–38657
ISBN   0–13–434100–7

© 1976, 1969, 1963, 1958, 1952
by Prentice-Hall, Inc.
Englewood Cliffs, New Jersey

10 9 8 7 6 5 4 3 2 1

Printed in the United States of America

PRENTICE-HALL INTERNATIONAL, INC., *London*
PRENTICE-HALL OF AUSTRALIA PTY. LIMITED, *Sydney*
PRENTICE-HALL OF CANADA, LTD., *Toronto*
PRENTICE-HALL OF INDIA PRIVATE LIMITED, *New Delhi*
PRENTICE-HALL OF JAPAN, INC., *Tokyo*
PRENTICE-HALL OF SOUTHEAST ASIA PTE. LTD., *Singapore*

Contents.

5   **Calculations from Formulas of Compounds.
Determining the Formula of a Compound.
Rounding off a Number. Significant Figures.**   *19*

6   **The Gas Laws.**   *37*

7   **Mole Relationships in Chemical Reactions.
I. Stoichiometry.**   *62*

Appendix Tables. *273*

Answers to Problems. *289*

Index. *297*

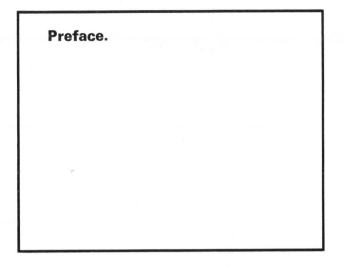

Preface.

In this fifth edition, we have tried as much as possible to preserve the self-teaching nature of previous editions, while introducing a much broader range of problems.

Features such as complete solutions for illustrative problems of most every type encountered, suggestions or clues for the solutions of more difficult problems, and answers to all the problems have also been retained.

Major new chapters involving the solution of many different types of problems in thermochemistry, thermodynamics, kinetics, and colligative properties of solutions have been added. Also, two new chapters of problems dealing with many aspects of electromagnetic radiation, the electronic structure of atoms, and the structure of molecules have been included. A detailed step by step development of the main reaction approximation for the solution of problems of simultaneous equilibria is included in Chapter 13, which now also includes many problems dealing with buffer solutions and other aspects of aqueous equilibria of biological interest. Included in Chapter 18 is a detailed step by step procedure for drawing Lewis structures of covalent molecules.

The purpose of preparing a fifth edition is to make available a self-teaching text from which the beginning chemistry student, who often has limited contact with an instructor, can learn to master the solution of the very broad range of problems encountered in a modern course in general chemistry.

The author wishes to thank Karen T. Boikess for her substantial assistance in the preparation of the manuscript, Kevin Lehmann for his assistance with the solutions to some of the problems, Joy Breslauer and Doris Mier for typing the manuscript, his teaching colleagues in general chemistry for their patience and expertise, and his students for providing stimulus and challenge.

ROBERT S. BOIKESS

How to Solve
GENERAL CHEMISTRY PROBLEMS

1

How to solve
a problem.

Every problem you encounter, whether in chemistry or elsewhere, is solved in essentially the same fashion. *First*, you size up the situation or read the problem carefully, and decide *what* you are supposed to do and what you have to do it with. *Second*, having determined what you are supposed to do and what you have to do it with, you figure out *how* to do it. *Last*, you go ahead and *solve it* according to plan. The first two steps represent the *analysis* of the problem. The third step represents the *arithmetical calculations*. Some problems are knottier than others, but they are all solved by these three fundamental steps.

To be more specific, when you go about solving any problem in this or any other book or in any test or examination:

1. Read the problem carefully. Note exactly what is given and what is sought. Recognize that most chemistry problems contain more information than is explicitly given. For example, if a problem specifies a given mass of water, it is also specifying a given number of moles, molecules, and atoms. Note any and all special conditions. Be sure you understand the meaning of all terms and units and that you are familiar with all chemical principles that are involved. Every problem in this book is designed to illustrate some principle, some relationship, some law, some definition, or some fact. If you understand the principle, relationship, law, definition, or fact you should have no difficulty solving the problem. The one big reason, the only reason in fact, why students have difficulty with chemistry problems is failure to understand, exactly and well, the chemical principles involved and the meaning and value of all terms and units that are used in the problem.

2. Plan, in detail, just how the problem is to be solved. Get into the habit of visualizing the entire solution before you execute a single step. Insist on knowing what you are going to do and why you are going to do it. Aim to learn to solve every problem in the most efficient manner; this generally means doing it the shortest way, with the fewest steps.

3. Specify definitely what each number represents and the units in which it is expressed when you actually carry out the mathematical operation of solving the problem. Don't just write

$$\frac{192}{32} = 6$$

Write

$$\frac{192 \text{ g of sulfur}}{32 \text{ g of sulfur per mole of sulfur}} = 6 \text{ moles of sulfur}$$

or whatever the case may be. Always divide and multiply the *units* as well as the *numbers*. This is one way to give exactness to your thought process and is a very good way to help avoid errors. You should jot down the unit or units in which your answer is to be expressed as the first step in the actual solution. For instance, if you are solving for the number of grams of oxygen in 200 g of silver oxide, you should jot down the fact that the answer will be "= ......g of oxygen." In reality, every problem is worked backward, since you first focus your attention on the units in which the answer is to be expressed and then plan the solution with these units in mind.

4. Having solved the problem, examine the answer to see if it is reasonable and sensible. The student who reported that 200 g of silver oxide contained 1380 g of oxygen should have known that such an answer was not sensible. When the slide rule is used, errors due to incorrect location of the decimal point are very likely to creep in unless you get into the habit of checking the answer to see if it is of the right order of magnitude.

5. If you do not understand how to solve a problem have it explained to you at the very earliest possible time. To be able to solve the later problems you must understand the earlier ones. After a problem has been explained to you, fix the explanation in your mind by working other similar problems at once, or at least within a few hours, while the explanation is still fresh in your mind. Test yourself to be sure that you can apply your understanding of the solution.

Units
of
measurement.

It will be assumed in this book that every student is familiar, through laboratory experience, with the common units of measure in the metric system and that he has a fair idea of the volume represented by 1 liter, 100 ml, and 1 ml, the mass represented by 10 g, 100 g, or 1 kg, and the length represented by 760 mm, 10 cm, and 1 m, etc. Also, it will be assumed that he is familiar with the Celsius (centigrade) thermometer scale.

Is should be recalled that the metric system employs decimal notations in which the prefix *micro-* means one millionth, *milli-* means one thousandth, *centi-* means one hundredth, and *deci-* means one tenth, while *kilo-* means one thousand times, and *mega-* means one million times.

Conversions of metric units (grams, liters, milliliters, cubic centimeters, centimeters, etc.) to other units (pounds, quarts, inches, feet, etc.) are not often required. The following table will serve where such conversions are called for.

Conversion units

$$1 \text{ meter (m)} = 10 \text{ decimeters (dm)} = 100 \text{ centimeters (cm)}$$
$$= 1000 \text{ millimeters (mm)} = 1,000,000 \text{ micrometers } (\mu\text{m})$$
$$= 39.37 \text{ inches (in.)} = 1.09 \text{ yards (yd)}$$

1 kilogram (kg) = 1000 grams (g) = 1,000,000 milligrams (mg)
= 2.2046 pounds (lb)

1 gram (g) = 1000 milligrams (mg)

1 milligram (mg) = 0.001 gram (g)

1 pound (lb) = 453.6 grams (g)

1 liter (l) = 1000 milliliters (ml) = 1000 cubic centimeters (cc)
= 0.264 U.S. gallons (gal) = 1.06 U.S. quarts (qt)

1 milliliter (ml) = 1000 μl = 1 cubic centimeter (cc)

1 cubic centimeter is the volume of about 20 drops of water

A new U.S. 5-cent piece has a mass of 5 g

Interconversion of Celsius (centigrade) and Fahrenheit temperature readings

The thermometers used in the laboratory are graduated in Celsius degrees, designated by the letter C. (It should be pointed out that the correct term is "Celsius degrees" or "degrees Celsius" rather than "centigrade degrees" or "degrees centigrade," in honor of the Swedish scientist, Anders Celsius, who devised the scale. Henceforth in this textbook the term "Celsius" rather than "centigrade" will be used. In any case the abbreviation is always C.) Most household thermometers are graduated in Fahrenheit degrees, designated by the letter F. The fixed points on both the Celsius and Fahrenheit temperature scales are the boiling point and freezing point of water. On the Celsius scale the freezing point of water is 0° and the boiling point is 100°; the space between the fixed points is divided into 100 units and the space above 100° and below 0° is divided into the same size units. On the Fahrenheit scale the freezing point of water is 32° and the boiling point is 212°; the space between the fixed points is divided into 180 units and the space above 212° and below 32° is divided into the same size units. Since the space between the freezing point and boiling point of water is divided into 100° on the Celsius scale and 180° on the Fahrenheit scale, it follows that 100 Celsius degrees must represent the same temperature change as 180 Fahrenheit degrees. That means that 1 Celsius degree is equal to 1.8 Fahrenheit degrees; or expressing it in fractional form, 1 Celsius degree is equal to $\frac{9}{5}$ Fahrenheit degrees and 1 Fahrenheit degree is equal to $\frac{5}{9}$ of a Celsius degree.

With these facts in mind we see that, if we wish to find the Fahrenheit value of a certain number of Celsius degrees, C, we first multiply the Celsius reading by $\frac{9}{5}$; this gives us $\frac{9}{5}$ C. Since the reference temperature (the freezing point of water) on the F scale is 32° above zero we must add 32° to $\frac{2}{5}$ C in order to get the actual reading on the Fahrenheit scale.

$$\text{Fahrenheit temperature} = \frac{9}{5}\text{ Celsius temperature} + 32$$

or

(1) $$F = \frac{9}{5}C + 32$$

Equation (1) can be transposed to the form,

(2) $$C = \frac{5}{9}(F - 32)$$

Equation (2) tells us that, to find the value, in degrees Celsius, of a Fahrenheit temperature, we first subtract 32° from the Fahrenheit temperature (because the Fahrenheit freezing point reference is 32° above zero) and then take $\frac{5}{9}$ of that answer.

To illustrate the use of the above relationships:

(a) Convert 144°F to a Celsius reading.

In thinking our way through this problem we note that 144°F is (144 − 32) or 112° above the freezing point of water. Since 1 Fahrenheit degree is equal to $\frac{5}{9}$ of a Celsius degree, 112 Fahrenheit degrees must be equal to 112 × $\frac{5}{9}$ or 62.2 Celsius degrees. That means that 144°F is 62.2 Celsius degrees above the freezing point of water. Since the freezing point of water is 0°C, 62.2 Celsius degrees above the freezing point of water will be 62.2°C.

(b) Convert 80°C to a Fahrenheit reading.

In thinking our way through this problem we note that 80°C is 80 Celsius degrees above the freezing point of water. Since 1 degree C equals $\frac{9}{5}$ degrees F, 80°C will be $\frac{9}{5}$ × 80 or 144 Fahrenheit degrees above the freezing point of water. But the freezing point of water on the Fahrenheit scale is 32°. Therefore, we must add 32 to our 144 to get the actual Fahrenheit temperature, 176°F.

PROBLEMS

2.1 What temperature, in degrees Celsius, is represented by each of the following Fahrenheit temperatures?

(a) 72.0°F

Solution: See solutions of problems given above.

(b) −20.0°F

2.2 What temperature in degrees Fahrenheit is represented by each of the following Celsius temperatures?

(a) 12.0°C

(b) −50.0°C

2.3 At what temperature will the readings on the Fahrenheit and Celsius thermometers be the same?

2.4 Suppose you have designed a new thermometer called the X thermometer. On the X scale the boiling point of water is 130°X and the freezing point of water is 10°X. At what temperature will the readings on the Fahrenheit and X thermometers be the same?

2.5 On a new Jekyll temperature scale water freezes at 17°J and boils at 97°J. On another new temperature scale, the Hyde scale, water freezes at 0°H and boils at 120°H. If methyl alcohol boils at 84°H what is its boiling point on the Jekyll scale?

Exponents.

Chemical problems often involve numbers which are either very large or very small. Such numbers are most conveniently expressed in *exponential form*.

To illustrate:

The number, 100, is 10^2, called "10 squared" or "10 to the second power," which is 1×10^2; 1000 is 1×10^3 and 1,000,000 is 1×10^6.

The number, 2,000,000, is $2 \times 1,000,000$, which is 2×10^6.

The number, 324,000,000, is $3.24 \times 100,000,000$, which is 3.24×10^8; but it is also $32.4 \times 10,000,000$, which is 32.4×10^7, and $324 \times 1,000,000$, which is 324×10^6. In other words, 324,000,000 may be represented as either 3.24×10^8, 32.4×10^7, or 324×10^6. The first of these, in which there is only one digit to the left of the decimal point in the nonexponential factor, is the preferred form.

Note that, in the above examples in which we are dealing with numbers *larger* than 1, a decimal point is placed to the *right* of the first digit in the number; the resulting expression is then multiplied by 10 raised to a positive power equal to the number of terms to the *right* of this decimal point.

To illustrate:

$$602,000,000,000,000,000,000,000 = 6.02 \times 10^{23}$$

and

$$31,730,000 = 3.173 \times 10^7$$

The number 0.0001 is one ten-thousandth, which is 1/10,000, which is 1/10⁴.

Keeping in mind that (a) $1 = 10^0$, (b) the fraction, $1/10^4$, means 1 divided by 10^4, and (c) in division of exponential numbers the exponent of the denominator is subtracted from the exponent of the numerator, then

$$\frac{1}{10^4} = \frac{10^0}{10^4} = 10^{0-4} = 10^{-4} = 1 \times 10^{-4}$$

Likewise,

$$0.00002 = 2 \times \frac{1}{100,000} = 2 \times \frac{1}{10^5} = 2 \times 10^{-5}$$

and

$$0.00000038 \text{ is } 3.8 \times \frac{1}{10,000,000} = 3.8 \times \frac{1}{10^7} = 3.8 \times 10^{-7}$$

Note that, in dealing with numbers *less* than 1, a decimal point is placed to the *right* of the first term to the right of the zeros and the resulting expression is then multiplied by 10 raised to a *negative* power equal to the number of terms to the *left* of this decimal point.

Thus

$$0.00000257 = 2.57 \times 10^{-6}$$

and

$$0.000016 = 1.6 \times 10^{-5}$$

Just as 3.24×10^8, 32.4×10^7, and 324×10^6 all represent the same number, so 2.57×10^{-6}, 25.7×10^{-7} and 257×10^{-8} are all the same number, and 48×10^{-6} is equivalent to 4.8×10^{-5}. Here again the expression with one digit to the left of the decimal point is preferred.

Note that in changing 48×10^{-6} to its equal, 4.8×10^{-5}, and in changing 32.4×10^{-7} to 3.24×10^{-6} we move the decimal point one place to the left and compensate for this by raising the exponent by one. Likewise, in changing 0.23×10^{-4} to 2.3×10^{-5} we move the decimal point one place to the right and compensate for this by lowering the exponent by one. If we were to move the decimal point two places to the right we would lower the exponent by two, and so on. Since, in each example we multiply one term by one or more factors of 10 and divide the other term by the same number of factors of 10, the value of the number is not changed.

Thus

$$2.36 \times 10^{-5} = 23.6 \times 10^{-6} = 236 \times 10^{-7} = 0.236 \times 10^{-4}$$

and

$$4.92 \times 10^5 = 49.2 \times 10^4 = 492 \times 10^3 = 0.492 \times 10^6$$

The use of exponents makes it quite easy to determine the correct number of digits in the answer to an operation involving multiplication and division of many numbers. Thus, if the expression

$$\frac{417,000 \times 0.0036 \times 15,300,000}{0.000021 \times 293 \times 183,000}$$

is changed to the form

$$\frac{4.17 \times 10^5 \times 3.6 \times 10^{-3} \times 1.53 \times 10^7}{2.1 \times 10^{-5} \times 2.93 \times 10^2 \times 1.83 \times 10^5}$$

it can be determined at a glance that the answer is approximately 2×10^7. Likewise, when the expression

$$\frac{0.0045 \times 0.082 \times 600}{204 \times 23}$$

is changed to the form

$$\frac{4.5 \times 10^{-3} \times 8.2 \times 10^{-2} \times 6 \times 10^2}{2.04 \times 10^2 \times 2.3 \times 10^1}$$

it can be seen that the answer is approximately

$$48 \times 10^{-6} \quad \text{or} \quad 4.8 \times 10^{-5}$$

PROBLEMS

3.1 Express each of the following numbers in exponential form:

(a) 21,000,000,000

(b) 760

(c) 0.0027

(d) 0.0000018

(e) 0.10

3.2 Carry out each of the following operations; first write each number in exponential form:

(a) $\dfrac{136,000 \times 0.000322 \times 273}{0.082 \times 4200 \times 129.2}$

(b) $\dfrac{120 \times 309 \times 800}{273 \times 600}$

3.3 Solve each of the following:

(a) $\dfrac{1.76 \times 10^{-3}}{8.0 \times 10^2}$

In solving this problem it may be helpful to express the numerator as 17.6×10^{-4} rather than leave it as 1.76×10^{-3}. This will give the answer

2.2 × 10⁻⁶ rather than 0.22 × 10⁻⁵. Although these two numbers have the same value, the preferred expression is 2.2 × 10⁻⁶.

(b) $\dfrac{0.0234 \times 10^{-3}}{3.6 \times 10^{-4}}$

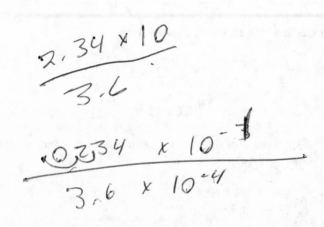

Atomic weight.
Gram-atomic weight.
Gram atom.
Mole.
The Avogadro number.

Atomic weight

The *atomic weight* of an element is a number which tells us how the mass of an *average atom* of that element compares with the mass of a standard reference atom. In the modern system of atomic weights, adopted in 1961, the atom of carbon-12 (^{12}C) is the reference standard and is assigned a value of exactly 12. The atomic weight of gold, listed in the table in the Appendix, is 196.967. This means that the mass of an *average atom* of gold is to the mass of an atom of ^{12}C as 196.967 is to 12. Likewise, the mass of an average atom of calcium (whose atomic weight is 40.08) is to the mass of an atom of ^{12}C as 40.08 is to 12, and so on down the list of elements. Since all atomic weights are referred to the same standard it must follow that these atomic weights are the relative masses of the average atoms of all the elements. That is, the mass of the average gold atom is to the mass of the average calcium atom as 196.967 is to 40.08, and the mass of the average calcium atom is 40.08/196.967 times the mass of the average gold atom, and so on.

The terms *atomic weight* and *atomic mass*, as commonly used, both refer to the same thing—namely, the mass of the atom. Strictly speaking, the term atomic *weight* is incorrect; the correct term is atomic *mass*. The weight of a substance is a measure of the pull of gravity on that substance; its value varies with its altitude. The mass of a substance represents the number of standard units of mass required to counterbalance the substance; its value is independent of the position of the substance and, hence, does not change as long as the substance maintains its identity. The term *atomic weight* is used by most scientists and is generally accepted as referring, correctly, to the *mass* of the *average* atom of the particular element. When the mass of an atom of a specific isotope of an element is under discussion the term *atomic mass* is generally used.

Gram-atomic weight

Atomic weight, as defined above, is simply a number. To give it more precise meaning we can say that the ^{12}C atom has a mass of 12 *atomic mass units* (amu) and not be concerned about what these units represent. The atomic weight of calcium then becomes 40.08 amu, of gold, 196.967 amu, and so on.

But we would like to use these atomic weights for solving problems. The amu, which is a convenient unit of mass for single atoms, is much too small to use for macroscopic quantities. The most commonly employed unit of mass (weight) is the gram, and it is convenient to find the conversion factor to change amu to grams. This factor has been found by many experiments to be 6.023×10^{23}. Thus, 1 g is equal to 6.023×10^{23} amu. If the mass of the ^{12}C atom is 12 amu, the mass of 6.023×10^{23} ^{12}C atoms is 12 g. If the atomic weight of sulfur is 32.064 amu then the mass of 6.023×10^{23} sulfur atoms is 32.064 g; and if the atomic weight of hydrogen is 1.00797 amu then the mass of 6.023×10^{23} hydrogen atoms is 1.00797 g. *The atomic weight of an element expressed in units of grams is a quantity called the gram-atomic weight of the element.*

One gram-atomic weight of an element always contains 6.023×10^{23} *atoms of that element.* This fundamental principle of chemistry is a direct consequence of the conversion of amu to grams, and the expression of all atomic weights relative to the same standard. This is illustrated by the following argument: The atomic weight of helium is 4.0026 and of sulfur is 32.064. That means that 1 atom of helium is 4.0026/32.064 times as heavy as 1 atom of sulfur. Therefore, 10 atoms of helium must be 4.0026/32.064 times as heavy as 10 atoms of sulfur and 6.023×10^{23} atoms of helium must be 4.0026/32.064 times as heavy as 6.023×10^{23} atoms of sulfur. But 6.023×10^{23} atoms of helium have a mass of 4.0026 g. Therefore, 6.023×10^{23} atoms of sulfur must have a mass of 32.064 g. Since 4.0026 g is one gram-atomic weight of helium and 32.064 g is one gram-atomic weight of sulfur, and since the above argument can be applied to each of the other 102 elements, it can be concluded that one gram-atomic weight of any element consists of 6.023×10^{23} atoms.

Gram atom

The term *gram-atomic weight* is abbreviated to *gram atom*. "*One gram atom*" of an element means one gram-atomic weight of that element.

Mole

The quantity, 6.023×10^{23} atoms, is referred to as one *gram mole* of atoms or, simply, one *mole*. It follows from what has been stated above that *one mole of atoms means one gram atom or one gram-atomic weight*. As we proceed, we shall learn that *mole* is a very widely used term. It is applied to molecules, ions, clusters of atoms, and electrons as well as to atoms. *One mole of a species is* 6.023×10^{23} *units of that species.*

The Avogadro number

The number, 6.023×10^{23}, which represents the number of units in one mole of a substance, is referred to as the *Avogadro number*, in honor of the Italian scientist who first postulated its existence. It is denoted by the letter N.

Review

Since, as has been noted above, the term *gram atom* is an abbreviation for gram-atomic weight, since one gram-atomic weight of an element is 6.023×10^{23} atoms, since one mole of atoms means 6.023×10^{23} atoms, and since the number, 6.023×10^{23}, is the Avogadro number, N, it follows that, when applied to a particular element, the terms

> one mole of atoms
>
> one gram atom
>
> one gram-atomic weight
>
> 6.023×10^{23} atoms
>
> N atoms

all mean the same thing.

Specifically, 32.064 g of sulfur is

> one mole of sulfur atoms
>
> one gram atom of sulfur
>
> one gram-atomic weight of sulfur
>
> 6.023×10^{23} atoms of sulfur
>
> N atoms of sulfur

Symbol

The name of an element is commonly represented by an abbreviation, referred to as the *symbol* for that element. Thus the symbols for sodium and oxygen are, respectively, Na and O.

   When the symbol for an element is used in any chemical formula or equation it is understood that it refers to one gram atom of the element. Since, as has already been noted, one gram atom means *one mole* of atoms, it follows that *the symbol for an element, when used in any chemical formula or equation, refers to one mole of atoms of that element.* **This is a very important fact to remember;** it is a basic fact in the solution of most chemical problems.

Pound-atomic weight, ton-atomic weight, pound atom, ton atom, pound mole, ton mole

It has been pointed out that, because the mass of 1 average atom of sulfur is to the mass of 1 average atom of helium as 32.064 is to 4.0026, the mass of any number, X, of atoms of sulfur is to the mass of the same number, X, of atoms of helium as 32.064 is to 4.0026. If we take 6.023×10^{23} atoms of helium they will have a mass of 4.0026 g; therefore, if we take 6.023×10^{23} atoms of sulfur they will have a mass of 32.064 g. Suppose X atoms of helium have a mass of 4.0026 *pounds*. If X atoms of helium have a mass of 4.0026 lb, X atoms of sulfur will then have a mass of 32.064 lb, X atoms of hydrogen will have a mass of 1.00797 lb, X atoms of sodium will have a mass of 22.9898 lb, and so on down the list of elements. Likewise, if Y atoms of helium have a mass of 4.0026 *tons*, Y atoms of hydrogen will have a mass of 1.00797 tons, Y atoms of sodium will have a mass of 22.9898 tons, Y atoms of sulfur will have a mass of 32.064 tons, and so on. The quantity, 4.0026 pounds of helium, is one *pound-atomic weight*, one *pound atom*, or one *pound mole* of He. The quantity, 32.064 tons of sulfur, is one ton-atomic weight, one ton atom, or one ton mole of S. In other words, just as we have gram-atomic weights, so we can have *pound-atomic weights, ton-atomic weights, pound atoms, ton atoms, pound moles* and *ton moles*. A pound-atomic weight is a mass in pounds of an element equal to its atomic weight, and a ton-atomic weight is a mass in tons of an element equal to its atomic weight. One pound mole of S is 32.064 pounds and one ton mole of Ag is 107.870 tons. Since the atomic weight table tells us how the masses of the average atoms of the elements *compare* with each other, the numerical values of the atomic weights will be the same regardless of the unit in which they are being expressed.

   It should be noted, however, that the number of individual atoms in a pound mole of an element will not be 6.023×10^{23} atoms. The Avogadro number, 6.023×10^{23}, represents the number of atoms in one *gram atom*, that is, in one *gram mole*. Since 1 ton = 2000 lb and 1 lb = 453.6 g, one pound mole of an element will contain $453.6 \times 6.023 \times 10^{23}$ atoms, and one ton mole will contain $2000 \times 453.6 \times 6.023 \times 10^{23}$ atoms.

Unless specifically stated otherwise *it shall always be understood that the* word **mole** *means* **gram mole. This is an important fact to remember.**

PROBLEMS

4.1   What fraction of a mole of zinc atoms is 12.00 g of Zn?

Solution: The atomic weight of zinc is 65.37. Therefore, 65.37 g is 1 mole of Zn, and 12.00 g is 12.00/65.37 mole of Zn.

The calculation, in detail, is

$$12.00 \text{ g of Zn} \div 65.37 \text{ g of Zn/mole of Zn} = \frac{12.00}{65.37} \text{ mole of Zn}$$

Since any fraction represents a process of division, the calculation can take the form

$$\frac{12.00 \text{ g of Zn}}{65.37 \text{ g of Zn/1 mole of Zn}} = \frac{12.00}{65.37} \text{ mole of Zn}$$

Note that g of Zn in the numerator cancels g of Zn in the denominator. The answer is then in moles of Zn.

The solution can also be carried out as follows: We are given 12.00 g of zinc. We want to know how many moles of zinc this represents. We reason that, if we multiply

$$\text{g of Zn} \times \frac{\text{moles of Zn}}{\text{g of Zn}}$$

then g of Zn in the numerator and denominator will cancel and the answer will be in units of moles of Zn. Since 1 mole of Zn has a mass of 65.37 g the actual calculation will be

$$12.00 \text{ g of Zn} \times \frac{1 \text{ mole of Zn}}{65.37 \text{ g of Zn}} = \frac{12.00}{65.37} \text{ mole of Zn}$$

Note that in this and subsequent problems there are three quantities involved: a mass of substance, the number of moles represented by this mass, and the mass of one mole of this substance. These quantities are related by

$$\text{number of moles} = \frac{\text{mass of substance}}{\text{mass of one mole}}$$

Given any two of these quantities we can find the third.

4.2   A mass of 30.42 g of calcium is what fraction of a mole of calcium atoms?

4.3   A mass of 120.0 g of helium is how many moles of He atoms?

4.4   Calculate the mass of 2.32 moles of carbon atoms.

Solution: The atomic weight of carbon is 12.011. Therefore, 1 mole of C has a mass of 12.011 g.

2.32 moles $\times$ 12.011 g/mole = 27.9 g (g/mole means "grams per mole")

Note that moles in the numerator and denominator cancel. The answer is in grams.

4.5   Calculate the mass of 4.72 moles of fluorine atoms.

4.6   Calculate the mass of 0.140 mole of sodium atoms.

4.7   Calculate the mass of 0.821 gram atom of manganese.

4.8   Calculate the mass of 3.20 ton moles of Si.

4.9   How many atoms are there in 20.00 g of boron?

Solution: We know that there are 6.023×10^{23} atoms in 1 mole of B. If we know how many *moles* of B there are in 20 g of B we can then multiply moles of B by 6.023×10^{23} atoms per mole. The atomic weight of boron is 10.811. That means that there are 10.811 g in 1 mole of B. Therefore

(1) $$\frac{20 \text{ g of B.}}{10.811 \text{ g of B/mole of B}} = \frac{20}{10.811} \text{ moles of B}$$

(2) $$\frac{20}{10.811} \text{ moles of B} \times 6.023 \times 10^{23} \text{ atoms/mole of B}$$

$$= \frac{20}{10.811} \times 6.023 \times 10^{23} \text{ atoms}$$

$$= 11 \times 10^{23} \text{ atoms}$$

$$= 1.1 \times 10^{24} \text{ atoms}$$

The entire calculation can be combined in one operation:

$$\frac{20 \text{ g}}{10.811 \text{ g/mole}} \times 6.023 \times 10^{23} \text{ atoms/mole} = 1.1 \times 10^{24} \text{ atoms}$$

Note that grams cancel grams and moles cancel moles.

4.10   An amount of 4.63×10^{21} atoms is how many moles of Sn?

Solution: 6.023×10^{23} atoms is 1 *mole* of Sn. Therefore, 4.63×10^{21} atoms is

$$\frac{4.63 \times 10^{21}}{6.023 \times 10^{23}} \text{ mole of Sn}$$

The detailed calculation would be

$$\frac{4.63 \times 10^{21} \text{ atoms}}{6.023 \times 10^{23} \text{ atoms/mole}} = \frac{4.63 \times 10^{21}}{6.023 \times 10^{23}} \text{ mole}$$

$$= 0.772 \times 10^{-2} \text{ mole}$$

$$= 7.72 \times 10^{-3} \text{ mole}$$

4.11 How many grams of copper will contain 3.22×10^{24} atoms of copper?

Solution: The atomic weight of copper is 63.54. This means that there are 63.54 g of Cu in 1 mole of Cu. If we know how many *moles* of Cu are represented by 3.22×10^{24} atoms we can multiply these moles by 63.54 g of Cu per mole.

We know that one mole contains 6.023×10^{23} atoms; therefore, 3.22×10^{24} atoms is

$$\frac{3.22 \times 10^{24}}{6.023 \times 10^{23}} \text{ moles}$$

$$\frac{3.22 \times 10^{24}}{6.023 \times 10^{23}} \text{ moles} \times 63.54 \text{ g/mole} = 339 \text{ g}$$

The entire calculation can be carried out in one operation:

$$\frac{3.22 \times 10^{24} \text{ atoms}}{6.023 \times 10^{23} \text{ atoms/mole}} \times 63.54 \text{ g/mole} = 339 \text{ g}$$

4.12 How many atoms are there in 120 g of magnesium?

4.13 Calculate the mass in grams of 8.00×10^{23} atoms of iron.

4.14 How many grams of chromium will contain 4.00×10^{23} atoms of chromium?

4.15 How many atoms of sulfur will have a mass of 40.0 g?

4.16 What fraction of a mole of aluminum atoms will contain 4.11×10^{20} atoms?

4.17 How many atoms are there in 125 pound moles of nickel?

Solution: Since 1 lb = 453.6 g, 125 lb moles = 125×453.6 gram moles. One gram mole = 6.023×10^{23} atoms. Therefore, the complete calculation is

125 pound moles $\times$ 453.6 gram moles/pound mole
$$\times \, 6.023 \times 10^{23} \text{ atoms/gram mole} = 3.41 \times 10^{28} \text{ atoms}$$

4.18 How many atoms are there in 0.0260 ton moles of chromium?

➜ **4.19*** A sample of chlorine consists of 80.00 mole percent of ^{35}Cl with atomic mass 35.00 and 20.00 mole percent of ^{37}Cl with atomic mass 37.00. Calculate the atomic weight of the chlorine in the sample.

Solution: The symbols ^{35}Cl and ^{37}Cl refer to the two isotopes of chlorine whose mass numbers are, respectively, 35 and 37 and whose atomic masses, according to the facts given in the problem, are 35.00 and 37.00, respectively. If the chlorine was pure ^{35}Cl its atomic weight

*The arrow preceding the number denotes the more sophisticated problems.

would be 35.00. If it was pure ^{37}Cl its atomic weight would be 37.00. Since 80.00 % of the atoms are ^{35}Cl and 20.00 % are ^{37}Cl the mass of the ^{35}Cl is 0.80 × 35.00 or 28.00 and the mass of the ^{37}Cl is 0.20 × 37.00 or 7.40. The atomic weight of the mixture is then 28.00 + 7.40 or 35.40.

➤ **4.20** Deuterium (D) is a naturally occurring heavy isotope of hydrogen which is crucial in nuclear fusion. Its atomic mass is 2.0141. The atomic mass of pure hydrogen (^1H) is 1.0078 and the atomic weight of naturally occurring hydrogen is 1.0080. What is the mole percent (relative abundance) of the two isotopes in naturally occurring hydrogen which is a mixture of ^1H and D only?

5

Calculations from formulas of compounds. Determining the formula of a compound. Rounding off a number. Significant figures.

When atoms combine to form compounds they always do so in definite proportions by weight. As a result, the composition of every pure compound is definite and constant. This is the *law of definite composition.*

The definite composition of a particular compound is represented by its *chemical formula.* To illustrate, it has been found by experiments that 22.9898 g of sodium will always combine with 35.453 g of chlorine to form 58.443 g of common salt. From the table of atomic weights we learn that 22.9898 g is 1 gram atom of sodium, while 35.453 g is 1 gram atom of chlorine. That means that sodium combines with chlorine in the *ratio* of 1 gram atom of sodium to 1 gram atom of chlorine to form the compound sodium chloride. We represent 1 gram atom of sodium by the symbol Na and 1 gram atom of chlorine by the symbol Cl. The formula for sodium chloride is therefore NaCl. This formula, NaCl, means that the compound, NaCl, is made up of sodium and chlorine combined in the *ratio* of 1 gram atom of sodium to 1 gram atom of chlorine. Since, as has already been emphasized, the symbol for an element refers to one mole of atoms of that element, the formula, NaCl, means that the compound, NaCl, is made up of sodium and chlorine combined in the ratio of 1 *mole* of Na atoms to 1 *mole* of Cl atoms. The experimentally determined formula for hydrogen sulfide is H_2S, which tells us that in this compound the hydrogen and sulfur are combined in the ratio of 2 moles of H atoms to 1 mole of S atoms. Similarly, *the chemical formula of every chemical compound that will be encountered repre-sents the experimentally-determined* **mole ratio** *in which the atoms of the elements in the compound are combined.*

The term **molecule** *refers to the smallest neutral unit in which a substance exists and displays its characteristics as a pure substance.* In many instances the chemical formula of a compound represents the actual number of atoms of each element in a single molecule of the compound. Thus a single mole-cule of carbon disulfide, CS_2, contains 1 atom of carbon in combination with 2 atoms of sulfur; the actual composition of each molecule is represented, correctly, by the *true chemical formula*, CS_2. Likewise, H_2S, CO, CO_2, HCl, NH_3, SO_2, CH_4, C_2H_6, and H_2O are true chemical formulas; they represent the actual number of atoms of each component element in a molecule of the compound.

A great many compounds, however, exist as ions, not as molecules in the sense that a molecule is a neutral particle of substance. Thus, every crystal, handful, and barrelful of solid sodium chloride and every cubic centimeter of melted NaCl is made up of many Na^+ and Cl^- ions but no neutral NaCl molecules. For every Na^+ ion there is one Cl^- ion; the total salt is neutral. The quantity of sodium chloride and the size of the crystal may vary, but the *ratio* of sodium to chlorine is constant and is correctly represented by the *empirical formula*, NaCl. The same is true of the hundreds of other salts, metal oxides, and metal hydroxides. Their formulas are empirical formulas, not true chemical formulas; each represents the ratio in which the elements combine but does not represent the number of indi-vidual atoms in a single molecule of the compound. Since, in most problems dealing with these kinds of substances the *ratio* in which elements combine is the important thing, the fact that we do not know the true chemical formula of a compound causes no real difficulty.

The sum of the atomic weights represented by the formula of a sub-stance is called its *formula weight*. If the true chemical formula of a substance is known the formula weight is also the *molecular weight*. Thus, the formula weights of NaCl and CS_2 are, 58.443 and 76.139, respectively; 76.139 is the molecular weight of CS_2.

The number of *grams* equal, numerically, to the formula weight is the *gram-formula weight*; expressed in pounds or tons it is the *pound-formula weight* or *ton-formula weight*. The formula weight of NaCl is 58.443; 58.443 *grams* is one *gram*-formula weight of NaCl, 58.443 *pounds* is one *pound*-formula weight, and 58.443 *tons* is one *ton*-formula weight.

The number of *grams* equal, numerically, to the molecular weight, is the *gram-molecular weight*; expressed in units of pounds or tons it is the *pound-molecular weight* or *ton-molecular weight*. The molecular weight of CH_4 is 16.043; 16.043 *grams* is one *gram*-molecular weight of CH_4, 16.043

pounds is one *pound*-molecular weight, and 16.043 *tons* is one *ton*-molecular weight.

Just as the symbol of an element, when used in a formula or an equation, represents one gram-atomic weight of that element, so the *formula of a compound, when used in an equation, represents one gram-formula weight of that compound. If the true chemical formula is known, this formula represents one gram-molecular weight.*

One gram-molecular weight of any substance contains 6.023 × 10²³ *molecules.* The reasoning that leads to this conclusion can be illustrated in the case of the compound whose true chemical formula is CS_2. This formula tells us that 1 atom of C combines with 2 atoms of S to form 1 molecule of CS_2. Therefore 6.023 × 10²³ atoms (1 mole) of C must combine with 2 × 6.023 × 10²³ atoms (2 moles) of S to yield 6.023 × 10²³ molecules of CS_2. Therefore, 1 gram-molecular weight of CS_2 must contain 6.023 × 10²³ individual molecules. In a similar manner it can be reasoned that 1 *gram-molecular weight of any substance whose true chemical formula is known contains* 6.023 × 10²³ *molecules.*

But 6.023 × 10²³ units is, by definition, one mole. Since the formula represents one gram-molecular weight, since one gram-molecular weight consists of 6.023 × 10²³ molecules, and since 6.023 × 10²³ molecules is 1 mole, *the chemical formula of a compound, when used in any equation, represents one mole of that substance.* **This is an important fact to remember.**

We will define one mole of NaCl as that quantity of NaCl which contains one mole of Na^+ ions and one mole of Cl^- ions; since the symbol for an element represents one mole of atoms of that element, the formula, NaCl, will, in fact, represent one mole of NaCl. Likewise, one mole of Na_2SO_4 is defined as that quantity of Na_2SO_4 which contains 2 moles of Na^+ ions and 1 mole of SO_4^{--} ions; the formula, Na_2SO_4, does, in fact, represent one mole of Na_2SO_4. The same line of reasoning can be applied to all compounds which, like NaCl and Na_2SO_4, eixst as ions rather than as discrete neutral molecules. Accordingly, the statement that the formula of a compound, when used in a chemical equation, represents one mole of that compound applies to all compounds even though they may not exist as discrete neutral molecules.

Use of the term "mole"

In the present-day use of the term *mole, the symbol or formula for any chemical substance,* whether it is an element, a compound, or an ion, when used in a formula or equation, *represents one mole of that substance.* So that there may be no question about the identity of the mole of substance, its symbol or formula should always be given. The following statements, each of which represents correct usage of the term *mole*, will illustrate this point:

The formula, $KClO_3$, tells us that 1 mole of $KClO_3$ contains 1 mole of K, 1 mole of Cl, and 3 moles of O.

One mole of $KClO_3$, when heated, will yield 1 mole of KCl and 1.5 moles of O_2.

One mole of Na_2SO_4 contains 2 moles of Na^+ ions and 1 mole of SO_4^{--} ions.

In the reaction represented by the equation, $C + O_2 = CO_2$, 1 mole of C atoms combines with 1 mole of O_2 molecules to form 1 mole of CO_2 molecules. A mole of CO_2 molecules can be considered to consist of 1 mole of C atoms and 2 moles of O atoms.

In the reaction, $Ag^+ + Cl^- = AgCl$, the Ag^+ ions and Cl^- ions combine in the ratio of 1 mole of Ag^+ ions to 1 mole of Cl^- ions to form 1 mole of AgCl.

If the true chemical formula is known, the term "one mole" means one *molecular weight*, expressed in the proper units of mass. If only the empirical formula is known, "one mole" means one *formula weight*.

Since the unit of mass commonly employed is the gram, the term *mole* will, unless otherwise stated, be understood to mean *gram mole* and represents either the gram-atomic weight, the gram-molecular weight, or the gram-formula weight. If the unit of mass employed is the pound or ton we will have a *pound mole* or *ton mole*. To illustrate, 1 mole of NaCl is 58.443 g and 1 mole of CS_2 is 76.139 g. One pound mole of CO_2 is 44.01 lb and 1 ton mole of H_2SO_4 is 98.10 tons.

Determining the formula of a compound

Attention has already been called, in one of the preceding paragraphs, to the experimental facts and scientific reasoning which lead to the conclusion that the empirical formula for the compound, sodium chloride, is NaCl. The formula of every compound that you will meet in chemistry has been determined experimentally in the same manner and by exactly the same kind of reasoning. Because it is imperative for the solution of all future problems that the significance and meaning of the chemical formula be clearly understood, the method used in the determination of a formula will now be discussed in detail.

Let us remember, first of all, that the chemical formula for a compound gives the ratio of the number of atoms of each of the elements in the compound. Since the ratio of the number of *atoms* of each element in a *molecule* of the compound is the same as the ratio of the number of *moles of atoms* of each element in a *mole* of the compound, what we are trying to do, when we determine the formula of a compound, is to find the number of moles of atoms of each element in one mole of the compound.

Suppose we want to determine the chemical formula for water. First we synthesize pure water (prepare it from pure oxygen and pure hydrogen). We find, by a careful experiment, that 15.999 g of oxygen combine with exactly 2.016 g of hydrogen to form exactly 18.015 g of water. We check our results by analyzing (breaking down) these 18.015 g of water, and we find that this amount of water yields exactly 15.999 g of oxygen and 2.016 g of hydrogen. The gram-atomic weight of oxygen is 15.999 g and the gram-atomic weight of hydrogen is 1.008 g. That means that 1 gram-atomic weight (1 gram atom) (1 mole) of O has combined with 2 gram-atomic weights (2 gram atoms) (2 moles) of H to form 1 gram-molecular weight of water. The chemical formula for water is therefore H_2O. The subscript to the right and below the H means that there are 2 atoms of hydrogen combined with 1 atom of oxygen. We obtained the 2, representing the number of atoms of H, by dividing the 2.016 g of hydrogen by 1.008 g (the gram-atomic weight of hydrogen). That is,

$$\frac{2.016 \text{ g of H}}{1.008 \text{ g per gram atom of H}} = 2 \text{ gram atoms of H}$$

$$\frac{15.999 \text{ g of O}}{15.999 \text{ g per gram atom of O}} = 1 \text{ gram atom of O}$$

Therefore, the chemical formula is H_2O.

Since 1 gram atom of an element is 1 mole of atoms of that element, the calculation given above can be represented as follows:

$$\frac{2.016 \text{ g of H}}{1.008 \text{ g per mole of H}} = 2 \text{ moles of H}$$

$$\frac{15.999 \text{ g of O}}{15.999 \text{ g per mole of O}} = 1 \text{ mole of O}$$

Summarizing what we did in getting the formula for water, we proceed as follows in determining the empirical formula for any chemical compound.

1. Determine the exact composition of the compound, that is, the mass of each element that combines, or the percent of each element in the compound, either by analysis or by synthesis.

2. Divide the mass of each element, or the percent of each element, by its gram-atomic weight. The simplest whole-number ratio between the quotients gives the empirical formula.

In Problems 5.40–5.48, which are designed to show how formulas are determined, the results of the experimental analysis or synthesis will be given. Only the subsequent calculations will be required.

Rounding off numbers

Up to this point the exact values of the various atomic weights have been used in making calculations. Since it is strongly recommended that a slide rule be used for all calculations, and since the average slide rule reading is not exact beyond three or four digits, nothing is gained by using such exact values. Accordingly, the "rounded off" values given in the table on the inside front cover will be used in all future calculations.

A number is "rounded off" by dropping digits starting from the right. In the case of atomic weights we will round off by dropping enough digits so that there is only one digit to the right of the decimal place. Thus 39.096 becomes 39.1 and 35.457 becomes 35.5.

The following rules govern the rounding-off process:

1. When the digit dropped is less than 5, the next digit to the left remains unchanged. Thus 69.72 becomes 69.7, and 12.011 becomes 12.0.

2. When the digit dropped is greater than 5, the value of the next digit to the left is increased by 1. Thus 65.38 becomes 65.4 and 35.457 becomes 35.5.

3. When the digit dropped is exactly 5, 1 is added to the digit on the left if that digit is odd but nothing is added if that digit is even. Thus 95.95 becomes 96.0 and 51.75 becomes 51.8, but 55.85 becomes 55.8 and 51.65 becomes 51.6.

Significant figures

Before we proceed further with calculations, it is desirable that we consider the question: To how many decimal places, if any, should we report the answer to a problem? In other words, we would like to know how many *significant figures* our answer should contain. A significant figure is one that is reasonably reliable. Suppose you have a yardstick which is divided into 36 one-inch units and suppose each inch is divided into tenths of an inch, but there are no smaller divisions. Now suppose you wish to measure the length of a table top using this yardstick. You can read the stick accurately to tenths of an inch. Thus, if the length of the table fell exactly on the twenty-eight and two-tenths mark, you could say the length was 28.2 in. All three of these numbers would be accurate, all three would be significant, and you would say that you had measured the length to *three significant figures*. Suppose, however, that the length of the table doesn't fall exactly on the 28.2 mark but falls somewhere between 28.2 and 28.3. You *estimate* that it falls two-tenths of the distnce between 28.2 and 28.3, and you report the length as 28.22. The last digit in this four-digit number is not exact because you had to estimate its value. So your answer still has only three absolutely significant figures, 28.2. If you were asked to report the length to the strictly significant figures only, you would report 28.2, not 28.22. However, experience has shown that estimates of the sort that you made are so close to being exact that they are considered to be significant and can be recorded as such. In other words, in the average careful measurement which involves taking a reading on a graduated scale of discernible length or width, the first estimated digit is considered to be significant and can be recorded. So, under ordinary circumstances, you would be justified in reporting the length of the table top as 28.22 in.

Suppose that this table whose top you have measured happens to stand end to end with a fine stainless-steel bench which you have just received from the National Bureau of Standards. The top of this bench has been carefully machined and has been measured by the Bureau of Standards with a very accurately graduated rule and is certified to be exactly 31.964 in. in length. The Bureau of Standards measurement is of such precision that all five digits in the number 31.964 are significant. Now you are asked to report the combined length of your table and the bench. The question is, will the combined length be reported as 60.184 in. (31.964 + 28.22) or 60.18 in.? The answer is 60.18 in. The rule is that the sum can have no more significant figures than the least significant of its parts. In other words, the sum is no more accurate than its least accurate part.

Now suppose you wish to calculate the area of the above table top. You have already found its length to be 28.22 in. You measure the width and report it, justifiably, as 20.16 in. To get the area in square inches you multiply 28.22 × 20.16 and get 568.9152. The question is, what figure shall you report? The answer is 568.9 sq in., and the rule is that the product shall contain no more significant figures than are present in the multiplier with the least number. In other words, the product can be no more exact than the least exact multiplier. By the same rule the product of 1.56 × 1.78 is reported as 2.78, not as 2.7768. Only three digits are significant, so 2.7768 has been rounded off to 2.78 in accordance with the rules given in the previous section.

In division also, we apply the same rule, namely, that the answer can be no more accurate than the least accurate of the terms involved in the operation. It follows, therefore, that the quotient obtained when 76.2 is divided by 47.24 is 1.61, not 1.613. The quotient obtained when *exactly* 200 is divided by *exactly* 3 is 66.66 . . . 6, because if the 200 is exactly 200 and the 3 is exactly 3, the numbers can be written 200.000 . . . 0 and 3.0000 . . . 0, respectively. In other words, a whole number has, in reality, an unlimited number of significant figures. The question of significance comes into the picture only when the number is the result, either directly or indirectly, of a physical measurement.

The number 0.00134, assuming that it does in fact represent, correctly, some measured value, has three significant figures, while the number 13.40 has four significant figures. Zeros to the left of a group of digits are not counted as significant figures but zeros to the right are. The zeros at the left serve only to locate the decimal point. The quotient obtained when 0.00134 is divided by 0.023 would be reported as 0.057, not 0.0573, because 0.023 has only two significant figures.

Since 0.00134 has three significant figures, 1.34×10^{-3}, which is equal to 0.00134, also has three significant figures. Likewise 1.5×10^{-20} has two significant figures, and the product of 2.32 and 1.5×10^{-20} will be 3.5×10^{-20}.

The number 6.023×10^{23} has four significant figures. It illustrates the fact that a more realistic picture of the degree of precision can be conveyed by writing large numbers in exponential form.

The question may arise as to why a number as small as 1.22×10^{-15} moles can have as many significant figures as the larger number, 235 g/mole. The answer is that, even though, in the first number, the unit 1×10^{-15}, is very small, we are certain that we have 1.22 such units (not 1.2 or 1.223). In the number, 235, the unit is large, but we are only certain that we have 235 of these units. In an exponential number the exponential term defines the unit of measure while the nonexponential term defines the number of these units; the latter determines the significant figures.

The upshot of all this discussion is that the answer obtained in multiplication or division of fractional or mixed numbers should never have any more digits than the number with the least number of significant digits. The answer obtained in addition or subtraction should never have any more digits to the right of the decimal point than does the number with the least digits to the right of the decimal point.

Certain refinements of the above rules must be considered in specific cases, but they need not concern us in this book.

The answers to all problems in this book are rounded off to the nearest significant figure. It should be stated, also, that all answers in this book have been obtained by slide-rule calculation and are, accordingly, subject to the normal chance of slight variation present in all slide-rule calculations. A student should never feel that he must duplicate the answer to the problem exactly. The correct method of solution is more important than the identically correct answer.

A slide rule should be used when solving problems. Longhand calculations are much too laborious and time-consuming. The increased availability of hand-held electronic calculators, however, is resulting in a decrease in the importance of the slide rule. If you own such a calculator and if you are certain that it will always be available to you, use it to perform the calculations. But remember that continued use of such calculators tends to dull the computational senses.

Calculations from the formula of a compound

PROBLEMS

5.1 Calculate the formula weight of KCl.

Solution: The formula weight is the sum of the atomic weights of the atoms represented by the formula. The atomic weight of K is 39.1. The atomic weight of Cl is 35.5.

$$39.1 + 35.5 = 74.6$$

5.2 Calculate the formula weight of $KClO_3$.

Solution: The formula shows that $KClO_3$ is made up of potassium, chlorine, and oxygen combined in the ratio of 1 atom of potassium to 1 atom of chlorine to 3 atoms of oxygen. The formula weight is the sum of the masses of the atoms represented by the formula.

$$
\begin{aligned}
1\ \text{K} &= 39.1 \\
1\ \text{Cl} &= 35.5 \\
3\ \text{O} &= \underline{48.0} \\
\text{formula weight} &= 122.6
\end{aligned}
$$

5.3 Calculate the formula weight of $Al_2(SO_4)_3$.

Solution: The chemical formula, $Al_2(SO_4)_3$, indicates that aluminum sulfate is made up of aluminum, sulfur, and oxygen combined in the ratio of 2 atoms of aluminum to 3 atoms of sulfur to 12 atoms of oxygen. The SO_4^{--} radical is enclosed in parentheses with a subscript 3 outside the parentheses. This means that the radical is taken three times. The subscript 2 applies only to the aluminum atom; the subscript 3 applies only to the SO_4 radical.

$$
\begin{aligned}
2\ \text{Al} &= 54.0 \\
3\ \text{S} &= 96.3 \\
12\ \text{O} &= \underline{192.0} \\
\text{formula weight} &= 342.3
\end{aligned}
$$

5.4   What is the mass of a mole of H_2SO_4?

Solution: A mole is the mass in grams of the elements represented by the formula of a substance. Therefore, to find the mass of a mole of H_2SO_4 we simply find the sum of the gram-atomic weights of its constituent atoms.

$$
\begin{aligned}
2\,H &= 2.0\text{ g} \\
1\,S &= 32.1\text{ g} \\
4\,O &= 64.0\text{ g} \\
\hline
\text{mole of } H_2SO_4 &= 98.1\text{ g}
\end{aligned}
$$

5.5   Calculate the mass of 0.0200 mole of $K_2Cr_2O_7$.

Solution: One mole of $K_2Cr_2O_7$ can be calculated (see Problem 5.4) to be 294.2 g.

$$0.0200 \text{ mole} \times 294.2 \text{ g/mole} = 5.88 \text{ g}$$

Note that moles cancel, giving the answer in grams.

5.6   What fraction of a mole of CH_4 (methane) is 7 g of CH_4?

Solution: One mole of CH_4 is 16.0 g.

$$\frac{7 \text{ g of } CH_4}{16 \text{ g of } CH_4/1 \text{ mole of } CH_4} = \frac{7}{16} \text{ mole of } CH_4$$

Note that g of CH_4 cancel, giving the answer as moles of CH_4.

5.7   An amount of 20 g of NH_3 (ammonia) is what fraction of a mole of NH_3?

5.8   An amount of 120 g of CO_2 (carbon dioxide) is how many moles of CO_2?

Solution:

$$\frac{120 \text{ g of } CO_2}{44 \text{ g of } CO_2/1 \text{ mole of } CO_2} = 2.7 \text{ moles of } CO_2$$

5.9   How many moles of $C_7H_5N_3O_6$ (TNT) are there in 384 g of $C_7H_5N_3O_6$?

X   **5.10**   How many moles of P are there in 2.4 moles of P_4O_{10}?

Solution: The formula P_4O_{10} shows that 1 mole of P_4O_{10} contains 4 moles of P.

$$2.4 \text{ moles of } P_4O_{10} \times \frac{4 \text{ moles of P}}{1 \text{ mole of } P_4O_{10}} = 9.6 \text{ moles of P}$$

Note that moles of P_4O_{10} cancel, giving the answer as moles of P.

5.11   How many moles of C are there in 0.106 mole of $C_{16}H_{18}N_2O_4S$ (penicillin)?

5.12 How many moles of H are there in 8.6 moles of N_2H_4 (hydrazine)?

5.13 How many moles of CCl_4 will contain 2.4 moles of Cl?

Solution:

$$\frac{2.4 \text{ moles of Cl}}{4 \text{ moles of Cl/1 mole of } CCl_4} = 0.60 \text{ mole of } CCl_4$$

Note that moles of Cl cancel, giving the answer as moles of CCl_4.

5.14 How many moles of C_3H_8 (propane) will contain 23 moles of H?

5.15 How many molecules are there in 2.70 moles of H_2S?

Solution: One mole of H_2S contains 6.023×10^{23} molecules.

$$2.70 \text{ moles of } H_2S \times \frac{6.023 \times 10^{23} \text{ molecules of } H_2S}{1 \text{ mole of } H_2S}$$
$$= 1.62 \times 10^{24} \text{ molecules}$$

Note that moles of H_2S cancel, giving the answer as molecules.

5.16 How many molecules are there in 0.0372 mole of $C_9H_8O_4$ (aspirin)?

5.17 How many moles of CH_4 will contain 4.31×10^{25} molecules of CH_4?

Solution:

$$\frac{4.31 \times 10^{25} \text{ molecules of } CH_4}{6.023 \times 10^{23} \text{ molecules of } CH_4/\text{mole of } CH_4} = 71.5 \text{ moles of } CH_4$$

5.18 How many moles of NH_3 will contain 5.16×10^{20} molecules of NH_3?

5.19 How many molecules of SO_2 are there in 200 g of SO_2?

Solution: The molecular weight of SO_2 is 64.1. Therefore, 200 g of SO_2 is 200/64.1 moles of SO_2.

$$\frac{200}{64.1} \text{ moles of } SO_2 \times 6.023 \times 10^{23} \text{ molecules/mole of } SO_2$$
$$= 1.88 \times 10^{24} \text{ molecules}$$

5.20 How many molecules of CH_4 are there in 1.25 g of CH_4?

5.21 How many grams of CO_2 will contain 5.10×10^{24} molecules of CO_2?

Solution: One mole of CO_2 has a mass of 44.0 g and contains 6.023×10^{23} molecules.

$$5.10 \times 10^{24} \text{ molecules of } CO_2 \times \frac{44.0 \text{ g of } CO_2}{6.023 \times 10^{23} \text{ molecules of } CO_2}$$
$$= 364 \text{ g of } CO_2$$

Note that molecules of CO_2 cancel molecules of CO_2.

5.22   Calculate the mass in grams of 9.00×10^{22} molecules of SO_3.

5.23   The mass of 2.60 moles of a compound is 312 g. Calculate the molecular weight of the compound.

Solution: The molecular weight is the mass in grams of 1 mole.

$$\frac{312 \text{ g}}{2.60 \text{ moles}} = 120 \text{ g/mole}$$

Therefore, molecular weight $= 120$

Note that any fraction, when solved, expresses the value per one unit.

Thus, in the above calculation, $\dfrac{312 \text{ g}}{2.60 \text{ moles}} = 120$ g per *one* mole.

5.24   A 6.2 mole sample of a compound has a mass of 105.4 g. Calculate the molecular weight of the compound.

5.25   How many grams of sulfur are there in 2.20 moles of H_2S?

Solution: One mole of H_2S contains 1 mole of S. One mole of S has a mass of 32.1 g.

$$2.20 \text{ moles of } H_2S \times \frac{1 \text{ mole of S}}{1 \text{ mole of } H_2S} \times \frac{32.1 \text{ g of S}}{1 \text{ mole of S}} = 70.6 \text{ g of S}$$

Since it is obvious that one mole of H_2S contains 32.1 g of S, the second factor in the above equation can be omitted to give

$$2.20 \text{ moles of } H_2S \times \frac{32.1 \text{ g of S}}{1 \text{ mole of } H_2S} = 70.6 \text{ g of S}$$

5.26   How many grams of sulfur are there in 1.67 moles of P_4S_3?

5.27   How many moles of SiO_2 (silica) will contain 50.0 g of oxygen?

5.28   How many moles of O are there in 182 g of $KClO_3$?

Solution:

$$\frac{182 \text{ g of } KClO_3}{122.6 \text{ g of } KClO_3/\text{mole of } KClO_3} \times \frac{3 \text{ moles of O}}{1 \text{ mole of } KClO_3}$$
$$= 4.46 \text{ moles of O}$$

5.29   How many moles of S are there in 0.0142 g of CS_2?

5.30   How many grams of Sb_2S_3 will contain 4.80 moles of S?

5.31   How many grams of phosphorus are there in 160 g of P_4O_{10}?

Solution: One mole of P_4O_{10} contains 4 moles of P. Moles of $P = 4 \times$ moles of P_4O_{10}.

$$\frac{160 \text{ g of } P_4O_{10}}{284 \text{ g of } P_4O_{10}/\text{mole of } P_4O_{10}} = \frac{160}{284} \text{ mole of } P_4O_{10}$$
$$\text{moles of } P = \frac{4 \times 160}{284}$$

$$\frac{4 \times 160}{284} \text{ mole of P} \times \frac{31.0 \text{ g of P}}{1 \text{ mole of P}} = 69.8 \text{ g of P}$$

The solution, in one operation, is

$$\frac{160 \text{ g of } P_4O_{10}}{284 \text{ g of } P_4O_{10}/\text{mole of } P_4O_{10}} \times \frac{4 \text{ moles of P}}{1 \text{ mole of } P_4O_{10}}$$
$$\times \frac{31.0 \text{ g of P}}{1 \text{ mole of P}} = 69.8 \text{ g of P}$$

5.32   How many grams of oxygen are there in 1.64 g of $K_2Cr_2O_7$?

5.33   How many grams of SO_3 will contain 2.00 g of oxygen?

5.34   How many tons of Fe_2O_3 will contain 12.0 tons of Fe?

5.35   Calculate the percent of carbon in CO_2.

Solution: By definition, percent means parts per 100 parts by mass. So all we need do in this problem is find how many grams of C there are in 100 g of CO_2; the answer will then be the percent of C in pure CO_2. A more general definition of percent, called fraction, is that it is the ratio of the number of parts by mass of the particular thing you want to find to the parts by mass of the whole thing. That is,

$$\text{fraction of C in } CO_2 = \frac{\text{mass of the C in } CO_2}{\text{mass of the } CO_2}$$

A mole of CO_2 has a mass of 44.0 g and contains 12.0 g of C. Therefore,

$$\text{fraction of C in } CO_2 = \frac{12.0 \text{ g of C}}{44.0 \text{ g of } CO_2} = 0.273$$

This gives a decimal (fractional) percent. To change this to the standard notation, we simply multiply by 100%. Doing the whole calculation in one operation

$$\text{percent of C in } CO_2 = \frac{12.0}{44.0} \times 100\% = 27.3\%$$

The general form of this relationship is:

percent of an element in a compound

$$= \frac{\text{mass of the element in the compound}}{\text{formula weight of the compound}} \times 100\%$$

5.36   Calculate the percent of oxygen in:

(a) $Fe_2(SO_4)_3$

(b) $(NH_4)_2CO_3$

5.37   For the compound $C_6H_5NO_2$ (nitrobenzene) calculate:

(a) moles of $C_6H_5NO_2$ in 200 g of $C_6H_5NO_2$

(b) grams of C in 5.00 moles of $C_6H_5NO_2$

(5) (6) (12.0)

(c) grams of C in 200 g of $C_6H_5NO_2$

(d) grams of C per 10.0 g of N

(e) moles of O in 150 g of $C_6H_5NO_2$

(f) moles of $C_6H_5NO_2$ which contain 5.0 g of N

(g) grams of $C_6H_5NO_2$ which contain 0.500 mole of C

(h) molecules of $C_6H_5NO_2$ in 3.00 g of $C_6H_5NO_2$

(i) atoms of C in 3.00 g of $C_6H_5NO_2$

(j) the percent of carbon

(k) atoms of N per atom of C

(l) grams of H per gram of C

(m) grams of C per mole of H

➤ 5.38   Dealer A sells NaClO bleach at 80 cents per pound of NaClO content. Dealer B sells the same identical bleach at $1.00 per pound of ClO content. Which dealer offers the better bargain?

➤ 5.39   The element M forms the chloride, MCl_4. This chloride contains 75.0% chlorine. Calculate the atomic weight of M, knowing that the atomic weight of chlorine is 35.5.

Solution: What must be the numerical value of the term

$$\frac{\% \text{ of M/atomic weight of M}}{\% \text{ of Cl/atomic weight of Cl}} ?$$

Calculation of the formula of a compound

PROBLEMS

5.40   It was found that 56 g of iron combined with 32 g of sulfur. Calculate the empirical formula of the compound that was formed.

Solution: The formula of a compound gives the number of moles of atoms of each element in one mole of the compound. The atomic weight of iron is 56 and of sulfur is 32. Therefore, 56 g is 1 mole of Fe and 32 g is 1 mole of S. Therefore, Fe and S are combined in the ratio of 1 mole of Fe to 1 mole of S. Since 1 mole of iron atoms is represented by the symbol, Fe, and 1 mole of sulfur atoms by the symbol, S, the empirical formula for the compound, iron sulfide, is FeS.

5.41   When heated, 433.22 g of a pure compound yielded 401.22 g of mercury and 32 g of oxygen. Calculate the empirical formula of the compound.

Solution: To find the number of moles of atoms of each element present, we will divide the quantity in grams of each element by the mass of 1 mole of atoms of the element, that is, by the atomic weight of that element.

$$\frac{401.22 \text{ g of Hg}}{200.61 \text{ g of Hg per mole of Hg}} = 2 \text{ moles of Hg}$$

$$\frac{32 \text{ g of O}}{16 \text{ g of O per mole of O}} = 2 \text{ moles of O}$$

The formula would thus appear to be Hg_2O_2. However, since we are interested in getting the simplest formula (the empirical formula), we will take the simplest ratio. Since 2 is to 2 as 1 is to 1, the empirical formula is HgO.

5.42 It was found that 10.0 g of a pure compound contains 3.65 g of K, 3.33 g of Cl, and 3.02 g of O. Calculate the empirical formula of the compound.

5.43 In the laboratory 2.38 g of copper combined with 1.19 g of sulfur. In a duplicate experiment 3.58 g of copper combined with 1.80 g of sulfur. Are these results in agreement with the law of definite composition?

5.44 When burned, 4.04 g of magnesium combined with 2.66 g of oxygen to form 6.70 g of magnesium oxide. Calculate the empirical formula of the oxide.

Solution: To find the number of moles of Mg in 6.70 g of magnesium oxide we will divide the mass in grams of the magnesium by the gram-atomic weight of Mg, and to find the number of moles of O we will divide the mass in grams of the oxygen by the gram-atomic weight of O.

$$\text{moles of Mg} = \frac{4.04 \text{ g of Mg}}{24.3 \text{ g per mole of Mg}} = 0.166 \text{ mole of Mg}$$

$$\text{moles of O} = \frac{2.66 \text{ g of O}}{16.0 \text{ g per mole of O}} = 0.166 \text{ mole of O}$$

Therefore, the magnesium and oxygen are combined in the ratio of 0.166 mole of Mg to 0.166 mole of O. But 0.166 is to 0.166 as 1 is to 1. Therefore, the simplest formula is MgO. The formula represents the simplest *whole-number* ratio of the moles. In this case the simplest ratio is 1 to 1.

5.45 A sample of 2.12 g of copper was heated in oxygen until no further change took place. The resulting oxide had a mass of 2.65 g. Calculate the empirical formula of the oxide.

5.46 A pure compound was found on analysis to contain 31.9% potassium, 28.9% chlorine, and 39.2% oxygen. Calculate its empirical formula.

Solution: To say that a compound contains 31.9% potassium, 28.9% chlorine, and 39.2% oxygen is equivalent to saying that 100 g of the compound contain 31.9 g of potassium, 28.9 g of chlorine, and 39.2 g of oxygen. Therefore, to find the number of moles of each element in a mole of the compound, we will divide the percent of each element by its atomic weight.

$$K = \frac{31.9 \text{ g of K}}{39.1 \text{ g/mole of K}} = 0.815 \text{ mole}$$

$$Cl = \frac{28.9 \text{ g of Cl}}{35.5 \text{ g/mole of Cl}} = 0.815 \text{ mole}$$

$$O = \frac{39.2 \text{ g of O}}{16 \text{ g/mole of O}} = 2.45 \text{ moles}$$

The mole ratio of K to Cl to O is 0.815 to 0.815 to 2.45. To simplify this, we divide all three of these numbers by the smallest.

$$K = \frac{0.815}{0.815} = 1 \qquad Cl = \frac{0.815}{0.815} = 1 \qquad O = \frac{2.45}{0.815} = 3$$

The simplest formula of the compound is, therefore, $KClO_3$.

5.47   A compound contains 29.1% sodium, 40.5% sulfur, and 30.4% oxygen. What is its empirical formula?

5.48   A compound was found on analysis to contain 21.6% magnesium, 27.9% phosphorus, and 50.5% oxygen. Calculate the empirical formula of the compound.

5.49   A compound was found to have the empirical formula CH_2 and a molecular weight of 71. Calculate the molecular formula.

Solution: Calculate the formula weight of CH_2: $1 \times 12 + 2 \times 1 = 14$. Calculate how many times the empirical formula goes into the molecular formula by dividing the empirical formula weight into the molecular weight $\frac{71}{14} = 5.07$ and round the answer off to the nearest integer, which is 5. Multiply each subscript in the empirical formula by this integer to get the molecular formula; $CH_2 \times 5 = C_5H_{10}$.

5.50   A compound was found to contain 55.8% carbon, 11.6% hydrogen, and 32.6% nitrogen. Its molecular weight is found to be 171. Calculate the molecular formula of the compound.

5.51   Mannose is a sugar which contains only carbon, hydrogen, and oxygen, and has a molecular weight of 180. A 2.36-g sample of mannose was found on analysis to contain 0.944 g of carbon and 0.158 g of hydrogen. Calculate the molecular formula.

Solution: Calculate the mass of oxygen in the sample by difference and then proceed as in Problems 5.46 and 5.49.

5.52 A 1.20-g sample of a compound which contains only carbon and hydrogen is burned completely in excess O_2 to yield 3.60 g of CO_2 and 1.96 g of H_2O. Calculate the empirical formula of the compound.

Solution: The mass of carbon in the sample is the same as the mass of carbon in CO_2; and the mass of hydrogen in the sample is the same as the mass of hydrogen in H_2O.

$$\text{g of C} = \frac{12 \text{ g of C}}{44 \text{ g of CO}_2} \times 3.60 \text{ g of CO}_2 = 0.982 \text{ g of C}$$

$$\text{g of H} = \frac{2 \text{ g of H}}{18 \text{ g of H}_2\text{O}} \times 1.96 \text{ g of H}_2\text{O} = 0.218 \text{ g of H}$$

Now proceed as in Problems 5.41 and 5.46.

5.53 A 2.36-g sample of a compound which contains only carbon, hydrogen, and oxygen is burned completely in excess O_2 to yield 5.76 g of CO_2 and 2.34 g of H_2O. Calculate the empirical formula of the compound.

Solution: As in Problem 5.52 the mass of carbon and hydrogen in the original sample can be calculated.

$$\text{g of C} = \frac{12 \text{ g of C}}{44 \text{ g of CO}_2} \times 5.76 \text{ g of CO}_2 = 1.57 \text{ g of C}$$

$$\text{g of H} = \frac{2 \text{ g of H}}{18 \text{ g of H}_2\text{O}} \times 2.34 \text{ g of H}_2\text{O} = 0.260 \text{ g of H}$$

Since the total mass of the sample is the sum of the masses of C, H, and O, the mass of oxygen in the sample can be obtained by difference.

$$\text{g of O} = \text{g of sample} - \text{g of C} - \text{g of H}$$
$$= 2.36 - 1.57 - 0.26 = 0.53 \text{ g of O}$$

Now proceed as in Problems 5.41 and 5.46.

5.54 A 1.78-g sample of a compound containing only carbon and hydrogen was burned in excess oxygen to yield 5.79 g of CO_2 and 1.80 g of H_2O. Calculate the empirical formula of the compound.

5.55 A 2.44-g sample of a compound containing only carbon, hydrogen, and nitrogen was burned in excess oxygen to yield 6.78 g of CO_2 and 1.35 g of H_2O. Calculate the empirical formula of the compound.

5.56 A 1.48-g sample of a compound containing only carbon, hydrogen, nitrogen, and chlorine was burned in excess O_2 to yield 2.21 g of CO_2 and 0.452 g of H_2O. Another sample of this compound, weighing 2.62 g, was found to contain 1.05 g of Cl. Calculate the empirical formula of the compound.

Solution: The g of Cl in the 1.48-g sample can be calculated from the g of Cl in the 2.62-g sample.

$$\text{g of Cl} = \frac{1.05 \text{ g of Cl}}{2.62 \text{ g of sample}} \times 1.48 \text{ g of sample} = 0.593 \text{ g of Cl}$$

The g of C and the g of H in the 1.48-g sample are calculated as in Problem 5.52. The g of N in the sample $= 1.48$ g $-$ g of C $-$ g of H $- 0.593$ g as in Problem 5.53. Now proceed as in Problems 5.41 and 5.46.

5.57 A 3.42-g sample of a compound containing only carbon, hydrogen, nitrogen, and oxygen was burned in excess O_2 to yield 2.47 g of CO_2 and 1.51 g of H_2O. Another sample of this compound weighing 5.26 g was found to contain 1.20 g of N. Calculate the empirical formula of the compound.

➡ **5.58** A 2.52-g sample of a compound containing only carbon, hydrogen, nitrogen, oxygen, and sulfur was burned in excess O_2 to yield 4.23 g of CO_2 and 1.01 g of H_2O. Another sample, which weighed 4.14 g, yielded 2.11 g of SO_3. Another sample, which weighed 5.66 g, yielded 2.27 g of HNO_3. Calculate the empirical formula of the compound.

➡ **5.59** A compound of molecular weight 177 contains only carbon, hydrogen, bromine, and oxygen. Analysis shows that a sample of the compound contains 8 times as much carbon as hydrogen, by weight. Calculate the molecular formula of the compound.

➡ **5.60** The elements X and Y form a compound which is 40% X and 60% Y by mass. The atomic weight of X is twice that of Y. What is the empirical formula of the compound?

Solution: Assign some value, 1 for instance, for the atomic weight of Y. The atomic weight of X will then be 2. Then proceed as in Problem 5.46.

➡ **5.61** Three pure compounds are formed when 1-gram portions of element X combine with, respectively, 0.472 g, 0.630 g, and 0.789 g of element Z. The first compound has the formula, X_2Z_3. What are the empirical formulas of the other two compounds? How does the atomic weight of X compare with that of Z?

Solution: Arbitrarily assign some convenient value, 0.50 for instance, for the atomic weight of X. Then proceed as in Problem 5.39.

6

The gas laws.

Boyle's law

We know from experience that if pressure is applied to any gas, to the air in a football, basketball, or tennis ball, for example, the volume of the gas will be decreased. As the pressure goes *up* the volume goes *down*; that is an *inverse proportion*. If we double the pressure, keeping the temperature constant, the volume will be reduced one half. We can state, therefore, that, at constant temperature, *the volume of a mass of gas is inversely proportional to the pressure.* This is *Boyle's law.* If we call P_1 the original or first pressure and V_1 the original or first volume, then if the pressure is increased to a second value, P_2, the volume will be decreased to a second value V_2. We can represent the change as follows:

The pressure goes up, from P_1 to P_2; the volume goes down, from V_1 to V_2. As already stated, this is an inverse proportion because as one increases the other decreases. That is,

$$\frac{P_2}{P_1} = \frac{V_1}{V_2}$$

or

$$\frac{V_1}{V_2} = \frac{P_2}{P_1}$$

This is the formula for Boyle's law. It tells us that, at constant temperature, *the volume of a mass of gas is inversely proportional to the pressure.*

Standard pressure

The pressure on a gas is usually expressed in units of millimeters of mercury or atmospheres. A pressure of 740 mm means the pressure which would be exerted by a column of liquid mercury 740 mm high. At sea level the average pressure of the atmosphere is 760 mm of mercury. A pressure of 760 mm is, therefore, called *standard pressure*. A pressure of 760 mm is referred to as *one atmosphere.*

The following problems will illustrate the application of Boyle's law.

PROBLEMS

6.1   A gas has a volume of 500 cc at a pressure of 700 mm of mercury. What volume will it occupy if the pressure is increased to 800 mm, the temperature remaining constant?

Solution: One way of solving will be to substitute in the Boyle's law equation

$$\frac{V_1}{V_2} = \frac{P_2}{P_1}$$

The first volume, $V_1 = 500$ cc.
The second volume, V_2, is what we want to find.
The first pressure, $P_1 = 700$ mm.
The second pressure, $P_2 = 800$ mm.
Substituting these values in the formula, we have

$$\frac{500 \text{ cc}}{V_2} = \frac{800 \text{ mm}}{700 \text{ mm}}$$

Solving,

$$V_2 = \frac{500 \text{ cc} \times 700 \text{ mm}}{800 \text{ mm}} = 437 \text{ cc}$$

Solving by substitution in the Boyle's law formula is an acceptable method as long as we understand the formula in the first place and are careful to keep our values straight.

A second way of solving will be to reason that since the volume is going to be changed by the change in pressure, the new volume will be the result of a pressure effect on the old volume. That is,

new volume = old volume × a pressure effect

The "pressure effect" will be the ratio of 800 mm to 700 mm. We know, since the pressure has gone up from 700 to 800, that the volume must have gone down. That is, the answer (the new volume) must be less than 500 cc. If the new volume is going to be less than 500, the "pressure effect" by which we multiply the old volume (500 cc) in the formula

new volume = old volume × pressure effect

must be less than 1. But the pressure effect is the ratio of 800 mm to 700 mm. Now 800/700 = 1.14, which is greater than 1, and 700/800 = 0.875, which is less than 1. (If the numerator is greater than the denominator the value of the ratio is greater than 1. If the numerator is smaller than the denominator the value of the ratio is less than 1.) Therefore, to get the correct new volume, we will multiply 500 cc by 700 mm/800 mm. That is,

$$\text{new volume} = 500 \text{ cc} \times \frac{700 \text{ mm}}{800 \text{ mm}} = 437 \text{ cc}$$

This, you will notice, is exactly the same combination of terms as we obtain in the final step of our solution by substitution in the formula. Of the two methods, the solution by reasoning is the more desirable one.

6.2 The volume of a gas is 800 liters at 750 mm and 20°C. What volume in liters will it occupy at 710 mm and 20°C?

6.3 The volume of a gas is 20 cu ft at 600 mm and 0°C. What volume in cu ft will it occupy at 1000 mm and 0°C?

6.4 A mass of hydrogen gas has a volume of 1200 liters at a pressure of one atmosphere. To what value in atmospheres must the pressure be changed if the volume is to be reduced to 2.00 liters?

6.5 A steel cylinder in which compressed helium gas is kept has an interior volume of 1.20 cu ft. When it was opened and the gas was allowed to escape into a dry storage tank at a pressure of 800 mm, the gas occupied a volume of 420 cu ft. Under what pressure in mm was the helium gas stored in the cylinder?

The effect of change of temperature

When a gas is heated it expands. That is, when the temperature goes up the volume increases. This is *direct proportion*, so we can say that *the volume of a gas varies directly as the temperature.* When the temperaturure increases the volume increases. When the temperature decreases the volume decreases.

Absolute zero

Suppose we were to take exactly 273 cc of some gas, helium for example, at exactly 0°C and at some constant pressure, say 750 mm. If we cool this gas, keeping the pressure constant, and note how the volume changes, the results obtained will be as summarized in the following table:

When the Temperature is	The Volume of the Gas is
0°C	273 cc
−1°C	272 cc
−2°C	271 cc
−3°C	270 cc
−4°C	269 cc
−5°C	268 cc
−10°C	263 cc
−20°C	253 cc
−100°C	173 cc
−200°C	73 cc
−250°C	23 cc
−270°C	3 cc

Notice that the volume has decreased by $\frac{1}{273}$ of the volume at 0°C for each degree drop in temperature. If the volume were to keep on shrinking at the same rate, then at −273°, the volume would be zero. Since cooling below −273°C would, on the basis of these data, give a volume less than zero, and since it is reasonable to assume that there is no such thing as negative volume, it can be concluded that −273°C is the lowest temperature theoretically possible. Therefore, −273°C is designated as *absolute zero*. (The exact value is −273.15°C. The value, −273°, is sufficiently accurate for all calculations in this book.) It is represented by the notation, 0°K. K means degrees Kelvin. Notice that absolute zero is 273° below Celsius zero. Zero degrees Celsius is therefore 273°K, and 273°C is 546°K. In other words, to change Celsius temperature to absolute temperature expressed in degrees Kelvin we simply add 273° to the Celsius reading.

Charles' law

If we had taken the 273 cc of helium gas at 0°C and heated it we would have found that, at 273°C, the volume would have increased to 546 cc. That is, at 0°C the volume is 273 cc while at 273°C it is 546 cc, or just twice as great. But 0°C is the same as 273°K (absolute) and 273°C is the same as 546°K.

When we double the absolute temperature (from 273°K to 546°K) we double the volume (from 273 cc to 546 cc). That means that *the volume of a mass of gas, at constant pressure, is directly proportional to the absolute temperature. This is Charles' law.* The formula for Charles' law is

$$\frac{V_1}{V_2} = \frac{T_1}{T_2}$$

Standard temperature

Zero degrees Celsius is referred to as *standard temperature.* Standard temperature (0°C) and standard pressure (760 mm) are commonly referred to by the notation STP. The following problems illustrate the application of Charles' law.

PROBLEMS

6.6 A gas occupies a volume of 200 cc at 0°C and 760 mm. What volume will it occupy at 100°C and 760 mm?

Solution: Since the pressure is constant, this is a problem involving temperature change only. We can solve by substituting in the Charles' law formula,

$$\frac{V_1}{V_2} = \frac{T_1}{T_2}$$

$V_1 = 200$ cc; $T_1 = 273°K(0 + 273°)$; $T_2 = 373°K(100 + 273°)$.

$$\frac{200 \text{ cc}}{V_2} = \frac{273°K}{373°K}$$

$$V_2 = \frac{200 \text{ cc} \times 373°K}{273°K} = 273 \text{ cc}$$

This problem can also be solved by applying the following logical reasoning: since the change of absolute temperature will change the volume, we can say that

new volume = old volume × a temperature effect

The *temperature effect* is the ratio of the old and new *absolute* temperatures. In this particular problem the temperature goes up. Therefore, the volume must increase. To get a larger volume, the old volume must be multiplied by a factor greater than 1. That means that the larger of the two absolute temperatures must go in the numerator. So we write

$$\text{new volume} = 200 \text{ cc} \times \frac{373°K}{273°K} = 273 \text{ cc}$$

6,7 The volume of a gas is 600 cc at 12°C. What volume will it occupy at 0°C, pressure remaining constant?

6.8 The volume of a gas is 20.0 cu ft at −20°C and 750 mm. What volume will it occupy at 20°C and 750 mm?

6.9 A mass of helium gas occupies a volume of 100 liters at 20°C. If the volume occupied by the gas is tripled, to what must the temperature be changed in order to keep the pressure constant?

Change of pressure and temperature

By combining the formulas for Boyle's law and Charles' law we get the general formula expressing the change of volume with change of pressure and change of temperature.

(1)
$$\frac{V_1}{V_2} = \frac{T_1 \times P_2}{T_2 \times P_1}$$

Equation (1) can be transposed to give Equation (2).

(2)
$$\frac{V_1 P_1}{T_1} = \frac{V_2 P_2}{T_2}$$

This equation is very useful because it gives us the relationship between any two sets of conditions of a given quantity of gas. Each set of conditions is specified by three variables (P, V, and T), so that this equation involves six variables. Given any five of these variables the sixth can be calculated. For example, if we know the volume of a quantity of gas at one temperature and pressure we can calculate its volume at any other temperature and pressure. Or if we know the temperature of a quantity of gas which occupies a certain volume at a certain pressure we can calculate its temperature when it occupies any volume at any pressure, and so on.

It should be noted that no gas obeys Boyle's and Charles' laws perfectly over all ranges of temperature and pressure; that is, no gas is "perfect." The higher the temperature and the lower the pressure the more nearly "perfect" every gas is in its response to changes of temperature and pressure. It will be assumed in the problems in this book that all gases are perfect and that Boyle's and Charles' laws are obeyed at the temperatures and pressures encountered.

PROBLEMS

6.10 The volume of a gas is 200 liters at 12°C and 750 mm. What volume will it occupy at 40°C and 720 mm?

Solution:

$$12°C = 285°K$$
$$40°C = 313°K$$

new volume = old volume × temperature effect × pressure effect

Since the temperature increases, the volume must thereby be increased. Therefore the temperature effect is greater than 1, so the larger absolute temperature goes in the numerator. The pressure decreases, therefore the volume must be increased, so the larger pressure must go in the numerator of the pressure effect factor. This gives us the relation

$$\text{new volume} = 200 \text{ liters} \times \frac{313°K}{285°K} \times \frac{750 \text{ mm}}{720 \text{ mm}} = 229 \text{ liters}$$

We can also solve by substituting in the general formula

$$\frac{V_1 P_1}{T_1} = \frac{V_2 P_2}{T_2}$$

$$\frac{200 \text{ liters} \times 750 \text{ mm}}{285°K} = \frac{V_2 \times 720 \text{ mm}}{313°K}$$

$$V_2 = 200 \text{ liters} \times \frac{313°K \times 750 \text{ mm}}{285°K \times 720 \text{ mm}} = 229 \text{ liters}$$

6.11   The volume of a dry gas is 50.0 liters at 20°C and 742 mm. What volume will it occupy at STP?

6.12   At what temperature will a mass of gas whose volume is 150 liters at 12°C and 750 mm occupy a volume of 200 liters at a pressure of 730 mm?

6.13   A gas in a 10.0-liter steel cylinder is under a pressure of 4.00 atm at 22.0°C. If the temperature is raised to 600°C what will be the pressure on the gas?

Note: From the combined gas law formula

$$\frac{V_1 P_1}{T_1} = \frac{V_2 P_2}{T_2}$$

it follows that, when the volume is kept constant, the pressure exerted by a given mass of gas is directly proportional to the absolute temperature.

6.14   A mass of helium gas contained in a 700-ml vessel at 710 mm pressure and 22°C is transferred to a 1000-ml vessel at 110°C. What is the pressure in the 1000-ml vessel?

➡ **6.15**   A cylinder contains helium gas at a pressure of 1470 lb per sq in. When a quantity of helium gas which occupies a volume of 4 liters at a pressure of 14.7 lb per sq in. is withdrawn, the pressure in the tank drops to

1400 lb per sq in. (Temperature remains constant.) Calculate the volume of the tank.

Solution: Let X = the volume, in liters, of the cylinder. Since the pressure drops from 1470 to 1400 when the quantity of gas is removed, this quantity must have exerted a pressure of 70 lb when confined to a volume of X liters at the constant temperature. Since we know that this quantity of gas exerted a pressure of 14.7 lb when confined to a volume of 4.00 liters we can solve for X by applying Boyle's law. Note that we are concerned only with the pressure exerted by the gas that was *removed*, since *it* is the gas which is confined in the 4.00-liter vessel.

Molar volume of a gas

If the volumes occupied at very high temperatures and very low pressures by *one-mole* (one-gram molecular weight) samples of a great many different gases are converted to volumes at STP by application of Boyle's and Charles' laws, the average value of these volumes at STP is 22.4 liters. This leads to the very important generalization that *one gram-molecular weight* (one mole) *of any gas occupies a volume of 22.4 liters at standard temperature and pressure* (0°C and 760 mm). The figure, 22.4 liters, is referred to as the *molar volume* or the *gram-molecular* volume. Since 1 mole of any gas occupies 22.4 liters at STP, it follows that the weight in grams of 22.4 liters of any gas at STP is the molecular weight of that gas.

Since one mole of any gas occupies a volume of 22.4 liters at STP, and since the gas laws apply equally to all gases, it follows that, *at the same temperature and pressure, equal volumes of all gases contain the same number of moles of gas.*

This leads to the conclusion that, *for different volumes of gases at the same temperature and pressure, the number of moles of a gas is directly proportional to the volume of that gas.* Thus, if at constant temperature and pressure, gas A occupies a volume of 100 liters and gas B occupies a volume of 50 liters, the number of moles of A is twice the number of moles of B.

It can also be concluded that *at a given temperature and in a given volume of gas, the pressure is directly proportional to the number of moles of gas.* Since the number of moles of gas in a given volume is the *molar concentration* of that gas, it can be stated that, *at a given temperature, the pressure exerted by a gas is directly proportional to its molar concentration.*

The fact that 1 mole of any gas occupies a volume of 22.4 liters at STP enables one, with the aid of Boyle's and Charles' laws, to calculate the mass of any volume of a gas and the volume of any mass of that gas under any conditions of temperature and pressure provided, of course, that its true chemical formula and, hence, its true molecular weight is known.

It should be noted that He, Ar, Ne, Kr, and Xe are monatomic gases; their molecular weights equal their atomic weights. The other elementary gases, H_2, O_2, N_2, F_2, Cl_2, Br_2, and I_2, form diatomic molecules.

PROBLEMS

6.16 What volume in liters will 2.71 moles of He gas occupy at STP?
Solution: The molar volume of a gas at STP is 22.4 liters. That means that 1 mole of He will occupy a volume of 22.4 liters at STP.

$$2.71 \text{ moles} \times \frac{22.4 \text{ liters}}{1 \text{ mole}} = 60.7 \text{ liters}$$

Note that "moles" cancel, leaving the answer in "liters."

6.17 What volume in liters will 0.362 moles of CO gas occupy at STP?

6.18 A volume of 50.0 liters of H_2S gas, measured at STP, is how many moles of H_2S?

$$\frac{50.0 \text{ liters}}{22.4 \text{ liters/mole}} = 2.23 \text{ moles}$$

6.19 How many moles are there in 18.3 liters of HF gas at STP?

6.20 What volume in liters will 100 g of CH_4 gas occupy at STP?
Solution: In solving this problem our first question is "How many *moles* of CH_4 are there?" We know that one mole of a gas occupies a volume of 22.4 liters at STP. Therefore, if we know how many moles of CH_4 gas there are in 100 g of CH_4, we can multiply this number of moles by 22.4 liters/mole.
The molecular weight of CH_4 is 16. Therefore, 100 g of CH_4 is 100/16 moles

$$\frac{100}{16} \text{ moles} \times \frac{22.4 \text{ liters}}{1 \text{ mole}} = 140 \text{ liters}$$

The complete calculation, in one operation, is:

$$\frac{100 \text{ g of } CH_4}{16.0 \text{ g of } CH_4/\text{mole of } CH_4} \times \frac{22.4 \text{ liters of } CH_4}{1 \text{ mole of } CH_4} = 140 \text{ liters of } CH_4$$

Note that "g of CH_4" and "mole of CH_4" cancel.

6.21 What volume in liters will 40.0 g of HCl gas occupy at STP?

6.22 Calculate the mass in grams of 40.0 liters of NO gas at standard conditions.
Solution: In solving this problem our first question is "How many *moles* of NO do we have?" If we know the number of moles, we can multiply this number of moles by the number of grams of NO in 1 mole.

22.4 liters of NO at STP is 1 mole. Therefore, 40.0 liters at STP is 40.0/22.4 moles of NO.

The molecular weight of NO is 30.0.

6.23   Calculate the mass in grams of 120 liters of CO_2 gas at STP.

6.24   How many grams of nitrogen are there in 100 liters of NO measured at STP?

Solution: In solving this problem we first ask "How many moles of NO are there?" Knowing the number of moles of NO we can easily calculate the number of grams of N, since 1 mole of NO contains 14.0 g of N. The formula, NO, tells us that there is 1 mole (1 gram-atom) of N in 1 mole of NO. One mole of N has a mass of 14.0 g. 100 liters of NO is 100/22.4 moles of NO and 100/22.4 moles of NO contain 100/22.4 moles of N.

$$\frac{100}{22.4} \text{ moles of N} \times \frac{14.0 \text{ g of N}}{1 \text{ mole of N}} = 62.5 \text{ g of N}$$

The solution, in one operation, is

$$\frac{100 \text{ liters of NO}}{22.4 \text{ liters of NO/mole of NO}} \times \frac{1 \text{ mole of N}}{1 \text{ mole of NO}}$$

$$\times \frac{14.0 \text{ g of N}}{1 \text{ mole of N}} = 62.5 \text{ g of N}$$

Note that all units except "g of N" cancel.

6.25   A volume of 65.0 liters of H_2S, measured at STP, contains how many grams of S?

6.26   How many liters of SO_2, measured at STP, will contain 50.0 g of S?

6.27   A volume of 2.00 liters of a gas, measured at STP, has a mass of 5.71 g. Calculate the approximate molecular weight of the gas.

Solution: The molecular weight of a gas is the mass in grams of 1 mole of the gas. One mole of a gas occupies a volume of 22.4 liters at STP. Therefore, the molecular weight of a gas is the mass in grams of 22.4 liters at STP.

We can reason that, since 2.00 liters have a mass of 5.71 g, the mass of 1 liter will be 5.71/2.00 g and the mass of 22.4 liters will be 22.4 × 5.71/2.00 g. The calculation, in one operation, is:

$$\frac{5.71 \text{ g}}{2.00 \text{ liters}} \times \frac{22.4 \text{ liters}}{1 \text{ mole}} = 64.0 \text{ g/mole}$$

Since the gas laws are not exact, the molecular weight calculated in this manner is not exact.

6.28 A volume of 6.82 liters of a gas, measured at STP, has a mass of 9.15 g. Calculate the approximate molecular weight of the gas.

6.29 What volume will 100 g of chlorine gas occupy at STP?

Solution: The formula for a molecule of chlorine gas is Cl_2. The atomic weight of chlorine is 35.5 and its molecular weight is 71.0. Therefore, 100 g of Cl_2 gas is 100/71.0 moles and occupies a volume of $100/71.0 \times 22.4$ liters at STP.

➤ **6.30** Flask A contains 20.0 liters of CH_4 gas. Flask B contains 30.0 liters of CO gas. Each volume is measured at the same temperature and pressure. If flask A contains 6.18 moles of CH_4, how many moles of CO are there in flask B?

Solution: At the same temperature and pressure, the number of moles of gas is directly proportional to the volume of the gas. Since the volume of B is 1.50 times the volume of A the number of moles of CO in B will be 1.50 times the moles of CH_4 in A.

➤ **6.31** If 25.0 g of CH_4 gas occupy a volume of 30.0 liters at a certain temperature and pressure, what volume in liters will 50.0 g of CO_2 gas occupy at the same temperature and pressure?

➤ **6.32** A volume of 40.0 liters of pure O_2 gas, measured at a certain temperature and pressure, was found to contain 40.0 g of oxygen. A volume of 60.0 liters of CH_4 gas, measured at the same temperature and pressure, will contain how many grams of CH_4?

The ideal gas law equation, $PV = nRT$

As has already been stated, the combined effect on the volume of a gas of change of temperature and pressure can be expressed by the equation

(1)
$$\frac{P_0 V_0}{T_0} = \frac{PV}{T}$$

This equation tells us that if we know the volume V_0, which a specific mass of gas occupies at pressure, P_0, and absolute temperature, T_0, we can calculate the volume, V, which this mass of gas occupies at some other pressure, P, and temperature, T.

As has been pointed out in the early part of this chapter we *do know* that a specific amount, *one mole*, of *an ideal gas* occupies a volume of 22.4 liters at 273°K and 760 mm (1 atm). Since we are assuming in our calculations that all gases are ideal, we can state that *one mole* of *any gas* occupies a volume of *22.4 liters* at 273°K and *760 mm*. That means that, in Equation (1) above, for one mole of any gas $V_0 = 22.4$ liters, $P_0 = 760$ mm, and $T_0 = 273°$. That means that, for one mole of any gas, $P_0 V_0/T_0$ is always equal to

(760 mm × 22.4 liters)/273 deg; that is, P_0V_0/T_0 is constant. If we calculate the value of this constant we find that, when P_0 is 760 mm,

$$\text{the constant} = \frac{22.4 \text{ liters}}{1 \text{ mole}} \times \frac{760 \text{ mm}}{273 \text{ deg}} = 62.4 \frac{\text{liters} \times \text{mm}}{\text{mole} \times \text{deg}}$$

When P_0 is 1.0 atm

$$\text{the constant} = \frac{22.4 \text{ liters}}{1 \text{ mole}} \times \frac{1.0 \text{ atm}}{273 \text{ deg}} = 0.082 \frac{\text{liters} \times \text{atm}}{\text{mole} \times \text{deg}}$$

If we call this constant R, then Equation (1) becomes

(2)
$$R = \frac{PV}{T}$$

It was stated that V_0 is the volume occupied by *one mole* at pressure P_0 and absolute temperature T_0. For n moles of gas the volume will be $n \times V_0$. Therefore, for n moles of gas,

(3)
$$\frac{PV}{T} = nR$$

and

(4)
$$PV = nRT$$

Equation (4) is the *ideal gas law equation*.

It is called the equation of state because for any quantity of an ideal gas it gives the relationship between pressure, volume, and temperature.

An alternate form of this equation can be written by substituting in the expression which defines a mole,

(5)
$$n = \frac{g}{MW}$$

where g is the mass of the sample in grams and MW is the molecular weight, giving,

(6)
$$PV = \frac{g}{MW} RT$$

When employing ideal gas equations to solve problems, it is essential that the correct units be used. Temperature must always be in °K and the units of pressure and volume must match the units of R. Thus, if $R = 0.082$ liter-atm/mole-deg then pressure must be in atm and volume must be in liters.

PROBLEMS

6.33 What volume will 45.0 g of CH_4 gas occupy at 27°C and 800 mm?

Solution: We will solve by substituting in the equation, $PV = nRT$.

$$n = \text{moles of CH}_4 = \frac{45.0 \text{ g}}{16.0 \text{ g/moles}} = \frac{45.0}{16.0} \text{ moles}$$

$$P = 800 \text{ mm}$$

$$T = 300°\text{K}$$

$$V = \frac{nRT}{P} = \frac{45.0}{16.0} \text{ moles} \times \frac{62.4 \dfrac{\text{liters} \times \text{mm}}{\text{mole} \times \text{deg}} \times 300 \text{ deg}}{800 \text{ mm}}$$

$$= 65.6 \text{ liters}$$

Note that moles, mm, and deg cancel.

6.34 The hydrogen gas in a 2.00-liter steel cylinder at 25°C was under a pressure of 4.00 atm. How many moles of H_2 were in the cylinder?

Solution:

$$PV = nRT$$

$$n = \frac{PV}{RT} = \frac{4.00 \text{ atm} \times 2.00 \text{ liters}}{0.082 \dfrac{\text{liters} \times \text{atm}}{\text{mole} \times \text{deg}} \times 298 \text{ deg}} = 0.33 \text{ mole}$$

Note that liters, atm, and deg cancel.

6.35 What pressure, in atmospheres, will 26.0 g of He gas exert when placed in a 3.24-liter steel cylinder at 200°C?

Solution:

$$PV = \frac{g}{\text{MW}} RT$$

$$P = \frac{gRT}{\text{MW}V} = \frac{26.0 \text{ g}}{\dfrac{4.0 \text{ g}}{1 \text{ mole}}} \times 0.082 \frac{\text{liters} \times \text{atm}}{\text{mole} \times \text{deg}} \times \frac{473 \text{ deg}}{3.24 \text{ liters}}$$

$$= 78 \text{ atm}$$

Note that g, liters, deg, and moles cancel.

6.36 Calculate the mass in grams of the pure H_2S gas contained in a 60.0-liter cylinder at 20°C under a pressure of 2.00 atm.

Solution:

$$PV = \frac{gRT}{\text{MW}}$$

$$g = \frac{PV\text{MW}}{RT} = \frac{2.00 \text{ atm} \times 60.0 \text{ liters} \times 34.1 \text{ g/mole}}{0.082 \dfrac{\text{liters} \times \text{atm}}{\text{mole} \times \text{deg}} \times 293 \text{ deg}}$$

$$= 170 \text{ g}$$

Note that mole, atm, liters, and deg cancel.

6.37   A 32.4-g sample of an ideal gas in a 12.0-liter container at 25°C exerted a pressure of 1.50 atm. Calculate its molecular weight.

Solution:

$$MW = \frac{gRT}{PV} = \frac{32.4 \text{ g} \times 0.082 \dfrac{\text{liters} \times \text{atm}}{\text{mole} \times \text{deg}} \times 298 \text{ deg}}{1.50 \text{ atm} \times 12.0 \text{ liters}}$$

$$= 44.0 \text{ g/mole}$$

6.38   Calculate the mass in grams of 200 liters of CO_2 gas at 20°C and 746 mm.

6.39   Calculate the volume in liters occupied by 200 g of CO gas at 22°C and 740 mm.

6.40   A 25.0-liter cylinder contains 14.2 moles of helium gas at 40°C. What is the pressure in atmospheres of the helium gas?

6.41   How many grams of carbon are there in 22.4 liters of CH_4 gas at 2.0 atm and 546°K?

6.42   A cylinder containing 85 g of steam at 200°C shows a pressure of 4.0 atm. What is the volume of the cylinder in liters?

6.43   A volume of 22.40 liters of H_2S gas measured at 273°C and 4.000 atm will contain how many molecules of H_2S?

6.44   A volume of 120 cc of a dry gaseous compound, measured at 22°C and 742 mm, has a mass of 0.820 g. Calculate the approximate molecular weight of the gas.

6.45   How many liters of C_2H_2 gas, measured at 25°C and 745 mm, will contain 10.0 g of carbon?

6.46   Twelve liters of dry nitrogen gas, measured at 22°C and 741 mm, have a mass of 13.55 g. Calculate the formula for a molecule of nitrogen gas.

→ **6.47**   At a given temperature 12.0 g of CO gas was placed in one evacuated container and 40.0 g of CH_4 gas was placed in a second evacuated container. The pressure of the CO in its container was 800 mm, of the CH_4 was 600 mm. Calculate the relative volumes of the two containers.

Solution: For each gas, $PV = nRT$. Since R and T are the same for both gases, $\dfrac{PV}{n}$ for $CH_4 = \dfrac{PV}{n}$ for CO. The ratio of the two volumes can then be calculated.

→ **6.48**   A 2.0-liter sample of helium gas measured at 27°C is under twice the pressure and contains 3 times as many molecules of gas as a sample of H_2 gas measured at 227°C. Calculate the volume in liters of the sample of H_2 gas.

➡ **6.49** If 0.200 g of H_2 is needed to inflate a balloon to a certain size at 20°C, how many grams will be needed to inflate it to the same size at 30°C? Assume elasticity of balloon is the same at 20°C and 30°C.

Densities of gases

The mass of a definite volume of gas is referred to as the *density* of the gas. The definite volume that is commonly used in expressing the density of a gas is either 22.4 liters or 1 liter. In the first instance the density of any gas is the mass in grams of 22.4 liters of that gas; it is expressed as so many grams per 22.4 liters. But the mass in grams of 22.4 liters at STP is the gram-molecular weight of the gas. It follows, therefore, that the density of a gas at STP in grams per 22.4 liters is equal numerically to its gram-molecular weight. Since the density of any gas at STP is equal, numerically, to its molecular weight, it follows that *the densities of two gases at the same temperature and pressure are to each other as their molecular weights.* The density of a gas in grams per liter at STP is obtained very simply by dividing the gram-molecular weight by 22.4 liters.

The density of a gas can also be expressed in units of moles per liter.

If the conditions are not STP; then the density of the gas in g/liter can be calculated if the pressure, temperature, and molecular weight of the gas are known, by using a variation of the ideal gas equation

(1) $$PV = \frac{gRT}{MW}$$

rearranging

(2) $$\frac{g}{V} = \frac{PMW}{RT} = d$$

The density of a gas can also be expressed in units of moles/liter. Then the equation:

(3) $$\frac{n}{V} = \frac{P}{RT}$$

can be used.

PROBLEMS

6.50 What is the density of C_2H_2 gas at STP?

Solution: The density of a gas is the mass of a unit volume of the gas. The volume normally selected is either the molar volume, 22.4 liters, or 1 liter. One mole of C_2H_2 at STP weighs 26.0 g and occupies 22.4 liters. Therefore, the density is 26.0 g per 22.4 liters or 1.16 g/liter.

6.51   Calculate the density of CO gas at STP in g/liter.

6.52   Calculate the density of SO_2 gas at 40°C and 730 mm.

Solution: The molecular weight of SO_2 is 64.0 g/mole. Substituting in Equation (2)

$$d = \frac{PMW}{RT} = \frac{730 \text{ mm} \times 64.0 \text{ g/mole}}{62.4 \dfrac{\text{liters} \times \text{mm}}{\text{mole} \times \text{deg}} \times 313 \text{ deg}} = 2.39 \frac{\text{g}}{\text{liter}}$$

6.53   Calculate the density of CH_4 gas at 120°C and 0.600 atm.

6.54   At a given temperature and pressure, which is heavier, 20.0 liters of C_2H_6 or 20.0 liters of O_2? How many times as heavy?

Solution: At the same temperature and pressure the densities of two gases are to each other as their molecular weights. At the same temperature and pressure equal volumes of two gases contain the same number of moles of gas.

6.55   If the density of He gas is 0.026 g/liter at a certain temperature, what is the density of Ne gas at the same temperature and pressure?

6.56   A gas which has a density of 2 moles per liter under a pressure of 44.8 atm will be at what temperature?

➡ **6.57**   A given mass of helium gas, which occupied a volume of 2 liters at STP, was allowed to expand to a volume of 4 liters by changing the temperature and pressure. What was its density at the new temperature and pressure?

Partial pressure: Dalton's law of partial pressures

The reason why a gas can occupy space and exert pressure is that the molecules are in motion. We have learned that one mole of any gaseous substance contains 6.023×10^{23} molecules. Since one mole of any gas occupies a volume of 22.4 liters at 0°C and 760 mm pressure, it follows that when 6.023×10^{23} molecules of a gas are confined to a volume of 22.4 liters at 0°C they will exert a total pressure of 760 mm. Now if only half that many molecules are placed in a 22.4-liter container at 0°C, they will exert only half that pressure, or 380 mm.

Suppose we place one half a mole of CH_4 gas and one half a mole of CO_2 gas together in a 22.4-liter container at 0°C. (CO_2 and CH_4 do not react with each other.) The half mole (3.011×10^{23} molecules) of CH_4 will exert a pressure of 380 mm, and the half mole of CO_2 gas will also exert a pressure of 380 mm. The two gases together will exert a pressure of 760 mm.

The pressure of 380 mm exerted by the half mole of CH_4 is referred to as the *partial pressure* of the CH_4, and the pressure of 380 mm exerted by the CO_2 is the *partial pressure* of the CO_2. One half, or 50 mole percent, of the molecules are CH_4 molecules and one half, or 50 mole percent, are CO_2 molecules. CH_4 provides half the molecules and exerts half the pressure; CO_2 provides the other half of the molecules and exerts the other half of the pressure. We can conclude from this that, *in a mixture of gases, the partial pressure of a gas is directly proportional to its mole percent* and that, *in a mixture of gases, the partial pressure of a gas in the mixture is equal to the product of its mole percent and the total pressure of the mixture of gases.* In a mixture of gases *the total pressure is the sum of the partial pressures of the gases in the mixture.*

In a mixture of CH_4 and CO_2 in which half the molecules are CH_4 and half are CO_2, the *mole fraction* of the CH_4 is 0.5 and the mole fraction of the CO_2 is also 0.5. *Mole fraction* is, therefore, the decimal equivalent of mole percent. That is

$$\text{mole fraction of } A = \frac{\text{moles of } A}{\text{total moles}}$$

In a mixture of 2 moles of CH_4 and 3 moles of CO_2 the mole fraction of CH_4 is 0.4 and the mole fraction of CO_2 is 0.6. If the total pressure of a mixture of 2 moles of CH_4 and 3 moles of CO_2 is 800 mm, the partial pressure of the CH_4 (P_{CH_4}) is found to be 320 mm and the partial pressure of the CO_2 (P_{CO_2}) is 480 mm. But 320 mm is 0.4×800 mm and 480 mm is 0.6×800 mm. That means that *in a mixture of gases, the partial pressure of a gas is equal to its mole fraction times the total pressure.*

$$P_A = \text{mole fraction of } A \times P_{\text{total}}$$

(P_A means partial pressure of A.)

From what has been stated above we can conclude that *the partial pressure of a gas is directly proportional to its mole fraction.* This is *Dalton's law of partial pressures.* This law is sometimes expressed as: *In a mixture of two gases, A and B, the partial pressure of A is to the partial pressure of B as the mole fraction of A is to the mole fraction of B.*

$$\frac{P_A}{P_B} = \frac{\text{mole fraction of } A}{\text{mole fraction of } B}$$

We can conclude that in a mixture of nonreacting gases, for any one of the gases:

$$\text{mole fraction} = \text{pressure fraction}$$

and

$$\text{mole percent} = \text{pressure percent}$$

This means that the relative number of moles of a gas in a mixture is equal to its relative partial pressure. Thus, if a mixture of 0.2 mole of He gas and 0.4 mole of Ne gas exerts a total pressure of 1500 mm, $\frac{1}{3}$ of the pressure (500 mm) will be exerted by the He and $\frac{2}{3}$ of the pressure (1000 mm) will be exerted by the Ne. Likewise, in a mixture of 0.2 mole of CH_4 and X moles of CO_2, if the partial pressure of the CH_4 is 18 mm and of the CO_2 is 45 mm, there must be $\frac{45}{18} \times 0.2$ mole or 0.5 mole of CO_2.

It should be emphasized that, *in a mixture of nonreacting ideal gases, each gas behaves exactly as it would if it alone were present. The other gases have no effect on its behavior. Its partial pressure is the pressure that it would exert if it alone occupied the volume.*

PROBLEMS

6.58 In a gaseous mixture of CH_4 and C_2H_6 there are twice as many moles of CH_4 as C_2H_6. The partial pressure of the CH_4 is 40 mm. What is the partial pressure of the C_2H_6?

Solution: The partial pressure of C_2H_6 is directly proportional to the number of moles of C_2H_6.

6.59 In a gaseous mixture of CH_4, C_2H_6, and CO_2, the partial pressure of the CH_4 is 50% of the total pressure. What is the mole fraction of the CH_4 in the mixture?

Solution: Mole fraction = pressure fraction.

6.60 In a mixture of gases, A, B, and C, the mole fraction of A is 0.25. What fraction of the total pressure is exerted by gas A?

6.61 A mixture of 2.0×10^{23} molecules of N_2 and 8.0×10^{23} molecules of CH_4 exerts a total pressure of 740 mm. What is the partial pressure of the N_2?

6.62 In a mixture of 0.200 moles of CO, 0.300 moles of CH_4, and 0.400 moles of CO_2 at 800 mm, what is the partial pressure of the CO?

6.63 If the 200 cc of H_2 gas contained in a cylinder under a pressure of 1200 mm are forced into a cylinder whose volume is 400 cc and which already contains 400 cc of CH_4 gas under a pressure of 800 mm, what will be the total pressure exerted by the mixture of H_2 and CH_4 gases in the 400 cc cylinder if the temperature is constant?

Solution: The total pressure will be the sum of the partial pressures.

6.64 The partial pressures of the four gases contained in a 6-liter cylinder at 1007°C were: $CO_2 = 63.1$ atm; $H_2 = 21.1$ atm; $CO = 84.2$ atm; $H_2O = 31.6$ atm. How many grams of CO_2 gas were there in the cylinder?

Solution: Since, in a mixture of gases, each gas exerts the pressure that it would exert if it alone occupied the volume, we can ignore the other gases and calculate, by the use of the equation, $PV = nRT$, how many grams of CO_2 gas would exert a pressure of 63.1 atm when placed in a 6-liter cylinder at 1007°C.

➡ **6.65** A reaction vessel contained 5 liters of a mixture of N_2 and O_2 gases at 25°C and 2 atm pressure. The oxygen in the mixture was completely removed by causing it to oxidize an excess of electrically heated zinc wire contained in the vessel to solid ZnO. (Solid ZnO is nonvolatile.) The pressure of the nitrogen gas that remained in the vessel, measured at 25°C, was 1.5 atm. What was the mole percent of oxygen in the original mixture; the mass percent?

Solution: $P_T = P_{O_2} + P_{N_2}$, $2 = P_{O_2} + 1.5$, $P_{O_2} = 0.5$ atm. When V and T are constant, the number of moles of a gas is directly proportional to its partial pressure. Therefore,

$$\frac{\text{moles of O}_2}{\text{moles of N}_2} = \frac{0.50 \text{ atm}}{1.5 \text{ atm}}. \qquad \text{Mole \% of O}_2 = \frac{0.50}{2.0} \times 100\% = 25\%.$$

To convert mole percent to mass percent convert moles of O_2 and N_2 to grams. Mass % of $O_2 = \dfrac{0.50 \times 32}{(0.50 \times 32) + (1.5 \times 28)} \times 100\% = 28\%$.
The fact that we do not know the exact number of moles of O_2 and N_2 is not important. The significant fact is that the number of moles of N_2 is three times the number of moles of O_2. Once we know the mole ratios in which substances are present, the mass ratios and mass percents can then be calculated.

➡ **6.66** In a gaseous mixture of equal grams of CH_4 and CO what is the ratio of moles of CH_4 to CO? What is the mole fraction of CH_4?

Solution: Let $X = $ g of $CH_4 = $ g of CO

Moles of $CH_4 = X/16$. Moles of CO $= X/28$

$$\frac{\text{moles of CH}_4}{\text{moles of CO}} = \frac{X/16}{X/28} = \frac{28}{16}$$

$$\text{Mole fraction of CH}_4 = \frac{\text{moles of CH}_4}{\text{moles of CH}_4 + \text{moles of CO}} = \frac{28}{44} = 0.64$$

Note that, in a mixture of *equal grams* of two gases, the number of *moles* of the gases are in the inverse ratio of their molecular weights. The fact that we do not know the exact number of grams or moles is not important. The significant thing is that we know the mole *ratios*. Mole fractions and mole percents represent *mole ratios*.

➡ **6.67**   A mixture of 50.0 g of oxygen gas and 50.0 g of CH_4 gas is placed in a container under a pressure of 600 mm. What is the partial pressure of the oxygen gas in the mixture?

Solution: The partial pressure of a gas is proportional to its mole fraction. Therefore, we must first find the number of moles of O_2 and CH_4 in the mixture.

$$\text{moles of } O_2 = \frac{50.0 \text{ g}}{32.0 \text{ g/mole}} = 1.56 \text{ moles of } O_2$$

$$\text{moles of } CH_4 = \frac{50.0 \text{ g}}{16.0 \text{ g/mole}} = 3.12 \text{ moles of } CH_4$$

$$\text{total moles} = 1.56 + 3.12 = 4.68 \text{ moles}$$

$$\text{mole fraction of } O_2 = \frac{1.56}{4.68}$$

$$\text{partial pressure of } O_2 = \text{mole fraction of } O_2 \times \text{total pressure}$$

$$= \frac{1.56}{4.68} \times 600 \text{ mm} = 200 \text{ mm}$$

➡ **6.68**   In a gaseous mixture of equal grams of C_2H_6 and CO_2 the partial pressure of the C_2H_6 is 22 mm. What is the partial pressure of the CO_2?

➡ **6.69**   How many grams of pure CO gas would have to be mixed with 40 g of pure CH_4 gas in order to give a mixture in which the partial pressure of the CO is equal to the partial pressure of the CH_4?

➡ **6.70**   Exactly 1.100 g of carbon dioxide were introduced into a 1-liter flask which contained some pure oxygen. The flask was warmed to 100°C and the pressure was found to be 815 mm. No chemical reaction occurred. Calculate the mass of oxygen in the flask.

➡ **6.71**   For the gaseous compound whose true chemical formula is C_3H_8 calculate:

(a) volume of 3.00 moles at STP

(b) volume of 32.0 g at 18°C and 752 mm

(c) moles in 140 liters of the gas at STP

(d) mass of 100 liters of the gas at 20°C and 700 mm

(e) density of the gas in g/liter at STP

(f) percent of carbon

(g) partial pressure of C_3H_8 in a gaseous mixture of equal masses of C_3H_8 and CH_4 at 750 mm

(h) density of the gas in g/liter at 80°C and 500 mm

Vapor pressure

When a dry gas, hydrogen for example, is collected over water at a given temperature, water evaporates and the molecules of water vapor (gaseous H_2O) *diffuse* into the space occupied by the hydrogen. This evaporation continues until the rate at which water molecules evaporate equals the rate at which they condense. The hydrogen gas is then said to be *saturated* with water vapor. The pressure exerted by this saturated water vapor is its *vapor pressure* at the given temperature. The vessel in which the hydrogen was collected now contains a mixture of hydrogen and water vapor, and each gas (hydrogen and water vapor) exerts its own partial pressure. The total pressure is the sum of the two partial pressures. It follows, therefore, that when a gas is collected over water only a part of the total pressure is being exerted by the gas itself; some of the pressure is being exerted by the water vapor. Since the outside pressure which is exerted on a confined mixture of gases counterbalances the sum of the pressures exerted by each gas, it follows that when a gas is collected over water, the actual pressure on the gas itself is the total outside pressure *minus* the vapor pressure of water at the particular temperature. That means that when Boyle's law is used in the calculation of the volume of a gas which is collected over water, the partial pressure of the water vapor must be subtracted from the total pressure under which the gas is collected to give the actual pressure on the gas itself. Table 1 in the Appendix gives the pressure of water vapor at various temperatures.

PROBLEMS

6.72 The volume of a dry gas is 600 cc at 25°C and 750 mm. What volume would this gas occupy if collected over water at 746 mm and 32°C?

Solution: Since the gas is to be collected over water, the vapor pressure of water at 32°C must be subtracted from the barometric pressure of 746 mm.

vapor pressure of water at 32°C = 35.4 mm (See Appendix Table 1)

actual new pressure = 746 − 35.4 = 710.6 mm

Using this pressure of 710.6 mm as the new pressure in place of the 746 mm, we can then solve in the manner shown in Problem 6.10.

$$\text{new volume} = 600 \text{ cc} \times \frac{750 \text{ mm}}{710.6 \text{ mm}} \times \frac{305°\text{K}}{298°\text{K}} = 648 \text{ cc}$$

6.73 The volume of a dry gas at 758 mm and 12°C is 100 cu ft. What volume will this gas occupy if stored over water at 22°C and a pressure of 740 mm?

6.74 A sample of 100 ml of dry gas, measured at 20°C and 750 mm, occupied a volume of 105 ml when collected over water at 25°C and 750 mm. Calculate the vapor pressure of water at 25°C.

6.75 A 1.20-liter volume of a gas was collected over water at 26°C and 1 atm. Calculate the number of moles of gas which was collected.

6.76 What volume in liters will 300 g of oxygen occupy when collected over water at 20°C and 735 mm?

6.77 The vapor pressure of water at 121°C is 2 atm. Some liquid water is injected into a sealed vessel containing air at 1 atm pressure and 121°C and is allowed to come to equilibrium with its vapor. What is the total pressure in the sealed container?

Solution: As long as there is some liquid water present and the temperature remains at 121°C the partial pressure of the water vapor in equilibrium with the liquid water will remain constant at 2 atm. It can not rise above 2 atm or fall below 2 atm. If we neglect the small solubility of air in water, the partial pressure of the air will remain at 1 atm. The total pressure will then be 3 atm.

➡ **6.78** Three 1-liter flasks, all at 27°C, are interconnected with stopcocks which are initially closed. The first flask contains 1 g of H_2O. The second flask contains O_2 at a pressure of 1 atm. The third flask contains 1 g of N_2.

(a) The temperature is kept constant at 27°C. At this temperature the vapor pressure of water is 0.0380 atm. The stopcocks are all opened. When equilibrium is reached, what is the pressure in the flasks?

(b) If the temperature of the whole system is raised to 100°C, what is the pressure?

Solution hint: Will there be any liquid H_2O present in the final system at 27°C? At 100°C?

➡ **6.79** At 37°C liquid A has a vapor pressure of 58.4 mm and liquid B a vapor pressure of 73.6 mm. To a sealed, evacuated container of volume 100 liters maintained at 37°C is added 0.20 mole of gaseous A and 0.50 mole of gaseous B.

What is the total pressure in the container when equilibrium is reached? (Each vapor is completely insoluble in the other liquid.)

Relative humidity

Air which contains a high concentration of water vapor is said to be *humid*. *Humidity* refers, therefore, to the water vapor content of a gas. If air is saturated with water vapor at a given temperature its *relative humidity* is 100%;

that is, it contains all (100%) the water vapor it can hold. If it contains only half the water vapor it can hold (is 50% saturated), its relative humidity is 50%. *Relative humidity* is, therefore, the ratio between the concentration of water vapor actually present and the concentration that would be present if the gas were saturated with water vapor. The simplest way to express the moisture content of a gas is in terms of the water vapor pressure; the relative humidity is then the ratio between the partial pressure of the water vapor and the equilibrium (saturated) vapor pressure at that temperature.

PROBLEMS

6.80   The partial pressure of the water vapor in the air in a room is 11.6 mm at 22°C. What is the relative humidity of the air in the room?

Solution: Relative humidity is the ratio between the partial pressure of the water vapor and the equilibrium (saturated) vapor pressure at that temperature. From Table 1 we learn that the vapor pressure of water at 22°C is 19.8 mm

$$\text{relative humidity} = \frac{11.6 \text{ mm}}{19.8 \text{ mm}} \times 100\% = 58.6\%$$

➡ **6.81**   A weather report gives the temperature as 22.5°C, barometric reading 750 mm, and relative humidity 75%. What is the mole fraction of water vapor in the atmosphere? (See Table 1.)

➡ **6.82**   At 24°C the vapor pressure of H_2O is 22.1 mm of Hg. If the relative humidity in a sealed container is 30% at 24°C and 57% when the temperature is lowered to 13°C, what is the vapor pressure of H_2O at 13°C?

Diffusion and effusion

If a gas is introduced into a container which already contains a gas, the new gas will move through the other gas to fill the container uniformly. This is an example of diffusion. If a container filled with gas has a small hole in one wall and a vacuum on the other side of the wall the gas will pass through the hole. This is called effusion. The rates of both these processes depend on the particular gases involved. For two gases the rates of diffusion (or effusion) are inversely proportional to the square roots of their densities. This generalization is known as Graham's law:

$$\frac{\text{the rate of diffusion of gas A}}{\text{the rate of diffusion of gas B}} = \frac{\sqrt{\text{density of B}}}{\sqrt{\text{density of A}}}$$

PROBLEMS

6.83 Gas A is nine times as dense as gas B. In a given diffusion apparatus and at a certain temperature and pressure, gas B diffuses 15 cm in 10 sec. In the same apparatus and at the same temperature and pressure, how fast will A diffuse?

Solution:

$$\frac{\text{rate of A}}{\text{rate of B}} = \frac{\sqrt{\text{density of B}}}{\sqrt{\text{density of A}}}$$

Substituting in this formula,

$$\frac{\text{rate of A}}{15 \text{ cm}/10 \text{ sec}} = \frac{\sqrt{1}}{\sqrt{9}} = \frac{1}{3}$$

$$\text{rate of A} = \frac{15 \text{ cm}/10 \text{ sec}}{3} = 5 \text{ cm in 10 sec}$$

6.84 The density of CH_4 is 16.0 g per 22.4 liters. The density of HBr is 81.0 g per 22.4 liters. If CH_4 diffuses 2.30 ft in 1 min in a certain diffusion apparatus, how fast will HBr diffuse in the same apparatus at the same temperature and pressure?

6.85 How do the rates of diffusion of HBr and SO_2 compare?
Solution:

$$\frac{\text{rate of A}}{\text{rate of B}} = \frac{\sqrt{\text{density of B}}}{\sqrt{\text{density of A}}}$$

Since the density of a gas is directly proportional to its molecular weight we can write:

$$\frac{\text{rate of A}}{\text{rate of B}} = \frac{\sqrt{\text{molecular weight of B}}}{\sqrt{\text{molecular weight of A}}}$$

6.86 In a given effusion apparatus 15.0 cc of HBr gas were found to effuse in 1 min. How many cc of CH_4 gas would effuse in 1 min in the same apparatus at the same temperature?

6.87 An unknown gas diffuses at the rate of 8 cc/sec in a piece of apparatus in which CH_4 gas diffuses at 12 cc/sec. Calculate the approximate molecular weight of the gas.

➡ **6.88** The average kinetic energy of gas molecules is calculated from the formula, $KE = \frac{1}{2} mv^2$, and is directly proportional to the absolute temperature. If H_2 gas molecules move at an average speed of 1.2 km per sec at 300°K, what will their average speed, in km per sec, be at 1200°K?

➡ **6.89** A gaseous compound contains C, H, and S. One molecule of the compound is known to contain 1 atom of S. A quantity of the compound is mixed, in a constant-volume reaction vessel, with the exact amount of oxygen gas required to oxidize it completely to CO_2, H_2O, and SO_2. The total pressure of the mixture of compound and oxygen, measured at 200°C, is 550 mm. The mixture is ignited, causing the oxidation reaction to occur. The total pressure of the gaseous mixture of CO_2, H_2O, and SO_2 (no liquid), measured at 200°C, is 600 mm. The partial pressure of the H_2O was found to be 1.5 times the partial pressure of the CO_2. Calculate the chemical formula of the compound.

➡ **6.90** A certain compound contains C, H, and O. The gaseous compound has a density of 2.08 g per liter at 127°C and 700 mm pressure.

A quantity of the gaseous compound is mixed with an excess of oxygen gas and placed in an evacuated steel vessel at 127°C, and its partial pressure is measured. An electric spark is then passed, causing all of the compound to be burned to give a gaseous mixture of CO_2, steam (H_2O), and excess O_2. There was no liquid. The sum of the partial pressures of the CO_2 and H_2O, measured at 127°C, was nine times as great as the partial pressure of the gaseous compound in the original mixture of compound and O_2, and the partial pressure of the H_2O was 1.25 times as great as the partial pressure of the CO_2. Calculate the chemical formula of the gaseous compound.

6.91 An unknown gas was found to diffuse 2.2 ft in the same time that methane gas (CH_4) diffused 3.0 ft. The unknown gas was found to contain 80% C and 20% H. Calculate the exact molecular weight of the gas.

6.92 In a given diffusion apparatus 15.0 cc of HBr gas diffused in 1 min. In the same apparatus and under the same conditions 33.7 cc of an unknown gas diffused in 1 min. The unknown gas contained 75% carbon and 25% hydrogen. Calculate its exact molecular weight.

7

Mole relationships in chemical reactions.
I. Stoichiometry.

The equation for a chemical reaction represents (a) the chemical formula for each reactant and each product, and (b) the relative number of moles of each reactant and product in the reaction. The equation is said to be *balanced* when the total number of atoms on the left-hand side equals the total number of atoms on the right. It will always be assumed that when we speak of an equation we mean a *balanced equation*.

The equation $2\,HgO = 2\,Hg + O_2$ tells us that when mercury(II) oxide, HgO, is heated, it decomposes to give mercury and oxygen, and it shows that 2 *moles* of HgO yield 2 *moles* of Hg and 1 *mole* of O_2. It tells us that, whether we heat a small amount of HgO or a large amount, the HgO that we heat and the Hg and O_2 that are formed are in the *ratio* of 2 *moles* of HgO to 2 *moles* of Hg and 1 *mole* of O_2. Furthermore, the equation emphasizes the fact that the total mass of the products is equal to the total mass of the reactants; that is, there are 2 moles of Hg and 2 moles of O on the left and the same number on the right.

We have learned in Chapters 4 and 6 that a mole of a particular substance is a specific mass of that substance and, if it is a gas at a definite temperature and pressure, a specific volume of that substance.

Therefore, having determined from the equation for a reaction the *mole relationships* of the substances involved, we can then calculate *mass* and *volume relationships*.

The process by which the balanced chemical equation is used as a basis for making calculations is called *stoichiometry*. In a common use of the term the *stoichiometry of a reaction* refers to the *mole relationships* represented by the equation for that reaction.

It is important to remember, always, that the *chemical formula* of a substance, when it appears in any chemical reaction, *refers to* **one mole** *of that substance.*

PROBLEMS

7.1 How many *moles* of O_2 will be obtained by heating 3.50 *moles* of $KClO_3$?

Solution: The equation for the liberation of O_2 from $KClO_3$ is

$$2 \text{ KClO}_3 = 2 \text{ KCl} + 3 \text{ O}_2$$
$$\text{2 moles} \qquad \text{2 moles} \quad \text{3 moles}$$

In order to be able to solve this problem at all we must *know* that heating liberates all of the oxygen from $KClO_3$. That is, we must *know the reaction* that occurs.

The equation tells us that 3 *moles* of O_2 are liberated from 2 *moles* of $KClO_3$.

That means that 1 *mole* of $KClO_3$ liberates 1.5 *moles* of oxygen.

Therefore, 3.50 *moles* of $KClO_3$ will liberate 3.50×1.5 or 5.25 *moles*.

It should be pointed out that the equation can be written in the form

$$\text{KClO}_3 = \text{KCl} + 1\tfrac{1}{2} \text{ O}_2$$

This tells us, at a glance, that the *moles* of O_2 liberated is 1.5 times the *moles* of $KClO_3$ heated, and 1.5 times 3.5 moles is 5.25 moles.

It should be emphasized that, if we know that $KClO_3$ liberates all of its oxygen when it is heated, and if we know that oxygen is a diatomic molecule, O_2, we do not actually need to *write* the equation at all. A glance at the formula will tell us that 1 *mole* of $KClO_3$ will liberate 1.5 *moles* of O_2; that is the important thing to know. This is not intended to convey the idea that it is not necessary to know the equation for the reaction on which the problem is based. We must *know* the equation, we must *know what happens*, but it is not always necessary to *write* the equation, particularly in a relatively simple problem.

The simplest, shortest, quickest solution is always desirable.

7.2 When antimony is burned in oxygen the following reaction occurs:

$$4 \text{ Sb} + 3 \text{ O}_2 = 2 \text{ Sb}_2\text{O}_3$$

How many *moles* of oxygen will be needed to burn 18 *moles* of antimony? How many grams of Sb_2O_3 will be formed?

$$4Sb + 3O_2 \rightarrow 2Sb_2O_3$$

Solution: A glance at the equation tells us that 3 *moles* of O_2 are needed for every 4 *moles* of Sb burned. That means that it takes $\frac{3}{4}$ *mole* of O_2 to burn 1 *mole* of Sb. Therefore, $18 \times \frac{3}{4}$ mole or 13.5 *moles* of O_2 will be needed to burn 18 *moles* of Sb.

To find the number of *grams* of Sb_2O_3 we first find the number of *moles* of Sb_2O_3. The equation tells us that the number of *moles* of Sb_2O_3 formed is $\frac{1}{2}$ the number of *moles* of Sb burned. Therefore, 9 moles of Sb_2O_3 will be formed.

$$9 \text{ moles of } Sb_2O_3 \times \frac{291.6 \text{ g of } Sb_2O_3}{1 \text{ mole of } Sb_2O_3} = 2628 \text{ g of } Sb_2O_3$$

7.3   When C_3H_8 is burned in O_2 gas, CO_2 and H_2O are formed as products. If 2.40 *moles* of C_3H_8 are burned in a plentiful supply of oxygen, how many *grams* of H_2O and how many *liters* of CO_2, measured at STP, will be formed?

Solution: The formula, C_3H_8 [*methane*], tells us that, since 1 molecule of C_3H_8 contains 3 atoms of C and 8 atoms of H, while 1 molecule of CO_2 contains 1 atom of C and 1 molecule of H_2O contains 2 atoms of H, *1 mole of C_3H_8 will yield 3 moles of CO_2 and 4 moles of H_2O.* Therefore, 2.40 moles of C_3H_8 will yield 2.40×3 or 7.20 moles of CO_2 and 2.40×4 or 9.60 moles of H_2O.

At STP 1 mole of CO_2 occupies a volume of 22.4 liters and 7.20 moles will occupy a volume of 7.20×22.4 liters or 161 liters.

Since the molecular weight of H_2O is 18, 1 mole of H_2O weighs 18 g and 9.60 moles will weigh 173 g.

Each calculation can be carried out in one operation.

$$2.40 \text{ moles of } C_3H_8 \times \frac{3 \text{ moles of } CO_2}{1 \text{ mole of } C_3H_8}$$

$$\times \frac{22.4 \text{ liters of } CO_2}{1 \text{ 1mole of } CO_2} = 161 \text{ liters of } CO_2$$

$$2.40 \text{ moles of } C_3H_8 \times \frac{4 \text{ moles of } H_2O}{1 \text{ mole of } C_3H_8}$$

$$\times \frac{18.0 \text{ g of } H_2O}{1 \text{ mole of } H_2O} = 173 \text{ g of } H_2O$$

Note that, in analyzing and solving this problem, we did not *write* the equation for the reaction. However we *thought* the significant part of the reaction when we observed that, since one *molecule* of C_3H_8 contains 3 atoms of C and 8 atoms of H, one *mole* of C_3H_8 will yield 3 *moles* of CO_2 and 4 *moles* of H_2O. This relationship was so obvious that it was not necessary to write it down. Had the problem asked us to calculate the number of moles, or grams, of O_2 required to burn the 2.40 moles of C_3H_8 we probably would have needed to *write* the equation for the reaction.

7.4 When C_4H_{10} is burned in excess oxygen the following reaction occurs.

$$2\ C_4H_{10} + 13\ O_2 = 8\ CO_2 + 10\ H_2O$$

How many *grams* of O_2 will be needed to burn 36.0 *grams* of C_4H_{10}? How many *grams* of CO_2 and how many *grams* of H_2O will be formed?

Solution: The equation tells us that 6.5 *moles* of O_2 are consumed for every *mole* of C_4H_{10} burned.

The equation, or the formula, C_4H_{10}, tells us that 1 *mole* of C_4H_{10} yields 4 *moles* of CO_2 and 5 *moles* of H_2O. The molecular weights of C_4H_{10}, O_2, CO_2, and H_2O are 58.0, 32.0, 44.0, and 18.0, respectively.

$$\frac{36.0 \text{ g of } C_4H_{10}}{58.0 \text{ g of } C_4H_{10}/\text{mole of } C_4H_{10}} \times \frac{6.5 \text{ mole of } O_2}{1 \text{ mole of } C_4H_{10}}$$

$$\times \frac{32.0 \text{ g of } O_2}{1 \text{ mole of } O_2} = 129 \text{ g of } O_2$$

$$\frac{36.0}{58.0} \text{ moles of } C_4H_{10} \times \frac{4 \text{ moles of } CO_2}{1 \text{ mole of } C_4H_{10}}$$

$$\times \frac{44.0 \text{ g of } CO_2}{1 \text{ mole of } CO_2} = 109 \text{ g of } CO_2$$

$$\frac{36.0}{58.0} \text{ moles of } C_4H_{10} \times \frac{5 \text{ moles of } H_2O}{1 \text{ mole of } C_4H_{10}}$$

$$\times \frac{18.0 \text{ g of } H_2O}{1 \text{ mole of } H_2O} = 55.9 \text{ g of } H_2O$$

Note that in solving this problem we first determined how many *moles* of C_4H_{10} were burned. Then we determined how many *moles* of O_2, CO_2, and H_2O, respectively, were involved. Then we converted moles of O_2, CO_2, and H_2O, respectively, to grams.

In general, this is the procedure that should be followed in all stoichiometric calculations. *The first questions should be: How many* **moles** *of reactant do we have? How many* **moles** *of product will be formed per* **mole** *of reactant?*

Having determined the number of *moles* we then convert to grams or liters, whatever the case may be.

7.5 How many *liters* of O_2 gas will be required to burn 50 *liters* of H_2 gas? The volumes of both gases are measured at STP. The equation for the reaction is $2\ H_2 + O_2 = 2\ H_2O$.

Solution: The equation tells us that 1 *mole* of O_2 reacts with 2 *moles* of H_2. At STP 1 mole of H_2 is 22.4 liters. Therefore, 50 liters is 50/22.4 moles of H_2.

[handwritten: 2.232 lmol H₂   1.116 lmol O₂]

$\frac{1}{2}$ of $\frac{50}{22.4}$ moles, or $\frac{50}{2 \times 22.4}$ moles of O_2 will be required

$$\frac{50}{2 \times 22.4} \text{ moles of } O_2 \times \frac{22.4 \text{ liters of } O_2}{1 \text{ mole of } O_2} = 25 \text{ liters of } O_2$$

We see that the volume of O_2 required is $\frac{1}{2}$ the volume of H_2 just as the number of moles of O_2 is $\frac{1}{2}$ the number of moles of H_2. In other words, at STP *the volumes of the two gases are to each other as the number of moles.* Since both gases will respond equally to changes of temperature and pressure it follows that, *when measured at the same temperature and pressure, the volumes of gases involved in a reaction are to each other as the number of moles of the gases in the equation for the reaction.* This very important generalization is an obvious consequence of the fact, pointed out in Chapter 6, that, *at constant temperature and pressure, the volume occupied by a gas is directly proportional to the number of moles of that gas.* The formula for the perfect gas law,

$$n = \frac{PV}{RT}$$

points out the fact that, if P and T are constant, the number of moles, n, is directly proportional to the volume, V.

7.6   How many liters of CO_2 gas, measured at 200°C and 1.20 atm pressure, will be formed when 40.0 g of carbon are burned? The equation is $C + O_2 = CO_2$.

Solution: moles of CO_2 formed = *moles* of C burned.

[handwritten: 3.33 mol C]

$$\frac{40.0 \text{ g of C}}{12.0 \text{ g/mole}} = \frac{40.0}{12.0} \text{ moles of C burned}$$

Therefore, 40.0/12.0 moles of CO_2 will be formed.

To find the volume occupied by 40.0/12.0 moles of CO_2 at 200°C and 1.2 atm we will use the ideal gas equation, $PV = nRT$.

$$V = \frac{nRT}{P} = \frac{40.0}{12.0} \text{ moles} \times 0.082 \frac{\text{liters} \times \text{atm}}{\text{mole} \times \text{deg}} \times \frac{473 \text{ deg}}{1.20 \text{ atm}} = 108 \text{ liters}$$

The entire calculation, in one operation, is

$$\frac{40.0 \text{ g of C}}{12.0 \text{ of C/mole of C}} \times \frac{1 \text{ mole of } CO_2}{1 \text{ mole of C}} \times 0.082 \frac{\text{liters} \times \text{atm}}{\text{mole} \times \text{deg}}$$

[handwritten: 3.3306 mol]

$$\times \frac{473 \text{ deg}}{1.20 \text{ atm}} = 108 \text{ liters}$$

[handwritten: 394.1467]

7.7   How many grams of copper will be formed when the hydrogen gas liberated when 41.6 g of aluminum are treated with excess HCl is passed over excess CuO?

Solution: The reactions that occur are:

$$2\,Al + 6\,HCl = 3\,H_2 + 2\,AlCl_3$$

$$H_2 + CuO = Cu + H_2O$$

We note that 2 *moles* of Al liberate 3 *moles* of H_2; that means that 1 *mole* of Al liberates 1.5 *moles* of H_2.

One *mole* of H_2, when it reacts with CuO, will produce 1 *mole* of Cu.

That means that the 1.5 moles of H_2 that are liberated by 1 mole of Al will produce 1.5 moles of Cu.

In short, 1 mole of Al will liberate enough hydrogen to produce 1.5 moles of Cu. The Al that reacts and the Cu that is produced are in the *ratio* of 1 *mole* of Al to 1.5 *moles* of Cu. The 41.6 g of Al is 41.6/27.0 moles. Therefore, 1.5 × 41.6/27.0 moles of Cu will be produced.

$$\frac{1.5 \times 41.6}{27.0}\ \text{moles of Cu} \times \frac{63.5\ \text{g of Cu}}{1\ \text{mole of Cu}} = 147\ \text{g of Cu}$$

7.8   When C_2H_6 is burned in excess oxygen the following reaction occurs:

$$2\,C_2H_6 + 7\,O_2 = 4\,CO_2 + 6\,H_2O$$

How many moles of oxygen will be consumed when 1.20 moles of C_2H_6 are burned? How many moles of CO_2 and how many moles of H_2O will be produced?

7.9   An amount of 0.262 moles of the compound, As_2S_5, was subjected to a series of treatments by which all of the sulfur in the As_2S_5 was converted to $BaSO_4$ and all of the arsenic was converted to Ag_3AsO_4. How many moles of $BaSO_4$ and Ag_3AsO_4, respectively, were formed?

Solution: Inspection of the formulas of the reactant, As_2S_5, and the products, $BaSO_4$ and Ag_3AsO_4, tells us that 1 *mole* of As_2S_5 will yield 5 *moles* of $BaSO_4$ and 2 *moles* of Ag_3AsO_4. We may, if we wish, write down the skeleton equation

$$1\,As_2S_5 \rightarrow 2\,Ag_3AsO_4 + 5\,BaSO_4.$$

7.10   An amount of 3.16 moles of $KClO_3$ was heated until all of the oxygen was liberated. This oxygen was then all used to oxidize arsenic to As_2O_5. How many moles of As_2O_5 were formed?

Solution: An inspection of the formulas, $KClO_3$ and As_2O_5, tells us that 1 *mole* of $KClO_3$ will liberate enough oxygen to produce $\frac{3}{5}$ *mole* of As_2O_5.

7.11 A quantity of $FeCl_3$ was completely oxidized, all of the chlorine being liberated as Cl_2 gas. This Cl_2 gas was all used to convert Si to $SiCl_4$. A total of 6.36 moles of $SiCl_4$ was produced. How many moles of $FeCl_3$ were oxidized?

7.12 How many moles of Cl_2 will be required to liberate all of the bromine from 8.0 moles of $CrBr_3$? The reaction is

$$3\ Cl_2 + 2\ CrBr_3 = 3\ Br_2 + 2\ CrCl_3$$

7.13 How many grams of oxygen will be required to prepare 200 g of P_4O_{10} from elemental phosphorus?

Solution: The molecular weight of P_4O_{10} is 284.

200 g of P_4O_{10} is $\frac{200}{284}$ mole of P_4O_{10}.

Inspection of the formulas P_4O_{10} and O_2 tells us that 5 *moles* of O_2 will be needed to produce 1 *mole* of P_4O_{10}.

Therefore $5 \times \frac{200}{284}$ moles of O_2 will be needed.

The molecular weight of O_2 is 32.

Therefore $5 \times \frac{200}{284} \times 32 = 113$ g of O_2

The entire calculation, in one operation, is:

$$\frac{200 \text{ g of } P_4O_{10}}{284 \text{ g of } P_4O_{10}/1 \text{ mole of } P_4O_{10}} \times \frac{5 \text{ moles of } O_2}{1 \text{ mole of } P_4O_{10}} \times \frac{32 \text{ g } O_2}{1 \text{ mole of } O_2}$$

$$= 113 \text{ g of } O_2$$

7.14 How many grams of pure zinc must be treated with an excess of dilute sulfuric acid in order to liberate 5.00 g of hydrogen?

7.15 How many grams of tin would be formed if an excess of pure SnO were reduced with 1500 cc of dry hydrogen gas measured at 300°C and 740 mm?

Solution: $SnO + H_2 = Sn + H_2O$

Using $PV = nRT$, first calculate the number of *moles* of H_2.

7.16 Ammonia gas is oxidized by oxygen gas in the presence of a catalyst as follows:

500L   625L   750L   500L
$$4\ NH_3 + 5\ O_2 = 6\ H_2O + 4\ NO$$

How many liters of oxygen will be necessary to oxidize 500 liters of NH_3 gas? How many liters of NO and how many liters of steam will be formed? All gases are measured under the same conditions of temperature and pressure.

Solution: At constant temperature and pressure the *volumes* of gases are directly proportional to the number of *moles* of each gas in the equation for the reaction. Therefore, in this problem the volumes of NH_3, O_2, H_2O (steam), and NO will be in the ratio, respectively, of 4 to 5 to 6 to 4. Accordingly, $\frac{5}{4} \times 500$ or 625 liters of O_2 will be needed, and $\frac{6}{4} \times 500$ or 750 liters of H_2O and $\frac{4}{4} \times 500$ or 500 liters of NO will be formed.

7.17 How many cubic feet of oxygen gas will be required for the oxidation of 6000 cu ft of SO_2 gas in the "contact" process? How many cubic feet of SO_3 gas will be formed? All gases are measured under the same conditions of temperature and pressure.

Solution: At constant temperature and pressure the *volumes* of gases are directly proportional to the number of *moles* of each gas in the equation for the reaction. It makes no difference in what unit volumes are expressed as long as the same unit (liters, cubic feet, quarts, cubic centimeters) is used for each gas.

7.18 How many grams of potassium chlorate must be heated to give 60.0 g of oxygen?

Solution:

$$2 \, KClO_3 = 2 \, KCl + 3 \, O_2$$

or

$$KClO_3 = KCl + 1.5 \, O_2$$

1 mole of $KClO_3$ will yield 1.5 moles of O_2. That means that $1/1.5$ or 0.667 mole of $KClO_3$ will yield 1 mole of O_2. In other words, moles of $KClO_3$ needed $= 0.667 \times$ moles of O_2 produced. The formula weight of O_2 is 32, of $KClO_3$ is 122.6.

$$\text{Moles of } O_2 = \frac{60 \text{ g of } O_2}{32 \text{ g of } O_2/1 \text{ mole of } O_2}$$

$$\text{Moles of } KClO_3 \text{ needed} = 0.667 \times \tfrac{60}{32}$$

$$\text{grams of } KClO_3 \text{ needed} = 0.667 \times \tfrac{60}{32} \times 122.6 = 153 \text{ g}$$

The entire calculation, in one operation, is:

$$\frac{60 \text{ g of } O_2}{32 \text{ g of } O_2/1 \text{ mole of } O_2} \times \frac{1 \text{ mole of } KClO_3}{1.5 \text{ moles of } O_2} \times \frac{122.6 \text{ g of } KClO_3}{1 \text{ mole of } KClO_3}$$

$$= 153 \text{ g of } KClO_3$$

7.19 How many tons of sulfur must be burned to produce 12 tons of SO_2 gas?

Solution:

$$S + O_2 = SO_2$$

Moles of SO_2 = moles of S.

$$\frac{12 \text{ tons of } SO_2}{64 \text{ tons/ton mole of } SO_2} = \frac{12}{64} \text{ ton moles of } SO_2$$

Therefore, $\frac{12}{64}$ ton moles of S must be burned.

$$\frac{12}{64} \text{ ton moles of S} \times \frac{32 \text{ tons of S}}{\text{ton mole of S}} \quad 6.0 \text{ tons of S}$$

As in previous problems we can if we wish include all of the detailed steps in one operation. Because some of the steps are generally obvious from an inspection of the equation for the reaction or the formulas of the substances and can be done "in our head," we will, it is hoped, fall into the practice of writing only the essential steps, thereby making the solution as brief and efficient as possible. In this problem, for example, we should be able to decide at a glance that $\frac{12}{64}$ mole and, hence, $\frac{12}{64} \times 32$ tons of S will be needed.

Note, in Problems 7.18 and 7.19, that the solution is the same regardless of the units in which mass is expressed.

7.20   How many pounds of ZnO will be formed by the complete oxidation of 100 lb of pure Zn?

7.21   How many grams of copper oxide (CuO) can be formed by the oxygen liberated when 160 g of silver oxide are decomposed?

Solution: Note that moles of CuO formed = moles of Ag_2O decomposed.

7.22   A sample of pure MgO was first dissolved in hydrochloric acid to give a solution of $MgCl_2$ which was then converted to a precipitate of pure dry $Mg_2P_2O_7$ having a mass of 6.00 g. Calculate the mass in grams of the sample of MgO.

Solution: Note that moles of MgO used = 2 × moles of $Mg_2P_2O_7$ formed.

7.23   The formula weight of P_4S_3 is 220.0 The formula weight of Ag_3PO_4 is 419. A 13.2-g sample of P_4S_3 was first boiled with excess HNO_3 and, eventually, treated with excess $AgNO_3$. In the process all of the phosphorus in the P_4S_3 was converted to insoluble Ag_3PO_4. How many grams of Ag_3PO_4 were formed?

7.24   A sample of impure copper having a mass of 1.25 g was dissolved in nitric acid to yield $Cu(NO_3)_2$. It was subsequently converted, first to $Cu(OH)_2$, then to CuO, then to $CuCl_2$, and finally to $Cu_3(PO_4)_2$. There was no loss of copper in any step. The pure dry $Cu_3(PO_4)_2$ that was recovered had a mass of 2.00 g. Calculate the percent of pure copper in the impure sample.

Solution: The equations for the reactions that occur can be written in the following skeleton form:

$$3 \text{ Cu} \longrightarrow 3 \text{ Cu(NO}_3)_2 \longrightarrow 3 \text{ Cu(OH)}_2 \longrightarrow 3 \text{ CuO} \longrightarrow$$
$$3 \text{ CuCl}_2 \longrightarrow 1 \text{ Cu}_3(\text{PO}_4)_2$$

This tells us at a glance that moles of pure Cu = 3 × moles of $Cu_3(PO_4)_2$.

$$\text{Percent of pure Cu} = \frac{\text{mass of pure Cu}}{\text{mass of sample}} \times 100\%$$

7.25 A 5.00-g sample of a crude sulfide ore in which all the sulfur was present as As_2S_5 was analyzed as follows: The sample was digested with concentrated HNO_3 until all the sulfur was converted to sulfuric acid. The sulfate was then completely precipitated as $BaSO_4$. The recovered $BaSO_4$ weighed 0.752 g. Calculate the percent of As_2S_5 in the crude ore.

7.26 How many tons of lead will be obtained from 2000 tons of ore containing 21.0% PbS, the yield of lead being 94.0% of the theoretical amount?

7.27 A pure compound containing 63.3% manganese and 36.7% oxygen was heated until no more reaction took place, oxygen gas having been evolved. The solid product was a pure compound containing 72% manganese and 28% oxygen. Write a chemical equation to represent the reaction which took place.

Solution: Calculate the formulas of the initial and final compounds as in Problem 5.46.

7.28 After complete reduction of 0.800 g of a pure oxide of lead with excess hydrogen gas, there remained 0.725 g of lead. Write the chemical equation for the reaction that took place.

7.29 A compound contained 27.1% sodium, 16.5% nitrogen, and 56.4% oxygen. Five g of this compound were heated until no more reaction took place. A mass of 0.942 g of oxygen was given off. A pure chemical compound remained as a solid product. Write a chemical equation to represent the reaction which took place.

7.30 A mixture of 12.2 g of potassium and 22.2 g of bromine was heated until the reaction was completed. How many grams of KBr were formed?

Solution: From the formula KBr and the equation,

$$2\,K + Br_2 = 2\,KBr$$

we see that potassium and Br_2 combine in the ratio of 2 moles of K to 1 mole of Br_2. The weights we start with are not in this molar ratio; since 12.2 g of K is 12.2/39.1 or 0.312 mole of K and 22.2 g of Br_2 is 22.2/159.8 or 0.139 mole of Br_2. Since 0.139 mole of Br_2 require only 0.278 mole of K for complete reaction, the K is present in excess and Br_2 is said to be the limiting reagent because it limits the quantity of KBr which will form. We now ignore the K and note that the 0.139 mole of Br_2 will form 0.278 mole of KBr.

$$0.278 \text{ mole} \times 119 \text{ g of KBr per mole} = 33.1 \text{ g of KBr}$$

7.31 What mass in grams of AgCl will be formed when 35.4 g of NaCl and 99.8 g of $AgNO_3$ are mixed in water solution?

➡ **7.32** How many grams of Bi_2S_3 will form when 12.3 g of H_2S are mixed with 126 g of $Bi(NO_3)_3$ in water solution?

➡ **7.33** The nitrogen in $NaNO_3$ and $(NH_4)_2SO_4$ is all available to plants as fertilizer. Which is the more economical source of nitrogen, a fertilizer containing 30% $NaNO_3$ and costing \$3.00 per 100 lb or one containing 20% $(NH_4)_2SO_4$ and costing \$2.70 per 100 lb?

➡ **7.34** Exactly 3.00 moles of chromium are reacted with excess of element Q; all of the Cr is converted to Cr_2Q_3. The Cr_2Q_3 is then treated with excess of strontium metal; all of the Q in the Cr_2Q_3 is converted to SrQ. The SrQ is then reacted with excess of sodium metal; all of the SrQ is converted to Na_2Q; 782 g of Na_2Q are formed. What is the atomic weight of element Q? Atomic weight of Na = 23.0.

   Solution: $2\,Cr \longrightarrow 1\,Cr_2Q_3 \longrightarrow 3\,SrQ \longrightarrow 3\,Na_2Q$

   Therefore, moles of Na_2Q = 1.5 × moles of Cr.

   Moles of Na_2Q = 1.5 × 3.00 moles = 4.5 moles of Na_2Q.

   Knowing that the atomic weight of Na is 23.0 and that 782 g is 4.5 moles of Na_2Q, the atomic weight of Q can be calculated.

➡ **7.35** When 2.451 g of pure, dry MXO_3 are heated, 0.9600 g of oxygen gas is liberated. The other product is 1.491 g of solid MX. When this MX is treated with excess $AgNO_3$, all of it reacts with the $AgNO_3$ to form solid AgX; 2.869 g of AgX are formed. Knowing that the atomic weights of O and Ag are, respectively, 16.00 and 108.0, calculate the atomic weights of M and X.

   Solution: $MXO_3 = MX + 1.5\,O_2$

   $$MX + AgNO_3 = AgX + MNO_3$$

   Moles of O_2 = 0.9600 g/32.00 g per mole = 0.0300 mole.

   Moles of AgX = moles of MX = 1/1.5 × moles of O_2 = 1/1.5
   ·   × 0.0300 = 0.0200.

   Molecular weight of AgX = 2.869 g/0.0200 mole = 143.45 g/mole.

   Atomic weight of X = 143.45 − 108.0 = 35.45.

   Molecular weight of MX = 1.491 g/0.0200 mole = 74.55 g/mole.

   Atomic weight of M = 74.55 − 35.45 = 39.10.

➡ **7.36** A 5.68-gram sample of pure P_4O_{10} was completely converted to H_3PO_4 by dissolving it in water. This H_3PO_4 was completely converted to Ag_3PO_4 by treatment with excess $AgNO_3$. The Ag_3PO_4 was then completely converted to AgCl by treatment with excess HCl. The AgCl weighed 34.44 grams. The molecular weight of P_4O_{10} is known to be 284.0, and the atomic weight of chlorine is known to be 35.5. Calculate the atomic weight of Ag.

➔ **7.37** Exactly 2.000 g of NH_3 were neutralized by HCl, NH_4Cl being formed as a product. In a separate experiment exactly 20.00 g of $AgNO_3$ were formed by the action of excess HNO_3 on 12.70 g of silver. All of the NH_4Cl formed in the first experiment was exactly sufficient to react with all of the $AgNO_3$ formed in the second experiment. Knowing that the atomic weight of H is 1.008 and of O is 15.999, calculate the atomic weights of N and Ag.

Solution: $NH_3 + HCl = NH_4Cl$

$Ag \rightarrow AgNO_3$ $NH_4Cl + AgNO_3 = AgCl + NH_4NO_3$

Moles of NH_3 = moles of NH_4Cl = moles of $AgNO_3$ = moles of NO_3 = moles of Ag

Let N represent the atomic weight of nitrogen.

Moles of NH_3 = g of $NH_3 \div$ mol wt of $NH_3 = \dfrac{2.000 \text{ g}}{(N + 3.024)\text{g/mole}}$

Moles of NO_3 = g of $NO_3 \div$ mol wt of $NO_3 = \dfrac{7.30 \text{ g}}{(N + 48.00)\text{g/mole}}$

Since moles of NH_3 = moles of NO_3 we can solve for N.

Having calculated N we can then solve for the atomic weight of Ag, knowing that moles of Ag = moles of NH_3.

➔ **7.38** When 1.82 g of zirconium metal (at. wt. 91.22) react with excess HCl, 1.04 liters of dry H_2 gas, measured at 27°C and 720 mm, are liberated. Write a balanced equation for the reaction that occurs when Zr is treated with HCl.

➔ **7.39** Equal volumes of hydrogen and oxygen, both at room temperature and atmospheric pressure, were introduced into a completely evacuated reaction bomb. The bomb was sealed and was heated to 120°C; the pressure of the mixture of gases in the bomb was found to be 100 mm at 120°C. An electric arc inside the container was turned on, which caused the reaction, $2 H_2 + O_2 = 2 H_2O$, to take place. When the reaction was over the bomb was cooled until the temperature was again 120°C. An examination revealed that there was no liquid water in the bomb. What was the new pressure inside the bomb?

Solution: At the same temperature and pressure equal volumes of O_2 and H_2 contain equal numbers of moles of O_2 and H_2. Since, from the equation, $2 H_2 + O_2 = 2 H_2O$, we see that H_2 and O_2 combine in the ratio of 2 moles of H_2 with 1 mole of O_2 to form 2 moles of H_2O and since, at constant volume and temperature, the pressure is directly proportional to the number of moles, 50 mm worth of H_2 will combine with 25 mm worth of O_2 to form 50 mm worth of H_2O. There will be an excess of 25 mm worth of O_2, so the final pressure will be 75 mm.

➡ **7.40**   A quantity of C_2H_4, when burned to CO_2 and H_2O, yielded 120 liters of CO_2 measured at a certain temperature and pressure. Measured at the same temperature and pressure, a quantity of C_5H_{12}, when burned to CO_2 and H_2O, yielded 50.0 liters of CO_2. Calculate the ratio of the quantities, in grams, of the C_2H_4 and C_5H_{12} that were burned.

Solution: Since, at constant temperature and pressure, the numbers of moles of gases is directly proportional to their volumes, the moles of CO_2 from C_2H_4 and the moles of CO_2 from C_5H_{12} are in the ratio of 120 to 50. Since 1 mole of C_2H_4 yields 2 moles of CO_2 and 1 mole of C_5H_{12} yields 5 moles of CO_2, the C_2H_4 and C_5H_{12} must have been present in what mole ratio to yield this $\frac{120}{50}$ ratio of CO_2? Knowing the mole ratios the mass ratios can then be calculated.

➡ **7.41**   At a very high temperature 2 volumes of H_2S gas decomposed completely to give 3 volumes of a mixture of H_2 gas and sulfur vapor. Write the equation for the reaction.

➡ **7.42**   A compound X contains C and H. When 1 mole of X is burned to CO_2, and H_2O, the total mass of the CO_2 and H_2O is 160 g greater than the mass of the 1 mole of X. The mass of the CO_2 is 60 g greater than the mass of the H_2O. Calculate the chemical formula of the compound.

Solution: Let y = atoms of C and z = atoms of H in 1 molecule of X. We can then write the following balanced equation for the reaction that occurs:

$$1\ C_yH_z + (y + z/4)O_2 = y\ CO_2 + z/2\ H_2O$$

Knowing the atomic weights of C, H, and O we can then write the following expressions:

Mass of y moles of $CO_2 = 44y$

Mass of $z/2$ moles of $H_2O = 18z/2 = 9z$

Mass of $(y + z/4)$ moles of $O_2 = 32(y + z/4) = 32y + 8z$

On the basis of the facts given in the problem we can then write the following equations:

$$44y = 9z + 60$$
$$32y + 8z = 160$$

We can then solve for y and z.
Note in this problem and in similar problems that follow, to solve for two unknowns (the subscripts y and z in this problem) we must set up two independent equations involving these two unknowns.

➡ **7.43**   To burn 2 moles of a gaseous hydrocarbon to CO_2 and H_2O (steam) required 9 moles of pure oxygen gas. Complete combustion of a sample of the hydrocarbon yielded 0.135 g of H_2O and 0.330 g of CO_2. Calculate the molecular weight of the hydrocarbon.

Solution hint: See Problem 7.42. Let C_xH_y be the formula of the compound. How many *moles* of O_2 are needed to burn *1 mole* of C_xH_y? What is the numerical value of the ratio of the *moles* of H_2O formed to the *moles* of CO_2 formed? Can this ratio be expressed in terms of x and y? What other relationship involving x and y is provided by the facts given in the problem?

➡ **7.44** A gaseous compound contains only C and H. To burn 1 mole of the gaseous compound required 6.5 moles of O_2 gas. In the mixture of CO_2 and H_2O (both are gases) that was formed as a product of the burning the mole fraction of the H_2O was 0.556. Calculate the chemical formula of the compound.

➡ **7.45** A certain compound contains C, H, and O. One molecule of the compound is known to contain 2 atoms of O, and the number of atoms of H per molecule is 2 times the number of atoms of C per molecule. When a quantity of the compound is burned completely in O_2 to form CO_2 and H_2O as the only products, 0.375 mole of O_2 is consumed and 0.300 mole of H_2O is formed. Calculate the chemical formula of the compound.

➡ **7.46** A mixture of O_2 and N_2 contains 30.0 mole percent oxygen. To this mixture of gases is added just enough carbon so that, on heating, all the oxygen and all of the carbon are consumed and, in the reaction, half of the carbon is converted to CO and half to CO_2. What is the average molecular weight of the resulting mixture of CO, CO_2, and N_2?

Solution: The average molecular weight of a mixture of gases is calculated by dividing the total mass of the mixture by the total number of moles of gases in the mixture. This means that, in this particular problem, we must know the *relative number* of moles of CO, CO_2 and N_2 in the final mixture; we need not know the *actual number* of moles of each species.

In the original mixture of N_2 and O_2 the ratio of moles of N_2 to moles of $O_2 = 70.0/30.0$ or $\frac{7}{3}$. From the equations

$$C + O_2 = CO_2$$
$$C + 0.5\,O_2 = CO$$

and the fact that $\frac{1}{2}$ the C is converted to CO_2 and $\frac{1}{2}$ to CO, we see that 1.5 moles of O_2 will yield 1 mole of CO_2 and 1 mole of CO. That means that the 3 moles of O_2 in the above $\frac{7}{3}$ ratio will be replaced by 2 moles of CO_2 and 2 moles of CO to give a final *ratio* of 7 moles of N_2 to 2 moles of CO_2 to 2 moles of CO.

$$\text{Av mol wt} = \frac{(7 \times 28.0) + (2 \times 44.0) + (2 \times 28.0)}{11} = 30.9$$

7.47 A mixture of CS_2 and excess O_2 gas contains a total of 1.0 mole. When the mixture is ignited by a spark, all of the CS_2 is oxidized to CO_2 and SO_2. There is now a total of 0.80 mole present. Calculate the number of grams of CS_2 in the original mixture.

Solution: If we examine the equation

$$CS_2 + 3\,O_2 = CO_2 + 2\,SO_2$$

we note that 4 moles of reactants yield 3 moles of products. There is a decrease of 1 mole. Most significant is the fact that *the decrease in the number of moles is equal to the number of moles of CS_2 that react.* To solve this problem we simply calculate the decrease in number of *moles* of gases; that gives us the number of *moles* of CS_2 that were burned.

➜ **7.48** When solid $CrCl_3$ is heated with H_2 gas, reduction occurs; HCl is the only gaseous product, the other possible products ($CrCl_2$, $CrCl$, and Cr) being nonvolatile solids. A reaction bomb contained 0.2000 g of anhydrous $CrCl_3$ and 0.1218 mole of H_2. When the temperature was raised to 327°C, a reduction reaction took place. The total number of moles of gas (H_2 and HCl) when the reaction was completed was 0.1237 mole. Write the equation for the reaction that took place.

Solution: See Problem 7.47. Three reactions are possible:

(a) $CrCl_3$ (solid) $+ \frac{1}{2}\,H_2$ (gas) $= CrCl_2$ (solid) $+$ HCl (gas)

(b) $CrCl_3$ (solid) $+ H_2$ (gas) $= CrCl$ (solid) $+ 2\,$HCl (gas)

(c) $CrCl_3$ (solid) $+ 1.5\,H_2$ (gas) $= Cr$ (solid) $+ 3\,$HCl (gas)

For each of these reactions how does the increase in the number of *moles of gas* compare with the number of *moles of solid* $CrCl_3$ that react? In the experiment that is reported how does the increase in the number of moles of gas compare with the number of moles of $CrCl_3$ consumed?

➜ **7.49** When CO_2 gas, in a steel bomb at 427°C and a pressure of 10 atm, is heated to 1127°C, the pressure rises to 22.5 atm. The following reaction occurs: $2\,CO_2 = 2\,CO + O_2$. Calculate the mole percent of CO_2 decomposed.

Solution: In all previous problems the reactions involved have gone to completion. In Problems 7.49, 7.50, and 7.51 we observe reactions at some point before they have gone to completion and calculate the mole percent or fraction of reactant that has reacted at that point. In making the calculation we can assume that we have, in effect, stopped the reaction momentarily.

In order to calculate the mole percent of CO_2 decomposed we must know how many moles of CO_2 we had at the start and how many we have at the moment the reaction is observed. Using $PV = nRT$ we can, from the facts given, calculate how many moles of CO_2 we have *per liter* at the start and how many moles of $CO_2 + CO + O_2$ we have *per liter* at observation time. Although we do not know the volume of the reaction vessel we are justified in selecting one liter of it for our calculations because the temperature, pressure and concentrations of gases will be completely uniform throughout the vessel.

Having calculated moles per liter of CO_2 at the start and of the gas mixture at observation time, and knowing that the reaction that takes place is $2 CO_2 = 2 CO + O_2$, we can then set up an algebraic equation involving X (the number of moles of CO_2 decomposed) which will enable us, eventually, to calculate the mole percent of CO_2 decomposed.

➡ **7.50** When a sample of acetylene, C_2H_2, is treated with a catalyst, some of it is converted to benzene, C_6H_6, according to the equation, $3 C_2H_2 = C_6H_6$. The density of the gaseous mixture of C_2H_2 and C_6H_6 at 27°C and a pressure of 0.44 atm is 0.760 g/liter. What fraction of the C_2H_2 originally present has changed to C_6H_6?

Solution hint: See Problem 7.49. The law of conservation of matter dictates that the mass per liter of C_2H_2 at the start must equal the mass per liter of the mixture of gases at observation time. This will enable us to calculate the number of moles of C_2H_2 per liter at the start.

➡ **7.51** A mixture of equal grams of C_4H_{10} and O_2 gases, when heated in a reaction vessel at 400°C, reacts slowly to form CO_2 gas and steam (H_2O). At the end of 1 minute the total pressure of the mixture of gases (C_4H_{10}, O_2, CO_2, and H_2O) was 2.96 atm and the density of the mixture was 2.000 g/liter. What mole percent of the C_4H_{10} had been oxidized at the end of the 1-minute interval?

7.52 A certain hydrocarbon gas was mixed, in a steel reaction bomb, with the exact amount of O_2 gas required to burn it completely to CO_2 and H_2O (steam). The mixture was ignited by a spark, which caused all of the hydrocarbon to react with all of the O_2 to form CO_2 and H_2O. The pressures, before and after ignition, measured at the same temperature, were the same. The partial pressures of the CO_2 and steam were the same. Calculate the exact molecular weight of the hydrocarbon.

8

Mole relationships in
chemical reactions.
II. Stoichiometry of mixtures.

It is frequently necessary to determine experimentally the absolute or relative amounts of the substances present in a *mixture* of two or more species. This can generally be accomplished by treating the mixture with an appropriate reagent and then determining the absolute or relative amounts of the products formed.

The problems in this chapter are concerned largely with the sort of calculations that may be encountered in the above type of situation. The approach used in solving these problems is essentially the same as that used in solving the problems in Chapter 7. Strict attention to and recognition of the *mole* relationships is the key to the solution of each problem.

PROBLEMS

8.1 A mixture of C and S, when burned, yielded a mixture of CO_2 and SO_2 in which the number of moles of the two gases were equal. Calculate the mole percent of C in the original mixture.

Solution:

$$C + O_2 = CO_2$$
$$S + O_2 = SO_2$$

(When 2 (or more) substances in a mixture react with another substance, as when the C and S in a mixture both react with O_2, *two separate equations should always be written*. The overall reaction should not be written as

$$C + S + 2\,O_2 = CO_2 + SO_2$$

This equation would be correct only if the C and S were present in a 1-to-1 mole ratio.)

The mathematical relationships involving the C and S are:

(1) moles of CO_2 formed = moles of C burned

(2) moles of SO_2 formed = moles of S burned

Since moles of C = moles of CO_2 and moles of S = moles of SO_2, and since moles of CO_2 = moles of SO_2, it follows that moles of C = moles of S.

That is, the mixture consists of 50 mole percent C and 50 mole percent S.

8.2   A sample of C was burned to CO_2 and a sample of S was burned to SO_2. Twice as many moles of CO_2 as SO_2 were formed. What was the mass percent of C of the total C and S burned?

8.3   A sample of C was burned to CO_2 and a sample of S was burned to SO_2. The combined mass of the C and S was 10.0 g. The moles of the CO_2 and SO_2 were equal. How many grams of carbon were burned?

8.4   A 10-g mixture of H_2S and CS_2 was burned in oxygen to form a mixture of H_2O, SO_2, and CO_2. The dried mixture on being separated into its pure components yielded 0.275 mole of SO_2 and 0.0774 mole of CO_2. How many grams of H_2S were there in the original mixture?

Solution: From the equations

$$H_2S + 1.5\,O_2 = H_2O + SO_2$$
$$CS_2 + 3\,O_2 = CO_2 + 2\,SO_2$$

we note that CO_2 and SO_2 are formed from CS_2 in the ratio of 1 mole of CO_2 to 2 moles of SO_2. Since 0.0774 mole of CO_2 are present, and this CO_2 was derived from the CS_2, 2(0.0774) = 0.155 mole of SO_2 must also have come from the CS_2. Therefore, (0.275 − 0.155) = 0.120 mole of SO_2 must have come from the H_2S.

8.5   A 12-g mixture of carbon and sulfur, when burned in air, yielded a mixture of CO_2 and SO_2 in which the moles of the CO_2 was one half the moles of the SO_2. How many grams of carbon were there in the mixture?

Solution: Let $X = $ g of C and $Y = $ g of S. Then $X + Y = 12$

$$C + O_2 = CO_2$$
$$S + O_2 = SO_2$$

We note that:

(1) moles of $CO_2 = $ moles of C burned
(2) moles of $SO_2 = $ moles of S burned
(3) $X/12 = $ moles of C $= $ moles of CO_2
(4) $Y/32 = $ moles of S $= $ moles of SO_2

Since moles of $CO_2 = \frac{1}{2}$ moles of SO_2, that is,

$$\frac{X}{12} = \frac{1}{2}\left(\frac{Y}{32}\right)$$

We now have two equations in two unknowns. The first is derived from the mass relationship; the total mass is the sum of the masses of the components. The second is derived from the mole relationship and the fact that moles $= $ g/MW. Many mixture problems can be solved by setting up and solving two such equations. In this case solving gives $X = 1.9$ g of carbon.

8.6   A mixture of CS_2 and CH_4 was burned to a mixture of CO_2, SO_2, and H_2O at 450°C. The partial pressures of the CO_2, SO_2, and H_2O in the mixture were 175 mm, 200 mm, and 150 mm, respectively. What was the mass percent of CH_4 in the original mixture?

Solution: Since the reaction takes place in one reaction vessel the three gases, CO_2, H_2O, and SO_2, are at the same temperature and occupy the same volume. Therefore, the number of moles of the three gases are in the ratios of their partial pressures.

$$CS_2 + 3\,O_2 = CO_2 + 2\,SO_2$$
$$CH_4 + 2\,O_2 = CO_2 + 2\,H_2O$$

Since the partial pressures of the SO_2 and H_2O are, respectively, 200 mm and 150 mm they must be present in the *ratio* of 2.00 moles of SO_2 to 1.50 moles of H_2O. Therefore, as the equations testify, the CS_2 and CH_4 must have been present in the *ratio* of 2.00 moles of CS_2 to 1.50 moles of CH_4. Knowing the mole ratio, the mass percent of CH_4 can then be calculated.

8.7   A mixture contains solid carbon, gaseous CS_2, and gaseous CH_4. The mixture is oxidized completely to give a gaseous mixture of CO_2, SO_2, and steam (H_2O). The ratio of the partial pressures of the CO_2, H_2O, and SO_2 in this gaseous mixture is 5.00 for the CO_2 to 2.00 for the H_2O to 1.00 for the SO_2. What was the percent by mass of solid carbon in the original mixture?

8.8   A 10.0-g sample of a mixture of $CuSO_4 \cdot 5\,H_2O$ and $CaCO_3$ was heated, decomposing the carbonate and dehydrating the hydrate. If 5.00 cc of water vapor (steam) were produced for every 2.00 cc of CO_2, both volumes being measured at the same T and P, calculate the mass percent of $CaCO_3$ in the original mixture.

8.9   A mixture of sulfur and carbon weighing 2.0 g, when burned, gave a mixture of SO_2 and CO_2 weighing 6.0 g. How many grams of carbon were there in the original mixture?

Solution: In solving this particular problem we should first note that each of the following is true:

(1) $X =$ g of C $Y =$ g of S

(2) $X + Y = 2.0$

(3) $C + O_2 = CO_2$

(4) $S + O_2 = SO_2$

(5)   moles of CO_2 = moles of C

(6)   moles of SO_2 = moles of S

(7)   moles of C + moles of S = moles of O_2

(8)   moles of C $= \dfrac{X}{12}$

(9)   moles of S $= \dfrac{Y}{32}$

(10) moles of $CO_2 = \dfrac{X}{12}$

(11) moles of $SO_2 = \dfrac{Y}{32}$

(12) grams of CO_2 + grams of SO_2 = 6.0 g

(13) g of $(CO_2 + SO_2)$ − g of $(C + S)$ = g of O_2 = 4.0 g

(14) moles of $O_2 = \dfrac{4.0}{32}$

(15) g of CO_2 = moles of $CO_2 \times \dfrac{\text{g of } CO_2}{\text{mole of } CO_2} = \dfrac{X}{12} \times 44$

(16) g of SO_2 = moles of $SO_2 \times \dfrac{\text{g of } SO_2}{\text{mole of } SO_2} = \dfrac{Y}{32} \times 64$

(17) g of O_2 in CO_2 + g of O_2 in SO_2 = 4.0 g

(18) g of O_2 in $CO_2 = \dfrac{32}{12} \times$ g of C $= \dfrac{32}{12} X$

Since 1 mole of CO_2 contains 1 mole of C and 1 mole of O_2 it follows that, in CO_2

$$\frac{\text{mass of the } O_2}{\text{mass of the C}} = \frac{\text{mol weight of } O_2}{\text{mol weight of C}} = \frac{32}{12}$$

Therefore,

$$\text{mass of the } O_2 = \frac{32}{12} \times \text{mass of the C}$$

(19) g of O_2 in $SO_2 = \frac{32}{12} \times$ g of S $= \frac{32}{32} \times Y$

(20) g of $CO_2 = \frac{44}{12} \times$ g of C $= \frac{44}{12} X$

Since 1 mole of CO_2 contains 1 mole of C it follows that

$$\frac{\text{mass of the } CO_2}{\text{mass of the C}} = \frac{\text{mol weight of } CO_2}{\text{mol weight of C}} = \frac{44}{12}$$

Therefore,

$$\text{mass of the } CO_2 = \frac{44}{12} \times \text{mass of the C}$$

(21) g of $SO_2 = \frac{64}{32} \times$ g of S $= \frac{64}{32} Y$

To solve for the value of X we need two equations in two unknowns. One is given by equation (2). This is the mass equation. A second equation can be constructed by various combinations of the 21 relationships. In one way or another any second equation will incorporate a mole relationship.

Solution 1:   From Equations (7), (8), (9), and (14).

(7)   moles of C + moles of S = moles of O_2

$$\frac{X}{12} + \frac{Y}{32} = \frac{4.0}{32} \qquad (X = 1.2)$$

Solution 2:   From Equations (12), (15), and (16).

(12) g of CO_2 + g of $SO_2 = 6.0$ g

$$\frac{X}{12} \times 44 + \frac{Y}{32} \times 64 = 6.0 \text{ g} \qquad (X = 1.2)$$

Solution 3:   From Equations (17), (18), and (19).

(17) g of O_2 in CO_2 + g of O_2 in $SO_2 = 4.0$ g

$$\frac{32}{12} X + \frac{32}{32} Y = 4.0 \text{ g} \qquad (X = 1.2)$$

Solution 4:   From Equations (12), (20), and (21).

(12) g of CO_2 + g of $SO_2 = 6.0$ g

$$\frac{44}{12} X + \frac{64}{32} Y = 6.0 \text{ g} \qquad (X = 1.2)$$

Note that, although Solutions 2 and 4 involve the same terms, they represent different approaches to the problem.

In most mixture problems more than one method of solution is possible. It is, of course, not necessary to probe all of the methods, as has been done in this problem. The simplest, shortest, and most obvious method should be selected.

8.10   A mixture of Mg and Zn having a mass of 1.000 g, when burned in oxygen, gave a mixture of MgO and ZnO which had a mass of 1.409 g. How much Zn was there in the original mixture?

Solution: X = g of Zn. Y = g of Mg, $X + Y = 1.000$. By examining the formulas, MgO and ZnO, we see that 1 mole of Mg will yield 1 mole of MgO and 1 mole of Zn will yield 1 mole of ZnO. Since X g of Zn = $X/65.4$ mole of Zn and Y g of Mg = $Y/24.3$ mole of Mg; $X/65.4$ mole of ZnO and $Y/24.3$ mole of MgO have formed. The total mass of ZnO and MgO are known, so we convert these molar quantities to masses by multiplying by the appropriate molecular weight: g of ZnO = 81.4 $(X/65.4)$ and g of MgO = 40.3 $(Y/24.3)$. The sum of these masses is given to be 1.409 g. This gives the second equation,

$$\frac{81.4}{65.4}X + \frac{40.3}{24.3}Y = 1.409 \text{ g} \qquad (X = 0.569 \text{ g of Zn})$$

8.11   A mixture of NaBr and NaI has a mass of 1.620 g. When treated with excess $AgNO_3$, it yields a mixture of AgBr and AgI which has a mass of 2.822 g. How many grams of NaI were there in the original mixture?

Solution: X = g of NaI; Y = g of NaBr;

$$X + Y = 1.620 \text{ (the mass equation)}$$

The formula weights are:

$$\text{NaBr} = 102.9; \text{ NaI} = 149.9; \text{ AgBr} = 187.8; \text{ AgI} = 234.8$$

Since 1 mole of NaBr and 1 mole of AgBr each contain 1 mole of Br and since 1 mole of NaI and 1 mole of AgI each contain 1 mole of I, 1 mole of NaBr will yield 1 mole of AgBr and 1 mole of NaI will yield 1 mole of AgI.

It follows, therefore, that

$$\frac{\text{mass of the AgBr}}{\text{mass of the NaBr}} = \frac{\text{mol weight of AgBr}}{\text{mol weight of NaBr}} = \frac{187.8}{102.9}$$

$$\text{mass of AgBr} = \frac{187.8}{102.9} \times \text{mass of NaBr}$$

Likewise,

$$\text{mass of AgI} = \frac{234.8}{149.9} \times \text{mass of NaI}$$

$$\text{g of AgI} = \frac{234.8}{149.9} \times \text{g of NaI} = \frac{234.8}{149.9}X$$

$$\text{g of AgBr} = \frac{187.8}{102.9} \times \text{g of NaBr} = \frac{187.8}{102.9} Y$$

$$\text{g of AgI} + \text{g of AgBr} = 2.822 \text{ g}$$

$$\frac{234.8}{149.9} X + \frac{187.8}{102.9} Y = 2.822 \text{ (the molar relationship equation)}$$

$$X = 0.520 \text{ g of NaI}$$

8.12   A mixture of CO_2 and SO_2 has a mass of 2.952 g and contains a total of 5.300×10^{-2} moles. How many moles of CO_2 are there in the mixture?

Molecular weights: $CO_2 = 44.01$, $SO_2 = 64.06$.

Solution: $X = \text{g of } CO_2$; $Y = \text{g of } SO_2$

$$X + Y = 2.952 \text{ (the mass equation)}$$

$$\frac{X}{44.01} + \frac{Y}{64.06} = 5.300 \times 10^{-2} \text{ (the mole equation)}$$

$$X = 0.9682 \text{ g; moles of } CO_2 \ 44.01/0.9682 = 2.200 \times 10^{-2} \text{ moles}$$

$$44.01X + 64.06(5.300 \times 10^{-2} - X) = 2.952 \text{ g}$$

$$(X = 2.200 \times 10^{-2} \text{ moles of } CO_2)$$

8.13   A mixture of pure AgCl and pure AgBr contains 66.35% silver. What is the mass percent of bromine in the mixture?

Solution: We will assume that we have 100 g of mixture. The answer, in grams, will then equal, numerically, the mass percent. Let

$$X = \text{g of Br} \qquad Y = \text{g of Cl} \qquad \text{g of Ag} = 66.35$$

$$X + Y = 100 - 66.35 \text{ (the mass equation)}$$

$$\text{moles of AgBr} = \text{moles of Br} = \frac{X}{79.91}$$

$$\text{moles of AgCl} = \text{moles of Cl} = \frac{Y}{35.45}$$

$$\text{moles of (AgCl + AgBr)} = \text{moles of Ag} = \frac{66.35}{107.9}$$

So that we can write the mole relationship equation:

$$\frac{X}{79.91} + \frac{Y}{35.45} = \frac{66.35}{107.9} \qquad (X = 21.3 \text{ g} = 21.3\%)$$

Other variations on the second equation are possible.

8.14   A mixture of $BaCl_2$ and $CaCl_2$ contains 43.1% chlorine. Calculate the mass percent of barium in the mixture.

8.15 A mixture of pure Na_2SO_4 and Na_2CO_3 has a mass of 1.200 g and yields a mixture of $BaSO_4$ and $BaCO_3$ having a mass of 2.077 g. Calculate the mass percent of Na_2SO_4 in the original mixture.

8.16 A mixture of CO_2 and CS_2 contains 20.0 mass % carbon. How many grams of SO_2 will be formed by the complete oxidation of 10.0 g of the mixture to CO_2 and SO_2?

8.17 A mixture of NaCl and NaBr contains twice as many grams of NaCl as NaBr. When treated with excess $AgNO_3$ this mixture yields 100 g of a mixture of AgCl and AgBr. How many grams of NaCl were there in the original mixture?

➡ **8.18** A mixture of As_2S_3 and CuS having a mass of 8.00 g was roasted in air until completely oxidized to SO_2, As_2O_3, and CuO. The SO_2 gas was oxidized to sulfate which was then completely precipitated as $BaSO_4$; 21.5 g of $BaSO_4$ were formed. Calculate the number of grams of combined Cu in the mixture.

➡ **8.19** When 50.0 g of mercury and 50.0 g of iodine are heated together they are completely converted into a mixture of Hg_2I_2 and HgI_2. How many grams of Hg_2I_2 are there in the mixture?

➡ **8.20** A mixture of Al and Mg contained 3 times as many grams of Al as Mg. When the mixture was treated with excess HCl the hydrogen that was liberated reduced 119.25 g of CuO to Cu. How many grams of Al were in the mixture?

➡ **8.21** When a mixture of H_2S and CS_2 was burned in oxygen to give H_2O, CO_2, and SO_2, the mass in grams of the SO_2 that was formed was four times the mass of the CO_2. Calculate the mass percent of CS_2 in the mixture of H_2S and CS_2.

Solution:

$$H_2S + 1.5\ O_2 = H_2O + SO_2$$
$$CS_2 + 3\ O_2 = CO_2 + 2\ SO_2$$

Since no specific quantities are given and since the answer called for (percent of CS_2) represents a *relative* value, let us assume that 1 mole (44 g) of CO_2 is formed. The total amount of SO_2 formed will then be 176 g. Since in the burning of CS_2 2 moles of SO_2 are formed for each mole of CO_2 formed, 2×64 g or 128 g of SO_2 will be formed from the CS_2. The remaining 48 g of SO_2 ($176 - 128 = 48$) must have been formed from the H_2S. With this information the relative number of moles of CS_2 and H_2S and the mass percent of CS_2 in the original mixture can be calculated.

➡ **8.22** A mixture of C_2H_6 gas and CS_2 gas was placed in a reaction vessel at 200°C. An excess of O_2 gas was then added to the mixture. When the resulting mixture was ignited by a spark all of the C_2H_6 and CS_2 was oxidized to CO_2, SO_2, and H_2O. When the reaction was complete, the partial pressures of the CO_2 and SO_2, measured at 200°C, were 108 mm and 120 mm, respectively. Calculate the mass percent of C_2H_6 in the original mixture of C_2H_6 and CS_2.

Solution: See Problem 8.6.

➡ **8.23**   To a mixture of C_2H_6 and C_2H_6S gases contained in a constant-volume reaction vessel was added the exact amount of O_2 gas required to burn it completely to CO_2, SO_2, and H_2O. When a spark was passed, complete combustion of the gases to CO_2, SO_2, and steam (H_2O) occurred. The mole fraction of the SO_2 in the gaseous mixture of CO_2, SO_2, and H_2O was 0.1237. (There was no liquid present.) Calculate the mass percent of the C_2H_6 in the original mixture of C_2H_6 and C_2H_6S.

Solution hint: Let $X =$ the mole fraction of the C_2H_6 in the original mixture of C_2H_6 and C_2H_6S. The mole fraction of C_2H_6S in the mixture will then be $1 - X$.

➡ **8.24**   A gaseous mixture of equal grams of CH_4 and C_2H_6 plus excess O_2 was contained in a reaction vessel at 300°C. The partial pressure of the CH_4 in this mixture was 15 mm. The mixture was ignited, resulting in all of the CH_4 and C_2H_6 being completely oxidized, yielding a gaseous mixture of CO_2, H_2O (steam), and O_2 (no liquid water). The mole fraction of the CO_2 in this gaseous mixture was 0.20. What was the partial pressure of the oxygen gas in the original mixture?

➡ **8.25**   A gaseous mixture contained in a 1-liter reaction vessel at 127°C and 10 atm pressure consisted of CS_2 and CH_4 and an excess of oxygen. The mixture was ignited by a spark and complete oxidation to CO_2, SO_2, and H_2O (steam) occurred. After the reaction had occurred the pressure in the 1-liter vessel, measured at 527°C, was 17.1 atm. How many grams of CS_2 were in the mixture?

Solution: See Problem 7.47.

➡ **8.26**   A gaseous mixture of CH_4 and CS_2 was added to an excess of oxygen gas. The resulting mixture was placed in an evacuated, constant-volume reaction vessel. The temperature of the mixture was 320°C. The mixture was ignited, resulting in the CH_4 and CS_2 being completely oxidized to CO_2, H_2O, and SO_2. The temperature of the reaction vessel was brought back to the original value, 320°C. The total pressure of the mixture of gases (no liquid) in the vessel at 320°C was 79 mm and the partial pressure of the SO_2 was 8.0 mm. Calculate the total pressure of the original mixture of CH_4, CS_2, and O_2.

9

Thermochemistry.
Thermodynamics.

Thermochemistry

Almost every chemical change involves the evolution or absorption of heat. Processes which evolve heat are said to be exothermic, and the quantity of heat evolved is expressed as a negative quantity. Processes which absorb heat are said to be endothermic and the quantity of heat absorbed is expressed as a positive quantity. The magnitude of the heat change associated with a process depends on a number of factors, including the conditions of the process. Most commonly, chemical processes occur in open containers at atmospheric pressure; that is, the pressure remains constant during the process. The heat change at constant pressure is called the enthalpy change, ΔH, and represents the difference between the enthalpy H of the products and of the starting materials. Enthalpy is sometimes called heat content.

A great deal of chemical information can be systematized and correlated by the use of ΔH. The ΔH associated with any chemical reaction is called the heat of reaction. More specifically, we refer to the heat of combustion, which is the ΔH associated with the rapid reaction of a substance with oxygen and to the heat of formation which is the ΔH associated with the formation of one mole of a substance from its constituent elements. Since the value of the heat of a reaction depends on the conditions, it is often desirable to specify the conditions exactly. This is conveniently done by defining a standard state, which refers to $T = 25°C$, $P = 1$ atm and the state of all substances to be their most stable at this temperature and pressure. The enthalpy change under these conditions is designated as $\Delta H°$, and a table of $\Delta H_f°$, standard heats of formation can be found in the appendix. A quantity related to the heat of reaction is the bond energy. When chemical bonds form, heat is given off; therefore, heat is required to break a bond. The bond energy is the average heat required to break a given type of chemical bond in a compound in the gas phase into its constituent atoms also in the gas phase. Bond energies are usually expressed in units of kcal/mole or kJ/mole, and are always positive numbers. A table of bond energies can be found in the appendix. These bond energy values can be used to calculate many heats of reaction. The ΔH associated with various phase changes can also be designated by name. For example, the heat of vaporization refers to a liquid becoming a gas; the heat of sublimation refers to a solid becoming a gas; the heat of fusion refers to solid becoming liquid.

A quantity of heat can be expressed in units of calories (cal), kilocalories (kcal), joules (J) or kilojoules (kJ) where 1 cal = 4.184J and 1 kcal = 10^3 cal, and 1 calorie is approximately the quantity of heat required to raise the temperature of one gram of water by 1°C. The heat capacity of a substance is defined as the quantity of heat required to raise the temperature of one mole of the substance by 1°C. At constant pressure the heat capacity, $C_p = \Delta H / \Delta T$. The specific heat of a substance is defined in the same way as the heat capacity but refers to a 1 gram quantity of material. Thus, the specific heat of water is 1 cal/g-deg and its heat capacity is 18 cal/deg-mole.

PROBLEMS

9.1   The specific heat of water is 1 cal/g × deg C. How many calories of heat will be required to raise the temperature of 150 g of water 40°C?

Solution: Since 1 cal is required to raise the temperature of 1 g of water 1°C, 150 cal will be required to raise the temperature of 150 g of water 1°, and 40 × 150 cal or 6000 cal will be required to raise the temperature of 150 g of water 40°.

The calculation, in one operation, is

$$150 \text{ g of water} \times 40° \times \frac{1 \text{ cal}}{1 \text{ g of water} \times 1°} = 6000 \text{ cal}$$

Note that "grams of water" and "degrees" will cancel. The answer will then be in calories.

9.2   The specific heat of water is 1 cal/g × deg. How many grams of water can be heated from 20 to 60°C by 3200 cal of heat?

9.3   The heat capacity of methyl alcohol is 19.2 cal/deg-mole. How many calories of heat will be required to raise the temperature of 4000 g of methyl alcohol from 2 to 22°C?

9.4 The heat of combustion of a sample of coal is 6000 cal/g. How many grams of this coal would have to be burned in order to generate enough heat to raise the temperature of 1000 g of water from 10 to 34°C? The specific heat of water is 1 cal/g × deg.

Solution:

$$\frac{\text{calories required to heat the water}}{\text{calories evolved per g of coal burned}} = \text{g of coal burned}$$

$$\frac{1000 \text{ g of H}_2\text{O} \times 24° \times \dfrac{1 \text{ cal}}{1 \text{ g of H}_2\text{O} \times 1°}}{\dfrac{6000 \text{ cal}}{1 \text{ g of coal}}} = 4 \text{ g of coal}$$

9.5 A sample of 12 g of a certain grade of coal gave off enough heat to raise the temperature of 4000 g of water from 12 to 30°C. Calculate the heat of combustion of the coal in calories per gram. The specific heat of water is 1 cal/g × deg.

9.6 A sample of 2.5 g of sulfur, when burned to SO_2, raised the temperature of 1080 g of water from 22.5 to 27.5°C. Calculate the heat of formation of SO_2 in kilocalories per mole.

Solution: To form 1 mole of SO_2 one must burn 1 mole of S. Therefore, the heat of formation of SO_2 is the same as the heat of combustion of S.

9.7 When burned in oxygen, 10.00 g of phosphorus generated enough heat to raise the temperature of 2950 g of water from 18 to 38°C. Calculate the heat of formation of P_4O_{10} in kilocalories per mole.

Solution: Note that 4 moles of P yield 1 mole of P_4O_{10}.

9.8 The heat of fusion of ice is 79.7 cal/g. The heat of combustion of methyl alcohol is 170.9 kcal/mole. How many grams of methyl alcohol, CH_3OH, must be burned in order to generate enough heat to melt a block of ice which weighs 9200 g?

9.9 The heat of combustion of gasoline is about 3×10^4 kcal/gallon. The energy of complete fission of 100 kg of uranium is 1 megaton, and 1 megaton $= 3.35 \times 10^{13}$ kcal. How many gallons of gasoline must be burned to yield the same amount of energy possible from fission of 1 kg of uranium?

9.10 The heat of combustion of methane (CH_4), the chief constituent of natural gas is 213 kcal/mole. The heat capacity of iron is 6.31 cal/deg-mole; its melting point is 1530°C and its heat of fusion is 3560 cal/mole. Assuming no heat loss, calculate the number of grams of methane which must be burned to cause 50 moles of iron initially at 30°C to melt.

Solution: Divide the problem into two parts. First calculate the heat required to raise the temperature to the melting point and then the amount of heat which must be absorbed for melting to occur. Be sure to keep the units consistent since both cal and kcal are used.

9.11   Calculate $\Delta H°$, the standard heat of reaction, for $CaO_{(s)} +$ $CO_{2(g)} \rightarrow CaCO_{3(s)}$ using the values of $\Delta H_f°$ in Table 7 in the Appendix.

Solution: It can be shown that

$$\Delta H° = \sum \Delta H_f° \text{ (products)} - \sum \Delta H_f° \text{ (reactants)}$$

Substituting the values from the table

$$\Delta H° = \left[\left(-288.4 \frac{\text{kcal}}{\text{mole}} \times 1 \text{ mole}\right)\right] - \left[\left(-151.8 \frac{\text{kcal}}{\text{mole}} \times 1 \text{ mole}\right)\right.$$

$$\left. + \left(-94.05 \frac{\text{kcal}}{\text{mole}} \times 1 \text{ mole}\right)\right] = -42.55 \text{ kcal}$$

9.12   Calculate $\Delta H°$ for the reaction

$$2 C_2H_{6(g)} + 7 O_{2(g)} \rightarrow 4 CO_{2(g)} + 6 H_2O_{(l)}$$

Solution: Applying the method of Problem 9.11,

$$\Delta H° = \left[\left(-94.05 \frac{\text{kcal}}{\text{mole}} \times 4 \text{ moles}\right) + \left(-68.32 \frac{\text{kcal}}{\text{mole}} \times 6 \text{ moles}\right)\right]$$

$$- \left[\left(-20.24 \frac{\text{kcal}}{\text{mole}} \times 2 \text{ moles}\right) + \left(0 \frac{\text{kcal}}{\text{mole}} \times 7 \text{ moles}\right)\right]$$

$$= -745.7 \text{ kcal}$$

It should be noted that $\Delta H_f°$ of $O_{2(g)} = 0$. *The heat of formation of any element in its standard state is equal to zero* by definition. Thus, the ΔH_f of $Br_{2(l)}$ at 25° and 1 atm is equal to zero, but ΔH_f of $Br_{2(g)}$ or of Br is not equal to zero. The element must be in its most stable state at 25°C and 1 atm for ΔH_f to be equal to zero.

9.13   The heat of vaporization of $H_2O_{(l)}$ at 25°C is 10.53 kcal/mole. Using the result from Problem 9.12, calculate ΔH for the reaction $C_2H_{6(g)}$ $+ \frac{7}{2} O_{2(g)} \rightarrow 2 CO_{2(g)} + 3 H_2O_{(g)}$.

Solution: This equation differs in two ways from that in Problem 9.12. All the molar quantities are cut in half; therefore, we can begin by dividing the calculated $\Delta H°$ by 2, to give $-745.7/2 = -372.9$ kcal/ mole. Furthermore, one of the products is $3 H_2O_{(g)}$ rather than $3 H_2O_{(l)}$, so that the heats of reaction will differ by ΔH_{vap} of $H_2O \times 3$ moles of $H_2O = 10.53$ kcal/mole $\times$ 3 moles $= 31.6$ kcal. Several approaches can be used to work out the direction of the difference. We can reason that since heat must be supplied to evaporate water the reaction which forms $H_2O_{(g)}$ will be less exothermic (less negative) than the reaction which forms $H_2O_{(1)}$, some of the heat produced will be needed for the evaporation of the water. $\Delta H = -372.9 + 31.6 = -341.3$ kcal. Instead we can use *Hess' law of heat summation*, which states that the energy liberated in a reaction is independent of the path that is followed and is the algebraic sum of the heat changes in the steps of the reaction. We can bring about the reaction in this problem by a path involving two steps of known ΔH.

(1) $C_2H_{6(g)} + \frac{7}{2} O_{2(g)} \rightarrow 2\,CO_{2(g)} + 3\,H_2O_{(1)}$ $\qquad \Delta H = -372.9$ kcal
(2) $3\,H_2O_{(1)} \rightarrow 3\,H_2O_{(g)}$ $\qquad\qquad\qquad\qquad \Delta H = 31.6$ kcal

Algebraic combination of Equations (1) and (2) gives

$$C_2H_{6(g)} + \frac{7}{2} O_{2(g)} \rightarrow 2\,CO_{2(g)} + 3\,H_2O_{(g)}$$
$$\Delta H_1 + \Delta H_2 = -341.3 \text{ kcal}$$

9.14 Using the data given in Table 7 calculate $\Delta H°$ for the following reactions.

(a) $HCl_{(g)} + NH_{3(g)} \rightarrow NH_4Cl_{(s)}$
(b) $2\,NO_{2(g)} \rightarrow N_2O_{4(g)}$
(c) $C_2H_{4(g)} + 3\,O_{2(g)} \rightarrow 2\,CO_{2(g)} + 2\,H_2O_{(g)}$
(d) $C_2H_{2(g)} + 2\,H_{2(g)} \rightarrow C_2H_{6(g)}$

9.15 Using the data given in Table 7 calculate the heat of combustion of $C_3H_{8(g)}$ to $CO_{2(g)}$ and $H_2O_{(1)}$.

9.16 Given that $\Delta H°$ for the reaction $2\,SO_{2(g)} + O_{2(g)} \rightarrow 2\,SO_{3(g)}$ is 46.9 kcal and that $\Delta H_f°$ of $SO_{2(g)}$ is -71.0 kcal/mole, calculate $\Delta H_f°$ of $SO_{3(g)}$.

Solution: Let $x = \Delta H_f°$ of $SO_{3(g)}$ and substitute into the equation of Problem 9.11.

9.17 Using the data given in Table 7, calculate ΔH for the following reactions

(a) $2\,HBr_{(g)} \rightarrow H_{2(g)} + Br_{2(g)}$ given ΔH_{vap} of $Br_{2(1)} = 7.4$ kcal/mole
(b) $2\,Fe_{(s)} + 3\,H_2O_{(g)} \rightarrow Fe_2O_{3(s)} + 3\,H_{2(g)}$ given ΔH_{vap} of $H_2O_{(1)}$ $= 10.53$ kcal/mole

9.18 The $\Delta H_f°$ of $TiI_{3(s)}$ is -80.1 kcal/mole and $\Delta H°$ for the reaction $2\,Ti_{(s)} + 3\,I_{2(g)} \rightarrow 2\,TiI_{3(s)}$ is -204.7 kcal. Calculate the ΔH of sublimation of $I_{2(s)}$. Note I_2 is a solid at 25°C.

9.19 Given (1) $NO_{(g)} + NO_{2(g)} \rightarrow N_2O_{3(g)}$ $\qquad \Delta H_1° = -9.6$ kcal
$\qquad\qquad$ (2) $N_2O_{4(g)} \rightarrow 2\,NO_{2(g)}$ $\qquad\qquad \Delta H_2° = 5.6$ kcal

Calculate $\Delta H°$ for (3) $2\,N_2O_{3(g)} \rightarrow 2\,NO_{(g)} + N_2O_{4(g)}$

Solution: If some algebraic combination of Equations (1) and (2) gives Equation (3) Hess' law of heat summation can be applied. We can see that this should be the case since all the substances in Equation (3) appear in (1) or (2) and the substance not in equation (3), $NO_{2(g)}$ appears in both (1) and (2) and can be made to cancel. The appropriate combination can be deduced by inspection or by the following reasoning. Equation (3) has 2 moles of $N_2O_{3(g)}$ on the left, $N_2O_{3(g)}$ appears in Equation (1) on the right. Reverse Equation (1) and multiply it by a factor of 2, giving Equation (4): $2\,N_2O_{3(g)} \rightarrow 2\,NO_{(g)} + 2\,NO_{2(g)}$. Reversing an equation has the effect of changing the sign of its ΔH. Therefore, $\Delta H_4° = -2(\Delta H_1°)$. Next, Equation (3) has $N_2O_{4(g)}$ on the right; $N_2O_{4(g)}$ appears in Equation (2) on the left. Reverse Equation (2) and change the sign of $\Delta H_2°$ to give Equation (5): $2\,NO_{2(g)} \rightarrow N_2O_{4(g)}$, $\Delta H_5° = -\Delta H_2°$. Combine Equations (4) and (5) to give $2\,N_2O_{3(g)} + 2\,NO_{2(g)} \rightarrow 2\,NO_{(g)} + 2\,NO_{2(g)} + N_2O_{4(g)}$. Cancel like terms ($2\,NO_{2(g)}$) from both sides to give Equation (3). Since Equation (4) and Equation (5) = Equation (3), $\Delta H_4° + \Delta H_5° = \Delta H_3° = -2(-9.6) + (-5.6) = 13.6$ kcal.

9.20 Given the equations:

$$C_{(s)} + O_{2(g)} \rightarrow CO_{2(g)} \qquad \Delta H = -94.4 \text{ kcal}$$

$$C_{(s)} + \tfrac{1}{2}O_{2(g)} \rightarrow CO_{(g)} \qquad \Delta H = -28.2 \text{ kcal}$$

Calculate ΔH for the reaction:

$$CO_{(g)} + \tfrac{1}{2}O_{2(g)} \rightarrow CO_{2(g)}$$

➤ **9.21** Given the equations:

$$SO_{2(g)} + \tfrac{1}{2}O_{2(g)} \rightarrow SO_{3(g)} \qquad \Delta H = 21.4 \text{ kcal}$$

$$S_{(s)} + \tfrac{3}{2}O_{2(g)} \rightarrow SO_{3(g)} \qquad \Delta H = -48.8 \text{ kcal}$$

Calculate the heat of formation of $SO_{2(g)}$ under these nonstandard state conditions.

➤ **9.22** Calculate ΔH for the reaction, $H_{2(g)} + \tfrac{1}{2}O_{2(g)} \rightarrow H_2O_{(l)}$ which is not carried out in the standard state, from the following data for processes carried out at the same conditions:

$$C_{(s)} + 2\,H_2O_{(g)} \rightarrow CO_{2(g)} + 2\,H_{2(g)} \qquad \Delta H = 39.0 \text{ kcal}$$

$$C_{(s)} + \tfrac{1}{2}O_{2(g)} \rightarrow CO_{(g)} \qquad \Delta H = -29.0 \text{ kcal}$$

$$H_2O_{(g)} \rightarrow H_2O_{(l)} \qquad \Delta H = -9.7 \text{ kcal}$$

$$CO_{(g)} + \tfrac{1}{2}O_{2(g)} \rightarrow CO_{2(g)} \qquad \Delta H = -67.7 \text{ kcal}$$

➤ **9.23** Use Hess' law to prove that:

$$\Delta H° = \sum \Delta H_f° \text{ (products)} - \sum \Delta H_f° \text{ (reactants)}$$

Solution hint: Pick a general reaction like $AB + CD \rightarrow AC + BD$ where A, B, C, and D are different elements.

9.24 Using the table of bond energies, Table 8 in the Appendix, calculate $\Delta H°$ for the reaction

$$H_{2(g)} + I_{2(g)} \rightarrow 2\,HI_{(g)}$$

Solution: Since the bond energy is the heat which must be absorbed in order to break a bond, it is always a positive quantity and is usually given in kcal/mole. It is defined when all species are in the gas phase. The heat of reaction can be calculated by using the relationship:

$$\Delta H = \sum \text{bonds broken} - \sum \text{bonds made.}$$

The reaction of interest must therefore be analyzed in order to determine the bonds which break and the bonds which form. In the above reaction, the H–H bond in the reactant $H_{2(g)}$ is broken, the I–I bond in the reactant $I_{2(g)}$ is broken, and two H-I bonds are formed, one in each $HI_{(g)}$. Substituting

$$\Delta H = [104.2 + 36.1] - [2(71.5)] = -2.7\,\text{kcal}$$

9.25 Using the bond energies given in Table 8, calculate ΔH for the following reactions.

(a) $2\,HBr_{(g)} \rightarrow H_{2(g)} + Br_{2(g)}$

(b) $H_2 \rightarrow 2\,H$

(c) $CH_4 + Cl_2 \rightarrow CH_3Cl + HCl$

(d) $2\,NH_{3(g)} \rightarrow 3\,H_{2(g)} + N_{2(g)}$

(e) $4\,NH_{3(g)} + 5\,O_{2(g)} \rightarrow 4\,NO_{(g)} + 6\,H_2O_{(g)}$

9.26 Using the bond energies given in Table 8, calculate ΔH for the reaction

$$H_2C{=}CH_2 + Cl_2 \rightarrow Cl{-}CH_2{-}CH_2{-}Cl$$

➡ **9.27** The heat of formation of $HF_{(g)}$ is -64.3 kcal/mole; use the bond energy values in Table 8 to calculate the bond energy of the H-F bond.

Solution: Let $x =$ the bond energy and substitute into the relationship of 9.24 applied to the reaction for forming 1 mole of HF from its elements.

➡ **9.28** Given $\Delta H_f°$ for $CH_{4(g)}$ is -17.9 kcal/mole, use the bond energy values in Table 8 to calculate the heat of vaporization of graphite.

Solution hint: The heat of formation refers to the process $C_{(s)} + 2\,H_{2(g)} \rightarrow CH_{4(g)}$. The bond energy can be applied to the process $CH_{4(g)} \rightarrow 4\,H_{(g)} + C_{(g)}$ and the heat of vaporization refers to the process $C_{(s)} \rightarrow C_{(g)}$. Use Hess' law.

Thermodynamics

Thermodynamics is an experimentally derived system for dealing with the macroscopic properties of matter. One of its most important uses is that it enables us to predict whether or not a given process is favorable and to what extent it can occur. It says nothing about the time that may be required for the process to occur. Thermodynamic systems are described by assigning values to *state functions* which are properties of the system, and whose values depend only on the particular state we are specifying and not on how the state is reached. Pressure, volume, and temperature are examples of state functions. Processes involve a change in state from an initial to a final state along a certain path. These processes are described by specifying the values of the changes in the state functions (symbolized by a Δ, e.g., ΔT is change in temperature) and the values of other thermodynamic quantities called path functions. While the values of the changes in state functions are independent of the path followed and depend only on the initial and final states of the system, the values of path functions are determined by the exact nature of the path followed. State functions are designated by capital letters, path functions by lower case letters. Generally the initial and final states in processes we will deal with are equilibrium states. That is, if the system is in an equilibrium state it will not change with time unless it is disturbed in some manner by something in the surroundings. Although thermodynamics has a very wide range of applicability, we shall restrict ourselves to certain idealized systems which can be dealt with by relatively simple treatments.

First law of thermodynamics

The first law of thermodynamics is an experimentally derived generalization about the behavior of matter which can be stated as: the energy of the universe is constant. A mathematical statement of this law is:

(1) $$\Delta E = q - w$$

The change in energy in going from an initial to a final state is equal to the heat absorbed by the system minus the work done by the system. E, the energy of a system, is a state function which can be defined as the stored-up capacity to do mechanical work. The heat, q, is a path function; it is the transfer of energy due to a temperature difference. It has a positive sign when the system absorbs heat from the surroundings and a negative sign when the system gives heat to the surroundings. Work, w, is also a path function; it is the transfer of energy brought about by performing a displacement against an opposing force. It has a positive sign when the system does work on the surroundings and a negative sign when the surroundings do work on the system. Since work involves a displacement, when ordinary chemical systems undergo a change in state, work can be performed only if there is a change in volume of the system. It is important to remember that $w = 0$ for all constant-volume processes, and therefore $\Delta E = q_v$, where q_v is the heat absorbed at constant volume.

The thermodynamics of ideal gases is often sufficiently uncomplicated to allow relatively simple calculation. A system involving an ideal gas may be of several types. The ideal gas may be in a container which is completely insulated, that is, the walls of the container do not allow heat flow. All changes in state are said to be *adiabatic*, $q = 0$. The ideal gas may be in a sealed container which cannot change its volume. All changes in state will be at constant volume, therefore, $w = 0$. The ideal gas may be in a container which allows heat flow and contains a mechanical link with the surroundings; most commonly this is a cylinder fitted with a frictionless piston, held in place by atmospheric pressure and/or weights.

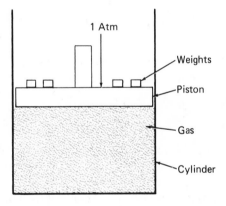

Since the piston in this system can move, volume changes can occur and work can be performed. This type of work is called pressure-volume (PV) work because it results from the displacement of the piston due to a volume change against an opposing pressure. If the volume of the gas in the cylinder increases ($\Delta V > 0$) the system is doing work on the surroundings ($w > 0$), and if the volume of the gas in the cylinder decreases ($\Delta V < 0$) the surroundings are doing work on the system ($w < 0$).

Calculation of the amount of work in processes of this type can readily be performed in certain situations. If the movement of the piston occurs against a constant opposing pressure then

(2) $$w = P_{opp}\Delta V.$$

For example, suppose that a gas in a cylinder has $P = 3$ atm and that a piston is held in place by 1 atm of external pressure and enough weights to total 3 atm. If the weights are then removed quickly the gas in the cylinder will expand against an opposing pressure of 1 atm from its initial state to a final state in which the pressure of the gas is 1 atm. Since the opposing external pressure during the entire expansion has been 1 atm, $w = (1 \text{ atm})\Delta V$. Note that the gas was initially in an equilibrium state, but removing the weights from the piston threw it out of equilibrium. Therefore, it spontaneously expanded until it finally reached a new equilibrium state. It is also theoretically possible to bring about this same change by a path in which the system is always in an equilibrium state. Such a process is said to be *reversible*. This can be imagined to occur by removing the weights from the piston in infinitely small increments. If this entire process occurs at a constant temperature, T in $°K$, then

(3) $$w = 2.3nRT \log V_2/V_1$$

where R is the gas constant, n is the number of moles of gas, and V_1 and V_2 are the initial and final volumes, respectively.

To summarize, we can calculate w for three types of processes:

1. constant volume: $w = 0$
2. volume change against a constant opposing pressure: $w = P_{opp}\Delta V$
3. reversible volume change: $w = 2.3nRT \log V_2/V_1$

Calculation of the value of q for certain processes involving ideal gases can also be accomplished. An *isothermal* process is one which occurs at constant temperature. For any isothermal process involving an ideal gas, by definition $\Delta E = 0$ and from the first law $q = w$. And so, if we can calculate w for the isothermal process, we have q as well. It is also possible to calculate q for processes in which the temperature changes, provided that either the volume or the pressure stays constant. For any temperature change at constant volume,

(4) $$q_V = nC_V\Delta T$$

where C_V is the heat capacity at constant volume and has the value $\frac{3}{2}R$ for one mole of an ideal gas and n is the number of moles of the gas. For any temperature change at constant pressure

(5) $$q_P = nC_P\Delta T$$

where q_P is the heat absorbed at constant pressure and where C_P is the heat capacity at constant pressure and has the value $\frac{5}{2}R$ for one mole of an ideal gas.

We have seen that $\Delta E = q_V$. It is also convenient to define another state function H, the enthalpy, such that $\Delta H = q_P$ since most chemical processes occur at constant pressure. It can be shown by use of the first law that

(6) $$\Delta H = \Delta E + \Delta(PV)$$

where $\Delta(PV) = P_2 V_2 - P_1 V_1$ for state 1, the initial state, and for state 2, the final state involved in the process of interest. For a constant-pressure process $\Delta(PV) = P\Delta V$.

Many of these ideas can also be applied to systems containing only liquids and solids by making the excellent approximation that the volume changes which solids and liquids undergo during ordinary processes are negligible. Therefore, $\Delta H \cong \Delta E$, $C_P \cong C_V$ and $w \cong 0$. For systems containing ideal gases as well as condensed phases only the volume changes involving the gases need be considered. In cases where an ideal gas is produced or consumed during the process again it is only the volume changes involving the gases that need to be considered. For example, if an isothermal process involves a change in the quantity of ideal gas it can readily be shown by substituting from the ideal gas equation into (6) that

(7) $$\Delta H = \Delta E + RT\Delta n$$

PROBLEMS

9.29 Calculate the change in energy for a system which absorbs 10 kcal of heat from the surroundings while performing 8 kcal of work in the surroundings.

Solution: The first law states that the change in energy is equal to the heat absorbed by the system minus the work it performs:

$$\Delta E = q - w = 10 - 8 = 2 \text{ kcal}$$

9.30 The energy of a system decreases by 17.2 kcal when it evolves 22.3 kcal of heat. How much work is performed?

9.31 A system containing 2.5 moles of an ideal gas absorbs 1700 cal/mole of heat. How much work must it perform in order for its energy to remain constant?

9.32 A 20-liter volume of gas in a cylinder with a frictionless piston is at a pressure of 3 atm. Enough weight is suddenly removed from the piston to lower the external pressure to 1.5 atm. The gas then expands isothermally until it reaches a new equilibrium state. Calculate ΔE, ΔH, q, and w for this change in state.

Solution: First it is necessary to calculate the final volume, which can be done by substitution into Boyle's law: $P_1 V_1 = P_2 V_2$, $(3)(20) = 1.5V_2$, $V_2 = 40$ liters. Since the process is isothermal $\Delta E = 0$. It is easily shown that $\Delta H = 0$ as well since at a given temperature

$P_1V_1 = P_2V_2$ and thus $\Delta(PV) = 0$. If $\Delta E = 0$ then $q = w$. To calculate the value of w we analyze the type of process. It is a volume change against a constant opposing pressure of 1.5 atm and thus we can use $w = P_{opp}\Delta V = 1.5(V_2 - V_1) = 1.5(40 - 20) = 30$ liter-atm.
Generally it is desirable to express thermodynamic quantities in units of calories or joules. The conversion factors are 1 liter-atm = 24.2 cal = 101 joules. Thus, $w = 30$ liter-atm $\times$ 24.2 cal/liter-atm = 726 cal and $q = 726$ cal.

9.33 One mole of an ideal gas occupying a volume of 7 liters expands isothermally and reversibly to a volume of 12 liters at 300°K. Calculate ΔE, ΔH, q, and w.

Solution: Since the process is isothermal $\Delta E = 0$, $\Delta H = 0$, and $q = w$. Since the process is also reversible $w = 2.3nRT \log V_2/V_1$. To calculate w, we substitute the data into this relationship taking care that suitable units are used. In order to obtain an answer in calories we use $R = 1.987$ cal/deg-mole. Thus, $w = 2.3(1)(1.987)(300)\log \frac{12}{7} = 319$ cal and $q = 319$ cal.

9.34 A 5.20-liter cylinder with a frictionless piston contains ideal gas at a pressure of 4.10 atm. Weight is quickly removed from the piston so that the external pressure is 1.20 atm. The gas then expands isothermally to a new equilibrium state. Calculate ΔE, ΔH, q, and w for this change in state.

9.35 A 17.8-liter cylinder with a frictionless piston contains ideal gas at a pressure of 1.20 atm. Weight is suddenly added to the piston so that the external pressure is 4.10 atm. The gas then contracts isothermally to a new equilibrium state. Calculate ΔE, ΔH, q, and w for this change in state.

9.36 Two moles of ideal gas reversibly expands isothermally at 25°C from a volume of 3.1 to a volume of 6.2 liters. Calculate ΔE, ΔH, q, and w for this change in state.

9.37 Three moles of ideal gas reversibly contracts isothermally at 25°C from a volume of 6.2 liters to a volume of 3.1 liters. Calculate ΔE, ΔH, q, and w for this change in state.

9.38 One mole of ideal gas at 27°C reversibly undergoes a pressure drop from 4.12 atm to 2.06 atm. Calculate ΔE, ΔH, q, and w for this change in state.

9.39 The temperature of 2 moles of ideal gas is raised from 25°C to 75°C at constant volume. Calculate ΔE, ΔH, q, and w for this change in state.

Solution: Since the volume is constant $w = 0$ and $\Delta E = q$. Substituting into (4) $q_V = nC_V\Delta T = (2)(\frac{3}{2})(1.987)(75 - 25) = 298$ cal $= \Delta E$. In order to calculate ΔH we substitute $\Delta(PV) = \Delta(nRT)$ into equation

(6) to obtain

(8) $\qquad \Delta H = \Delta E + \Delta(nRT) = \Delta E + nR\Delta T$

since for this change in state the number of moles of gas is constant. Thus, $\Delta H = 298 + (2)(1.987)(50) = 497$ cal.

9.40 The temperature of 2 moles of an ideal gas is raised from 25°C to 75°C at constant pressure. Calculate ΔE, ΔH, q, and w for this change in state.

Solution: Since the pressure is constant we can substitute into (5), $\Delta H = q_P = nC_P\Delta T = (2)(\frac{5}{2})(1.987)(75 - 25) = 497$ cal. Substitution into (8) allows calculation of ΔE

$$497 = \Delta E + (2)(1.987)(50)$$
$$\Delta E = 298 \text{ cal}$$

Since we know q and ΔE the first law allows calculation of w.

$$w = q - \Delta E = 497 - 298 = 199 \text{ cal}$$

9.41 The temperature of 2.50 moles of an ideal gas is raised from 273°K to 373°K at constant volume. Calculate ΔE, ΔH, q, and w.

9.42 The temperature of 2.50 moles of an ideal gas is raised from 373°K to 473°K at constant pressure. Calculate ΔE, ΔH, q, and w.

9.43 Calculate the work for the following four processes. One mole of a perfect gas at 300°K in a cylinder with a frictionless piston is at 4 atm of pressure caused by the pressure of the air and three weights of equal mass pushing on the piston. The pressure is dropped to 1 atm by (a) taking off all three weights at once, (b) taking off 2 weights, waiting, taking off the third, (c) taking off 1 weight, waiting, taking off the next 2 at once, or (d) taking off the weights one at a time.

9.44 The temperature of 2.25 moles of an ideal gas is lowered from 350°K to 175°K at constant pressure. Calculate ΔE, ΔH, q, and w.

9.45 The change in state: 1 mole Perfect Gas (400°K, 2 atm) → 1 mole Perfect Gas (300°K, 1 atm) is brought about by a number of two-step processes. For each two-step process calculate ΔE, ΔH, q, and w.

(a) Reversible expansion at 400°K to final volume; cooling at this volume to 300°K.

(b) Sudden pressure drop to 1 atm and isothermal expansion back to equilibrium; cooling at constant pressure of 1 atm to 300°K.

(c) Reversible cooling at constant pressure of 2 atm to 300°K; reversible isothermal expansion until pressure is 1 atm.

Solution hint: For multistep processes it is most convenient to treat each step separately and then sum the quantities for each step to get values for the entire process.

9.46 The heat of vaporization of water is 9720 cal/mole at 100°C. Calculate ΔE, ΔH, q, and w for the evaporation of 1000 g of water at 100°C and 1 atm.

Solution hint: Calculate ΔE by using (7), and w by using the first law.

➤ **9.47** Using C_P for $H_2O_{(l)} = 17.9$ cal/mole and C_P for $H_2O_{(g)} = 5.98$ cal/mole and the heat of vaporization given in 9.46, calculate ΔE, ΔH, q, and w for the change in state: 1 mole $H_2O_{(l)}$ at 80°C and 1 atm going to 1 mole $H_2O_{(g)}$ at 110°C and 1 atm.

Solution hint: Consider the process as three steps, warming $H_2O_{(l)}$ from 80°C to 100°C, evaporating $H_2O_{(l)}$ to $H_2O_{(g)}$, and warming $H_2O_{(g)}$ from 100°C to 110°C.

➤ **9.48** Using the data in the previous two problems and C_P for $H_2O_{(s)} = 8.8$ cal/mole and ΔH_{fusion} of $H_2O_{(s)} = 1430$ cal/mole; calculate ΔE, ΔH, q, and w for the change in state 1 mole $H_2O_{(l)} \rightarrow 1$ mole $H_2O_{(s)}$ at -10°C and 1 atm.

9.49 For the reaction $2\,NO_{2(g)} \rightarrow N_2O_{4(g)}$ at 300°K, ΔE is found to be -13.1 kcal. Calculate ΔH.

Second law of thermodynamics

The first law tells us that any process must satisfy the requirement of conservation of energy. However, real processes have another feature; a preferred spontaneous direction. For example, if a hot and a cold object are placed in contact, heat will flow from hot to cold until the temperature of each object is the same. The reverse process, heat flowing from one object to another when both are at the same temperature does not occur, even though it does not violate the first law. Intuitively we know that systems proceed spontaneously from a condition of nonequilibrium to the equilibrium state. Such a process is called an irreversible process. Once at equilibrium systems change no further unless disturbed by some external agent. The second law of thermodynamics deals with this phenomenon. It can be stated in many ways. One statement is: "There is a quantity S called entropy, which is a state function. In an irreversible process, the entropy of the universe increases. In a reversible process the entropy of the universe remains constant. At no time does the entropy of the universe decrease."

It can be shown that this new state function, S, is a measure of the disorder of the system; and therefore, it is possible to make qualitative predictions about changes in the entropy of a system, ΔS, when it undergoes a change in state, by evaluating the relative disorder of the initial and final states. For example, if the process involves an isothermal expansion the increased volume of the final state results in more disorder in this state and a positive ΔS. In order to calculate the value of ΔS for a system undergoing a change in state we use the relationship:

$$(9) \qquad \Delta S = \frac{q_{rev}}{T}$$

which can be used for any isothermal change in state. It should be noted that just as with any state function, ΔS for the system does not depend on the path of the process, but only on the initial and final states. However, in order to calculate ΔS we must calculate q for the hypothetical reversible process. For an ideal gas we can use (3) since $q_{rev} = w_{rev}$. Substituting into (9) gives ΔS for any isothermal volume change of an ideal gas

$$(10) \qquad \Delta S = 2.3nR \log V_2/V_1$$

Calculation of ΔS for processes in which the temperature changes but which occur at constant pressure or constant volume is also possible by using the derived relationships:

$$(11) \qquad \Delta S = 2.3nC_P \log T_2/T_1$$

$$(12) \qquad \text{or} \quad \Delta S = 2.3nC_V \log T_2/T_1$$

The second law deals with ΔS for the universe, so that, in order to judge whether a process can occur we will want to calculate not only ΔS for the system but also ΔS for the surroundings, since the sum of these two quantities gives ΔS for the universe. In order to calculate ΔS_{surr}, we must calculate the value of q for the specific path the system is following during the change in state. Then:

$$(13) \qquad \Delta S_{surr} = -q/T$$

where T is the temperature of the surroundings.

Using entropy as a criterion for spontaneity is somewhat awkward since we must make two calculations. In order to avoid this, still another state function is defined called the free energy, G:

$$(14) \qquad G = H - TS$$

$$\text{or} \qquad \Delta G = \Delta H - T\Delta S$$

The value of ΔG for the system alone is also a criterion for spontaneous change for processes which occur at constant temperature and pressure. If $\Delta G = 0$ for a process, the process is a reversible one; if $\Delta G < 0$ the process is spontaneous. Processes for which $\Delta G > 0$ are not spontaneous and the system can only undergo such a process under influence from the surroundings.

PROBLEMS

9.50 Calculate ΔS of the universe and ΔG of the system, for the change in state: 2 moles of an ideal gas at 300°K occupying a volume of 24.6

liters expands isothermally against an opposing pressure of 1 atm to reach equilibrium.

Solution: First we calculate the final volume of the ideal gas at equilibrium

$$V = \frac{nRT}{P} = \frac{(2)(.082)(300)}{1} = 49.2 \text{ liters}$$

In order to calculate ΔS of the universe, we calculate ΔS_{sys} and ΔS_{surr} separately and sum.

$$\Delta S_{sys} = \frac{q_{rev}}{T} = 2.3 \, nR \log \frac{V_2}{V_1} = (2.3)(2)(1.987) \log \frac{49.2}{24.6}$$

$$= 2.76 \text{ cal/deg}$$

Since ΔS_{surr} requires the actual heat of the process we calculate as in Problem 9.32.

$$q = w = P\Delta V = (1)(24.6) = 24.6 \text{ liter-atm}$$

$$= 595 \text{ cal}$$

$$\Delta S_{surr} = -q/T = -\tfrac{595}{300} = -1.98 \text{ cal/mole}$$

Then $\Delta S_{univ} = \Delta S_{sys} + \Delta S_{surr}$

$$= 2.76 + (-1.98) = 0.78 \text{ cal/deg}$$

As the process is isothermal ΔG for the system can be calculated using

$$\Delta G = \Delta H - T\Delta S$$

$$\Delta H = 0 \text{ (isothermal process of an ideal gas)}$$

$$\Delta G = 0 - (300)(2.76) = -828 \text{ cal}$$

Note that consistent with the second law the entropy of the universe has increased and ΔG for the system is negative since this is a process in which a system not at equilibrium proceeds to equilibrium.

9.51   Calculate ΔS of the universe and ΔG of the system for the change in state: 2.5 moles of an ideal gas expands isothermally at 300°K against a constant opposing pressure of 0.5 atm from a volume of 15 liters to equilibrium volume.

9.52   Calculate ΔS of the universe and ΔG of the system for the change in state: 2 moles of an ideal gas contracts isothermally at 0°C due to a pressure of 1.5 atm from a volume of 40 liters to equilibrium volume. Assume the surroundings are at 0°.

9.53   Calculate ΔS of the system and ΔG of the system for the change in state: 1 mole of an ideal gas at 1 atm is heated reversibly at constant pressure from 25°C to 100°C and is then cooled reversibly at constant volume to 25°C. The temperature of the surroundings is 25°C.

Solution hint: Changes in state for the system depend only on the initial and final states, not on the path; so that only the calculation of ΔS_{surr} requires dealing with the path.

9.54 The heat of vaporization of water at its boiling point is 9720 cal/mole. Calculate ΔS of vaporization at the boiling point at 1 atm.

Solution: Since the vaporization is a reversible process at constant T and P, $\Delta G = 0$ and $\Delta H = T\Delta S$. Substituting $9720 = (373)(\Delta S)$

$$\Delta S = 26.1 \text{ cal/deg-mole}$$

9.55 At its boiling point it is found that gold has ΔH of vaporization = 74.1 kcal/mole and ΔS of vaporization = 29.6 cal/deg-mole. Calculate the boiling point of gold.

9.56 Calculate ΔS of the system for the change in state in Problem 9.47.

➡ **9.57** Calculate ΔS of the universe and ΔG of the system for the change in state in Problem 9.48.

➡ **9.58** In the grid below for each process written on the left put into each box a + if the indicated parameter is positive, a − if it is negative, and a 0 if it is zero (I.G. = ideal gas).

Process	q	w	ΔE	ΔH	ΔS	ΔG
1 mole I.G. (25°C, 1 atm) ⟶ 1 mole I.G. (25°C, 0.5 atm)						
$H_2O_{(l)} \longrightarrow H_2O_{(g)}$ (25°C, 1 atm)			+	+		
$H_2O_{(l)} \longrightarrow H_2O_{(s)}$ (−10°C, 1 atm)		0				
1 mole I.G. (25°C, 10 liters) ⟶ 1 mole I.G. (100°C, 10 liters)						+
$H_{2(g)}$ (1 atm, 3000°K) ⟶ 2H (equilibrium pressure, 3000°K)		0				

Concentration of solutions.

In order to be able to deal with solutions quantitatively it is necessary to have methods of expressing the concentration, that is, the relative amounts of solute and solvent in a given solution. There are a number of ways of doing this.

(1) Mass–Mass expressions indicate the relative masses of solute and solvent, either directly or in terms of moles, in a given solution.

(a) mass percent or percent strength $= \dfrac{\text{mass of solute}}{\text{mass of solution}} \times 100$

(b) mass fraction $= \dfrac{\text{mass of solute}}{\text{mass of solution}}$

(c) mole percent $= \dfrac{\text{moles of solute}}{\text{total moles in the solution}} \times 100$

(d) mole fraction (abbreviated f or X) $= \dfrac{\text{moles of solute}}{\text{total moles in the solution}}$

(e) molality (abbreviated m) $= \dfrac{\text{moles of solute}}{\text{g of solvent}} \times 1000$

It should be noted that in all the mass–mass expressions except molality, the denominator contains a term involving the entire solution.

(2) Mass–Volume expressions are the most common method of expres-

sing concentration and indicate how many moles of solute are in a given volume of solution.

(a) molarity (abbreviated M) $= \dfrac{\text{moles of solute}}{\text{liters of solution}}$

(b) formality (abbreviated F) is closely related to molarity and will be discussed in Chapter 12.

It should be emphasized that in problems dealing with reactions that occur in solution *the mole relationships are treated exactly as they were treated in the chapter on stoichiometry*. The only new thing in this chapter is that instead of being given a certain number of moles or grams of a particular reactant, you will be given, for example, a certain number of milliliters of a solution of a certain molarity. You will first determine how many *moles* of reactant there are in the solution; having determined the number of *moles* the rest of the calculation follows the usual pattern for stoichiometric calculations.

(3) Volume–Volume expressions are used for solutions of two or more liquid components.

(a) volume percent $= \dfrac{\text{volume of solute}}{\text{volume of solution}} \times 100$

(b) proof $= 2 \times$ volume percent

Often solution of a problem requires converting from one concentration unit to another. In order to change between a mass–mass and a mass–volume expression it is necessary to know the density of the solution; that is the mass of a given volume. Density (abbreviated d) is usually defined as $d = m/v$ where m is mass in grams and volume is in cubic centimeters.

PROBLEMS

10.1 Twelve g of NaCl are dissolved in 68 g of water. Calculate the mass percent of NaCl in the solution.

Solution: Calculation of the mass percent requires the mass of solution which is equal to the mass of solute plus the mass of solvent. Twelve g of NaCl in 68 g of H_2O gives $12 + 68 = 80$ g of solution. Substituting into the expression for mass percent gives:

$$\text{mass percent} = \frac{12 \text{ g}}{80 \text{ g}} \times 100 = 15\%$$

It should be noted that in the calculation of a percent or a fraction any units can be used provided the numerator and the denominator have the same units.

10.2 How many grams of NaCl are there in 60 g of a 15% solution of NaCl in water?

Solution: 15% means that 0.15 of the mass of the solution is NaCl.

$$0.15 \times 60 \text{ g} = 9.0 \text{ g}$$

10.3 How many grams of sugar would have to be dissolved in 60 g of water to yield a 25% solution?

Solution: What we want is a solution in which the mass of the sugar is 0.25 of the mass of the solution.

The mass of the solution is the mass of the sugar plus the mass of the water.

Let X = mass of sugar

X + 60 g = mass of the solution

$$X = 0.25 (X + 60 \text{ g})$$

Solving, X = 20 g of sugar

10.4 How many grams of water and how many grams of salt would you use to prepare 80 g of a 5.0% solution?

Solution: In a 5.0% solution, the mass of the salt is 0.050 of the mass of the solution.

$$0.050 \times 80 \text{ g} = 4.0 \text{ g of salt}$$

The mass of the solution = mass of salt + mass of water.

$$80 \text{ g of solution} - 4.0 \text{ g of salt} = 76 \text{ g of water}$$

10.5 The mass of 15 cc of a solution is 12 g. Calculate the density of the solution.

Solution: Density means grams per cubic centimeter. Therefore, to find the density we will divide the mass in grams by the volume in cc. That is,

$$\text{density} = \frac{\text{g}}{\text{cc}} = \frac{12 \text{ g}}{15 \text{ cc}} = 0.80 \text{ g/cc}$$

10.6 The density of a solution is 1.80 g/cc. What volume will 360 g of the solution occupy?

Solution: To find the volume in cc occupied by 360 g of a solution we must divide the mass in grams (360 g) by the mass in grams of 1 cc. that is,

$$\frac{\text{g}}{\text{g/cc}} = \text{cc}$$

$$\frac{360 \text{ g}}{1.80 \text{ g/cc}} = 200 \text{ cc}$$

10.7 A 44.0% solution of H_2SO_4 has a density of 1.343 g/ml. How many grams of H_2SO_4 are there in 60 ml of this solution?

Solution: The mass in grams of 60 ml of this solution will be

60 ml of solution $\times$ 1.343 g/ml = 80.58 g of solution

0.440 g of H_2SO_4 per g of solution

$\times$ 80.58 g of solution = 35.46 g of H_2SO_4

The entire calculation, in one operation, is

60 ml $\times$ 1.343 g/ml $\times$ 0.440 g of H_2SO_4/g = 35.46 g of H_2SO_4

10.8   A 44.0% solution of H_2SO_4 has a density of 1.343. A volume of 25.0 cc of 44.0% H_2SO_4 solution was treated with an excess of Zn. What volume did the dry hydrogen gas which was liberated occupy at STP?

Solution: $Zn + H_2SO_4 = ZnSO_4 + H_2$. Therefore, 1 mole of H_2SO_4 will liberate 1 mole of H_2.

We will calculate how many moles of H_2SO_4 are present in the solution.

25.0 cc $\times$ 1.343 g/cc $\times$ 0.440 = 14.8 g of H_2SO_4

$$\frac{14.8 \text{ g of } H_2SO_4}{98.1 \text{ g/mole}} = 0.151 \text{ mole of } H_2SO_4$$

Therefore 0.151 mole of H_2 was evolved. One mole of H_2 occupies a volume of 22.4 liters at STP. Therefore,

$$0.151 \text{ mole} \times \frac{22.4 \text{ liters}}{1 \text{ mole}} = 3.38 \text{ liters of } H_2$$

The entire calculation, in one operation, is

$$\frac{25.0 \text{ cc} \times 1.343 \text{ g/cc} \times 0.440 \text{ g of } H_2SO_4/g}{98.1 \text{ g of } H_2SO_4/\text{mole of } H_2SO_4} \times \frac{1 \text{ mole of } H_2}{1 \text{ mole of } H_2SO_4}$$

$$\times \frac{22.4 \text{ liters of } H_2}{1 \text{ mole of } H_2} = 3.38 \text{ liters of } H_2$$

10.9   How many liters of dry HCl gas, measured at 25°C and 740 mm, can be prepared by combining chlorine gas with the hydrogen which will be liberated when 100 cc of a 20.0% solution of H_2SO_4 of density 1.14 is treated with an excess of aluminum?

10.10   Ten g of NH_4Cl are dissolved in 100 g of a 10% solution of NH_4Cl in water. Calculate the percent strength of the resulting solution.

10.11   You are given 100 g of a 10.0% solution of $NaNO_3$ in water. How many more grams of $NaNO_3$ would you have to dissolve in the 100 g of 10.0% solution to change it to a 20.0% solution?

10.12   Sixty g of a 12% solution of NaCl in water were mixed with 40 g of a 7.0% solution of NaCl in water. What was the percent strength of the resulting solution?

10.13 A solution is prepared by mixing 10 g of CH_3OH and 100 g of H_2O. Calculate the mole fraction of each component.

Solution: To calculate the mole fraction of a component of a solution we must know the number of moles of that component and the total number of moles of all the components in the solution. The moles of CH_3OH, $n_{CH_3OH} = \frac{10}{32} = 0.31$; the moles of H_2O, $n_{H_2O} = \frac{100}{18} = 5.5$. Then

$$f_{CH_3OH} = \frac{n_{CH_3OH}}{n_{CH_3OH} + n_{H_2O}} = \frac{0.31}{0.31 + 5.5} = 0.053$$

Once the fraction of all the components of a solution but one are known, the remaining fraction is most easily calculated by remembering that the sum of the fractions is 1. In this case $f_{CH_3OH} + f_{H_2O} = 1$ or $f_{H_2O} = 1 - 0.053 = 0.947$.

10.14 A solution is prepared at 25°C by mixing 15 g of Na_2SO_4 with 125 cc of H_2O. The density of H_2O at this temperature is 1.0 g/cc. What is the molality of Na_2SO_4 in the solution?

Solution: Molality refers to the number of moles of solute which are dissolved in 1000 g of solvent. Thus, there are 15 g/142 = 0.106 moles of Na_2SO_4 and there are 125 cc × 1.0 g/cc = 125 g of H_2O. Substituting into the expression for molality

$$m = \frac{0.106 \text{ moles}}{125 \text{ g}} \times 1000 = 0.84$$

10.15 Calculate the mole fraction and the molality of KCl in a solution prepared at 25°C from 17 g of KCl and 196 cc of H_2O.

10.16 How many grams of glucose $C_6H_{12}O_6$ must be dissolved in 150 cc of water to prepare a solution which is 0.2 *m* in glucose?

10.17 What is the mole fraction of glucose in the solution in Problem 10.16?

➡ **10.18** A solution of C_2H_6O in water has a mole fraction of $H_2O = 0.97$. Calculate the molality of the C_2H_6O.

➡ **10.19** To 223 ml of a solution containing 9.20% $C_2H_6O_2$ (ethylene glycol) by weight in water with a density of 1.12 g/ml is added 21.4 g of $C_3H_8O_3$ (glycerol). Calculate the mole fraction of glycerol in the resultant solution.

10.20 How many grams of NaOH will be required to prepare 1.0 liter of 1.0 *M* NaOH?

Solution: 1.0 *M* NaOH will contain 1.0 mole of NaOH dissolved in enough water to make 1 liter of solution. One mole of NaOH is 40 g of NaOH. Therefore 40 g will be required.

10.21 How many grams of K_2SO_4 will be required to prepare 1.00 liter of 0.500 *M* K_2SO_4?

Solution: One liter of 0.500 *M* K_2SO_4 will contain half a mole of K_2SO_4 dissolved in enough water to make a liter of solution. Half a mole of K_2SO_4 is 87.1 g. Therefore, 87.1 g of K_2SO_4 will be required.

10.22 How many grams of $Al_2(SO_4)_3$ will be required to prepare 300 ml of 0.200 *M* $Al_2(SO_4)_3$?

Solution: One liter of 0.200 *M* $Al_2(SO_4)_3$ will contain 0.200 mole of $Al_2(SO_4)_3$. 300 ml is 0.300 liter; 0.300 liter of 0.200 *M* $Al_2(SO_4)_3$ will contain 0.3×0.2 or 0.0600 mole of $Al_2(SO_4)_3$. One mole of $Al_2(SO_4)_3$ is 342 g. Therefore, 0.0600×342 g = 20.6 g.
What has been stated above can be summarized, briefly, as follows:

$$0.300 \text{ liter} \times \frac{0.200 \text{ mole}}{\text{liter}} \times \frac{342 \text{ g}}{\text{mole}} = 20.6 \text{ g}$$

Note that liters and moles cancel, leaving the answer in grams.

10.23 If 12 g of NaOH are dissolved in enough water to give 500 ml of solution, calculate the molarity of the solution.

Solution: To find the molarity means to find the number of moles of solute that are present in 1000 ml (1 liter) of solution.

$$1 \text{ mole of NaOH} = 40 \text{ g of NaOH}$$
Therefore,

$$12 \text{ g of NaOH} = \tfrac{12}{40} \text{ mole of NaOH} = 0.30 \text{ mole of NaOH}$$

The 0.30 mole of NaOH is present in 500 ml of solution. Since the molarity is the number of moles per 1000 ml, then,

$$1000 \text{ ml} \times \frac{0.30 \text{ mole}}{500 \text{ ml}} = 0.60 \text{ mole}$$
Therefore, the solution is 0.60 *M*.

10.24 A solution of $Cu(NO_3)_2$ contains 100 mg of the salt per milliliter. Calculate the molarity of the solution.

Solution: 100 mg = 0.1 g. 0.1 g in 1 ml would be the same concentration as 100 g in 1000 ml. Therefore, this is a solution containing 100 g of $Cu(NO_3)_2$ per liter. To find the molarity we must find the number of moles per liter. There are 187.5 g of $Cu(NO_3)_2$ in a mole. Therefore,

$$\frac{100 \text{ g of } Cu(NO_3)_2}{187.5 \text{ g/mole}} = 0.53 \text{ mole of } Cu(NO_3)_2$$

Therefore, the molarity is 0.53.

10.25 How many grams of KOH will be required to prepare 400 ml of 0.12 *M* KOH?

10.26   How many liters of 0.20 M Na_2CO_3 can be prepared from 140 g of Na_2CO_3?

10.27   A solution of NaCl contained 12 g of NaCl in 750 ml of solution. What was the molarity of the solution?

10.28   If 200 ml of 0.30 M Na_2SO_4 are evaporated to dryness, how many grams of dry Na_2SO_4 will be obtained?

10.29   10.0 cc of a 70.0% solution of sulfuric acid of density 1.61 were dissolved in enough water to give 25.0 cc of solution. What was the molarity of the final solution?

10.30   In 3.58 M H_2SO_4 there is 29.0% H_2SO_4. Calculate the density of 3.58 M H_2SO_4.

10.31   To what volume in ml must 44.20 ml of a 70.00% solution of sulfuric acid whose density is 1.610 be diluted to give 0.4000 M H_2SO_4?

10.32   A 1.00 M solution of CH_3OH in water has a density of 0.91 g/ml. Calculate the mole fraction, mass percent, and molality of the CH_3OH.

Solution: From the density, 1 liter of solution weighs (0.91) g/ml $\times$ 1000 ml = 910 g. It contains 1 mole = 32 g of CH_3OH and therefore $910 - 32 = 878$ g of H_2O.
Thus,

$$f_{CH_3OH} = \frac{n_{CH_3OH}}{n_{CH_3OH} + n_{H_2O}} = \frac{1}{1 + \frac{878}{18}} = 0.020$$

$$\text{mass percent} = \frac{32}{910} \times 100 = 3.52\%$$

$$\text{molality} = \frac{1 \text{ mole}}{(910 - 32)\, g} \times 1000 = 1.13\, m$$

10.33   A 1.30 M solution of KBr in water has a density of 1.16 g/ml. Calculate the mole fraction, mass percent, and molality of the KBr.

10.34   A solution which is 5.30% LiBr by mass has a density of 1.04 g/ml. Calculate its molarity.

10.35   How many moles of hydrogen will be liberated from 320 ml of 0.50 M H_2SO_4 by an excess of magnesium?

Solution:

$$Mg + H_2SO_4 = MgSO_4 + H_2$$

0.50 M H_2SO_4 contains 0.50 mole of H_2SO_4 per liter

$$0.320 \text{ liter of } H_2SO_4 \times \frac{0.50 \text{ mole of } H_2SO_4}{1 \text{ liter of } H_2SO_4} = 0.16 \text{ mole of } H_2SO_4$$

1 mole of H_2SO_4 liberates 1 mole of H_2
Therefore, 0.16 mole of H_2 will be liberated.

10.36 How many moles of hydrogen will be liberated from 400 ml of 0.40 M HCl by an excess of zinc?

Solution:

$$Zn + 2\,HCl = ZnCl_2 + H_2$$

2 moles of HCl yield 1 mole of H_2
1 mole of HCl yields 0.5 mole of H_2
0.40 M HCl contains 0.40 moles of HCl per liter

$$0.400 \text{ liter} \times \frac{0.40 \text{ mole}}{1 \text{ liter}} = 0.16 \text{ mole of HCl}$$

Therefore, 0.080 mole ($\frac{1}{2}$ of 0.16 mole) of H_2 will be liberated.

10.37 How many moles of hydrogen gas will be liberated when an excess of magnesium reacts with:

(a) 600 ml of 0.80 M H_2SO_4?
(b) 600 ml of 0.80 M HCl?

10.38 600 ml of 0.40 M HCl were treated with excess Mg. The evolved H_2 gas was all used to reduce CuO to Cu. How many grams of free copper were formed?

10.39 A 447-ml solution of Na_2CO_3 was warmed with an excess of sulfuric acid until all action ceased. Five liters of dry CO_2 gas, measured at standard conditions, were given off. Calculate the molarity of the sodium carbonate solution.

Solution: The analysis of this problem could run something like this: To find the molarity we must find the number of moles of Na_2CO_3 per liter of solution. If we knew how many moles of Na_2CO_3 there were in 447 ml, we could calculate the number in 1000 ml (1 liter). From the equation,

$$Na_2CO_3 + H_2SO_4 = Na_2SO_4 + H_2O + CO_2$$

or, more simply, from the formulas, Na_2CO_3 and CO_2, we can see that 1 mole of Na_2CO_3 yields 1 mole of CO_2. Therefore, if we knew how many moles of CO_2 were formed we would know how many moles of Na_2CO_3 were present in the 447 ml of solution. We know that 5 liters of CO_2 were evolved. Since 1 mole of CO_2 occupies a volume of 22.4 liters at STP, 5 liters of CO_2 is 5/22.4 mole. Therefore, 5/22.4 mole of Na_2CO_3 is present in the 447 ml (0.447 liter) of solution.

$$\frac{5}{22.4} \text{ mole} \div 0.447 \text{ liter} = 0.5 \text{ mole/liter}$$

Therefore, the solution is 0.5 M.

10.40 A beaker contained 130 ml of hydrochloric acid. The contents were treated with excess zinc. The result was that 7.13 liters of dry hydrogen gas, measured at 22°C and 738 mm, were obtained. Calculate the molarity of the acid.

➡ **10.41** A 17.4-ml sample of a 70.0% solution of sulfuric acid whose density is 1.61 was diluted to a volume of 100 ml and was then treated with a large excess of zinc. The evolved hydrogen gas was combined with chlorine gas to form HCl. This HCl gas was then dissolved in enough water to form 200 ml of hydrochloric acid. There was no loss of material in the reactions. Calculate the molarity of the hydrochloric acid.

10.42 How many milliliters of 0.250 M HCl will be required to neutralize 120 ml of 0.800 M KOH?

Solution: One mole of HCl will neutralize 1 mole of KOH. Therefore, moles of HCl required = moles of KOH present in the solution. 0.800 M KOH contains 0.800 mole of KOH per liter; 0.250 M HCl contains 0.250 mole of HCl per liter.

$$0.120 \text{ liter} \times \frac{0.800 \text{ mole of KOH}}{1 \text{ liter}} = \text{moles of KOH present}$$

$$X \text{ liters} \times \frac{0.250 \text{ mole of HCl}}{1 \text{ liter}} = \text{moles of HCl required}$$

Therefore, since moles of HCl required = moles of KOH present

$$X \text{ liters of HCl} \times \frac{0.250 \text{ mole}}{1 \text{ liter}} = 0.120 \text{ liter} \times \frac{0.800 \text{ mole}}{1 \text{ liter}}$$

$$X = 0.384 \text{ liter} = 384 \text{ ml}$$

10.43 If 12.0 g of NaOH were required to neutralize 82.0 ml of sulfuric acid, calculate the molarity of the acid.

10.44 25.0 ml of NaOH solution exactly neutralized 40 ml of 0.10 M H$_2$SO$_4$. Calculate the molarity of the NaOH.

10.45 How many milliliters of 0.250 M AgNO$_3$ will be required to precipitate the chloride from 80.0 ml of 0.400 M NaCl? How many grams of AgCl will be precipitated?

➡ **10.46** An 85-g sample of an antimony sulfide ore containing 40% by weight of Sb$_2$S$_3$ and 60% inert material is oxidized until all of the S in the Sb$_2$S$_3$ is converted to SO$_3$. This SO$_3$ is dissolved in enough water to give 200 ml of solution. How many ml of 0.400 M NaOH will be required to completely neutralize the contents of the 200 ml solution?

Solution hint: Sb$_2$S$_3$ → 3 SO$_3$ → 3 H$_2$SO$_4$ → 6 NaOH

moles of NaOH required = 6 × moles of Sb$_2$S$_3$

$$\text{ml of NaOH required} \times \frac{0.400 \text{ moles of NaOH}}{1000 \text{ ml of solution}}$$

$$= \text{moles of NaOH required}$$

➡ **10.47** A 16-g mixture of sodium and potassium, when allowed to react with water, gave a solution which neutralized 602.5 ml of 0.40 M H_2SO_4. How many grams of sodium were there in the mixture?

Solution hint: One mole of Na yields 1 mole of NaOH; 1 mole of K yields 1 mole of KOH. Moles of hydroxides = 2 × moles of H_2SO_4.

➡ **10.48** A 14.8-g mixture of Na_2CO_3 and $NaHCO_3$ was dissolved in enough water to make 400 ml of solution. When these 400 ml of solution were treated with excess 2.00 M H_2SO_4 and boiled to remove all dissolved gas, 3.73 liters of dry CO_2 gas measured at 740 mm and 22.0°C were obtained. Calculate the molarity of the Na_2CO_3 and of the $NaHCO_3$ in the 400 ml of solution.

Solution hint: Let X = molarity of Na_2CO_3 and Y = molarity of $NaHCO_3$.
moles of Na_2CO_3 in 400 ml of solution = 0.400 X
moles of $NaHCO_3$ in 400 ml of solution = 0.400 Y
One mole of Na_2CO_3 yields 1 mole of CO_2.
One mole of $NaHCO_3$ yields 1 mole of CO_2.

➡ **10.49** To a beaker containing 164 ml of a solution of $CuSO_4$ was added 10.00 g of magnesium metal. When reaction was complete, a mixture of Mg and Cu having a mass of 14.45 g remained in the beaker. Calculate the molarity of the original $CuSO_4$ solution.

Properties of solutions.

A number of the physical properties of a solution, to a first approximation, do not depend on the nature of the solute but only on the number of solute particles which are dissolved. These properties are called *colligative properties*, and some examples are vapor pressure, boiling point, melting point, and osmotic pressure. A study of these properties often reveals important information about the solute such as its molecular weight or its state in solution. The approach we will use here involves an approximation similar to the one we use in treating gases as ideal gases. Dilute solutions are treated as ideal solutions with the result that relatively simple relationships govern colligative properties.

It is found that at a given temperature the vapor pressure of a liquid solvent is always greatest where the liquid is pure and decreases as solute is dissolved in it. For dilute solutions, *Raoult's law*

$$P_A = P_A{}^\circ f_A$$

states that the vapor pressure of solvent A (P_A) above a solution is equal to the vapor pressure of the pure liquid ($P_A{}^\circ$) at the given temperature multiplied by the mole fraction (f_A) of the solvent in the solution. The same relationship also applies to solutions with two or more volatile components, for example water and ethanol. The vapor pressure of each volatile component above the solution at a certain temperature is given by its vapor pressure when pure at that temperature multiplied by its mole fraction in the solution.

Raoult's law may also be applied to the calculation of the melting point and boiling point of the solvent in a solution which contains only nonvolatile solutes. The boiling point of the solvent in such a solution is *higher* than it is when the solvent is pure and the freezing point is *lower* for the solvent in such a solution than when it is pure. This is referred to as *boiling point elevation* and *freezing point depression*. It can readily be demonstrated for the ideal solution that the magnitude of the boiling point elevation or freezing point depression depends on the number of nonvolatile solute particles and on the nature of the solvent. The exact relationship is: $\Delta T = K_b m$ for boiling point and $\Delta T = K_f m$ for freezing point, where ΔT is the difference between the boiling or freezing point of the pure solvent and the solvent in the solution expressed as a positive number, where m is the total molality of solute particles, and where K_b and K_f are the boiling point constant and freezing point constant. These constants are experimentally determined quantities characteristic of each solvent. For water it is found that $K_b = 0.52$ and $K_f = 1.86$ when the temperature is measured in °C.

Another property of solutions which depends on the number of solute particles is the osmotic pressure. This can be measured by placing the solution in a tube fitted at one end with a semipermeable membrane through which the solvent but not the solute can pass. If the tube is placed in a container of the pure solvent, the solvent will enter the tube causing the liquid level in the tube to rise to a certain height. The difference in height between the level of the pure solvent and the level of the solution which has risen in the tube can be used to calculate the osmotic pressure, π. The osmotic pressure is related to the number of moles of solute particles by a relationship which closely resembles the ideal gas equation:

$$\pi V = nRT$$

where V is the volume of solution, n is the number of moles of solute particles in that volume, T is the temperature, and $R = 0.082$ liter-atm/deg-mole, the ideal gas constant. Measurement of osmotic pressure is a useful way of finding the molecular weight of large molecules since very small osmotic pressures can be measured.

It is crucial to recognize that colligative properties depend on the number of moles of solute particles in the solution which is not necessarily the same as the number of moles of solute used in preparing the solution. When strong electrolytes such as salts, strong acids, and strong bases are dissolved in polar solvents such as water they dissociate into ions. Thus, when one mole of NaCl dissolves in water it produces two moles of particles: one mole of Na^+ ions and one mole of Cl^- ions. Similarly each mole of $Mg(OH)_2$ produces three moles of particles (one Mg^{++} and two OH^-), each mole of $Ca_3(PO_4)_2$ produces 5 moles of particles (three Ca^{++} and two PO_4^{---}). When weak electrolytes such as weak acids and weak bases dissolve in polar solvents they undergo partial dissociation to produce varying numbers of solute particles. Thus, it is possible by measuring some of the colligative properties of a solution whose composition is known to gain information about the behavior of substances when they dissolve.

PROBLEMS

11.1   A quantity of 60.0 g of a nonelectrolyte dissolved in 1000 g of H_2O lowered the freezing point 1.02°C. Calculate the approximate molecular weight of the nonelectrolyte.

Solution: One mole of nonelectrolyte in 1000 g of water would have depressed the freezing point 1.86°C. Since a depression of 1.02° was observed, 1.02/1.86 mole of nonelectrolyte must have been dissolved.

$$\frac{1.02}{1.86} \text{ mole} = 60 \text{ g}$$

$$1 \text{ mole} = \frac{1.86}{1.02} \times 60 \text{ g} = 109 \text{ g}$$

The molecular weight is approximately 109.

11.2   When 4.20 g of a nonelectrolyte were dissolved in 40.0 g of water, a solution which froze at −1.52°C was obtained. Calculate the approximate molecular weight of the nonelectrolyte.

Solution: The molecular weight is the mass of solute that will depress the freezing point 1.86°C when dissolved in 1000 g of solvent. Therefore, we will first find the concentration in grams of solute per 1000 g of water.

$$1000 \text{ g of water} \times \frac{4.20 \text{ g of solute}}{40.0 \text{ g of water}} = 105 \text{ g of solute}$$

A solution containing 105 g of solute in 1000 g of water is the same *concentration* as one containing 4.20 g of solute in 40.0 g of water. Therefore, 105 g of solute will depress the freezing point of 1000 g of solvent exactly the same number of degrees that 4.20 g of solute will depress the freezing point of 40.0 g of solvent, namely, 1.52°C. Continue as in Problem 11.1.

11.3   A 20-g quantity of $C_6H_{10}O_5$, a nonelectrolyte, is dissolved in 250 g of H_2O. Calculate the boiling point of the solution at 760 mm.

Solution: First we calculate the molality of the solution. Since the molecular weight of $C_6H_{10}O_5$ is 162,

$$m = \frac{20/162}{250} \times 1000 = 0.494.$$

ΔT is the boiling point of the solution in °C minus 100°C, the boiling point of water. Thus

$$\Delta T = \text{B.P.}_{\text{solution}} - 100°C = 0.52(m) = (0.52)(0.494)$$

$$\text{B.P.}_{\text{solution}} = (0.52)(0.494) + 100$$

$$= 100.26°C$$

11.4 When 6.00 g of a nonelectrolyte were dissolved in 54.0 g of H_2O, a solution was obtained which boiled at 100.41°C at 760 mm. Calculate the approximate molecular weight of the nonelectrolyte.

11.5 6.00 g of C_2H_5OH, a nonelectrolyte, were dissolved in 300 g of H_2O. Calculate the freezing point of the solution.

11.6 How many grams of $C_3H_5(OH)_3$, a nonelectrolyte, must be dissolved in 600 g of H_2O to give a solution which will freeze at −4.00°C?

11.7 When 12 g of a nonelectrolyte were dissolved in 300 g of water, a solution which froze at −1.62°C was obtained. What was the approximate molecular weight of the nonelectrolyte?

11.8 When 5.12 g of the nonionizing solute, naphthalene $(C_{10}H_8)$, are dissolved in 100 g of CCl_4 (carbon tetrachloride), the boiling point of the CCl_4 is raised 2°. What is the boiling-point constant for CCl_4?

11.9 A compound was found, on analysis, to consist of 50.00% oxygen, 37.50% carbon, and 12.50% hydrogen. When dissolved in 100.0 cc of water, 1.666 g of the compound gave a nonconducting solution which froze at −1.00°C. Calculate the exact molecular weight of the compound.

11.10 A solution is prepared by dissolving 2.37 g of NaCl in 100 g of water. What is the boiling point of the solution?

Solution: This problem is treated like Problem 11.3 except that we must include the fact that each mole of NaCl gives two moles of solute particles when it dissolves in water. In order to calculate the boiling point we use

$$m = 2 \times \frac{2.37}{58.4} \Big/ 100 \times 1000.$$

11.11 A solution is prepared by dissolving 1.27 g of an electrolyte whose molecular weight is 310 in 50 g of water, and the freezing point of the solution is found to be −0.762°C. How many moles of solute particles are formed in solution from each mole of solute?

Solution: First calculate the freezing point depression assuming no dissociation

$$\Delta T = K_f m = (1.86)\frac{(1.27/310 \times 1000)}{50} = 0.152$$

Then divide this value into the observed value: $0.762/0.152 = 5.01$ suggesting that each mole of solute has dissociated into 5 moles of solute particles.

11.12 A solution of 6.27 g of K_2SO_4 in 50 g of water is prepared. What is its boiling point?

11.13 How many grams of $Mg(NO_3)_2$ are dissolved in 100 g of a water *solution* which boils at $101.22°C$?

Solution hint: First calculate the molality of the $Mg(NO_3)_2$ taking into account the complete dissociation; then calculate the weight of the salt in 100 g of solution.

11.14 A solution of a salt whose molecular weight is 142.1 is prepared from 2.84 g of salt and 500 g of water. Its freezing point is observed to be $-0.223°C$. How many moles of solute particles are formed in solution from each mole of salt which is dissolved?

11.15 When 1.16 g of acetic acid, $C_2H_4O_2$, is dissolved in 50 g of benzene the freezing point of the solution is found to be depressed by $1.024°C$. The value of K_f for benzene is 5.12. Describe the state of acetic acid in benzene solution.

➡ **11.16** When H_2SO_3, a weak acid, is dissolved in water it undergoes partial dissociation according to the equation $H_2SO_3 \rightleftharpoons H^+ + HSO_3^-$. A solution is prepared which contains 1.64 g of H_2SO_3 in 100 g of water. The freezing point of this solution is found to be $-0.649°C$. Calculate the fraction of H_2SO_3 which dissociates.

Solution: First proceed as in Problem 11.11 to calculate the freezing point depression assuming no dissociation occurs.

$$\Delta T = (1.86)\left(\frac{1.64/82.1}{100} \times 1000\right) = 0.372°C$$

Divide this calculated ΔT into the observed ΔT: $0.649/0.372 = 1.74$. This verifies that dissociation was not complete since if it were we would find that the observed ΔT is twice the calculated one.

To calculate the fraction dissociated let $X =$ fraction dissociated. From the equation given for the reaction, for each mole of H_2SO_3 which dissociates, two moles of particles are formed. If X dissociates, $2X$ particles are formed and $1 - X$ particles of H_2SO_3 remain. Thus the total number of particles is $2X + (1 - X) = X + 1$. We have also calculated that this number of particles gives 1.74 times as much of a freezing point depression as if there were no dissociation. Thus $X + 1 = 1.74$, $X = 0.74$, which is the fraction of H_2SO_3 which has dissociated.

➡ **11.17** When HNO_2 is dissolved in water it partially dissociates according to the equation $HNO_2 \rightleftarrows H^+ + NO_2^-$. A solution is prepared which contains 7.050 g of HNO_2 in 1000 g of water. Its freezing point is found to be $-0.2929°C$. Calculate the fraction of HNO_2 which has dissociated.

➡ **11.18** It is found that when a solution containing 39.2 g of H_2SO_4 in 1000 g of water is prepared, the H_2SO_4 dissociates completely into H^+ and HSO_4^- and that 17.8 mole percent of the HSO_4^- further dissociates into $H^+ + SO_4^=$. Calculate the freezing point of the solution.

11.19 A solution of 42.2 g of glucose, $C_6H_{12}O_6$, a nonelectrolyte, in 200 g of water is prepared at 25°C. What is the vapor pressure of water above the solution?

Solution: Assuming ideal solution behavior we can substitute into the Raoult's law expression; using the value of $P°_{H_2O}$ from Table 1 in the appendix.

$$P_{H_2O} = P°_{H_2O} f_{H_2O} = (23.6)\left(\frac{\frac{200}{18}}{\frac{200}{18} + \frac{42.2}{180}}\right) = 23.1 \text{ mm}$$

11.20 A solution is prepared from 27.1 g of K_2SO_4 and 1000 g of water. Calculate the vapor pressure of water above the solution at 100°C.

Solution: Since pure water boils at 100°C, its vapor pressure is 760 mm. The vapor pressure above the solution will depend on the mole fraction of water based on the number of moles of solute in the solution. Since one mole of K_2SO_4 gives 3 moles of solute particles, 27.1 g gives $3 \times 27.1/174 = 0.467$ moles of solute particles.

$$f_{H_2O} = \frac{\frac{1000}{18}}{\frac{1000}{18} + 0.467} = 0.992$$

and $P = (760)(0.992) = 754$ mm.

Note that the solution does not boil at 100°C, since its vapor pressure is less than 760 mm.

11.21 A solution of 50 g of $C_2H_4O_2$, ethylene glycol, a nonelectrolyte, and 50 g of water is prepared. Calculate the vapor pressure of water above this solution at 40°C.

11.22 A solution of 75 g of NH_4Cl (a salt) in 150 g of water is prepared. Calculate the vapor pressure of water above the solution at 35°C.

11.23 A solution is prepared at 70°C from 80 g of C_2H_6O, ethanol, and 100 g of water. Calculate the vapor pressure of each component above the solution and the mole fraction of ethanol in the vapor. The vapor pressure of pure ethanol at this temperature is 543 mm and of pure water is 234 mm.

Solution: For a solution with more than one volatile component Raoult's law can be applied to each separately. First we calculate the mole fraction of each component:

$$f_{H_2O} = \frac{\frac{100}{18}}{\frac{100}{18} + \frac{80}{46}} = 0.762$$

$$f_{C_2H_6O} = 1 - f_{H_2O} = 0.238$$

$$P_{H_2O} = P°_{H_2O}f_{H_2O} = (234)(0.762) = 178 \text{ mm}$$

$$P_{C_2H_6O} = P°_{C_2H_6O}f_{C_2H_6O} = (543)(0.238) = 129 \text{ mm}$$

Since for a given set of conditions moles are directly proportional to pressure the $f_{C_2H_6O}$ in the vapor can be calculated directly from the pressures.

$$f_{C_2H_6O} = \frac{129}{129 + 178} = 0.420$$

11.24 Calculate the vapor pressure of C_2H_6O, ethanol, and of H_2O, and the mole fraction of H_2O in the vapor above a solution prepared at 30°C by mixing 100 g of ethanol and 100 g of water. The vapor pressure of pure ethanol at this temperature is 78.0 mm.

➡ 11.25 The vapor above the solution in Problem 11.24 is separated and condensed. Calculate the mole fraction of water in the vapor above this condensate. The vapor above this condensate is again separated and condensed. Calculate the mole fraction of water in the vapor above the second condensate. This problem describes a process called fractional distillation.

➡ 11.26 A solution prepared at 30°C from 50 g of C_6H_6, benzene, and 50 g of C_7H_8, toluene, has a total vapor pressure of 80.8 mm. A solution prepared at 30°C from 75 g of C_7H_8 and 25 g of C_6H_6 has a total vapor pressure of 59.9 mm. Calculate the vapor pressure of pure C_6H_6 and pure C_7H_8 at 30°C.

11.27 A 1.00-liter volume of a solution of 10.0 g of hemoglobin, the oxygen-carrying substance in red blood corpuscles, in water is found to have an osmotic pressure of 2.80 mm of Hg at 27°C. Calculate the molecular weight of hemoglobin.

Solution: The osmotic pressure depends on the number of moles of solute particles in a given volume of solution. Assuming hemoglobin does not dissociate, there are 10.0/MW moles in 1.00 liters of solution. Substituting into the osmotic pressure relationship gives:

$$n = \frac{\pi V}{RT} = \frac{10.0}{MW} = 2.80 \text{ mm} \times \frac{1}{760} \frac{\text{atm}}{\text{mm}} \times 1.00 \text{ liter}$$

$$\times \frac{1}{0.082} \frac{\text{deg-mole}}{\text{liter-atm}} \times \frac{1}{273 + 27} \text{deg}$$

$$MW = 66800$$

11.28 A water solution of 1 g of catalase, an enzyme found in the liver, has a volume of 100 ml at 27°C. Its osmotic pressure is found to be 0.745 mm. Calculate the molecular weight of catalase.

11.29 Calculate the osmotic pressure of a solution which contains 10.0 g of NaCl in 150 ml volume of solution at 20°C.

11.30 How many grams of KCl must be dissolved in water to make 1.00 liter of a solution with an osmotic pressure of 1.00 atm at 25°C?

➡ **11.31** The ocean is approximately a 3 % solution of NaCl in water with a density of 1.04 g/ml. Calculate its osmotic pressure at 20°C, due to the NaCl, assuming no other salts are dissolved.

12

Chemical equilibrium.
Equilibrium constants.

In a great many reactions the products are capable of reacting with each other and begin to do so as soon as they are formed. A familiar example is the gas phase reaction between carbon monoxide and steam where the products, CO_2 and H_2, react with each other as indicated by the lower arrow:

$$CO + H_2O \rightleftarrows CO_2 + H_2$$

This type of reaction is called an *incomplete reaction* or *reversible reaction*. Incomplete reactions eventually reach a *state of equilibrium*. At equilibrium the rate of the reaction to the right is exactly equal to the rate of the reaction to the left.

At the start of the above reaction the CO and H_2O are present in high concentration and, therefore, react with each other at maximum speed, but they are gradually used as the reaction progresses so the speed to the right gradually decreases.

At the beginning of the reaction, CO_2 and H_2 are present in very low concentration and, therefore, react with each other at a very low rate, but they increase in concentration as the reaction progresses and the speed to the left gradually increases. Finally the speed to the left exactly equals the speed to the right. That means that the CO, H_2O, CO_2, and H_2 are being used up as fast as they are being formed. There is no further change in the concentration of any reactant. The reaction just keeps on going "round and round" without any apparent change. A state of equilibrium has been reached.

If we were to start with a mixture of CO_2 and H_2, rather than CO and H_2O, a state of equilibrium would again be reached. In this case, at the start of the reaction CO_2 and H_2 are present in high concentration, while CO and H_2O are present in very low concentration. As the reaction proceeds, the speed to the left gradually decreases as CO_2 and H_2 are used up while the speed to the right increases as more CO and H_2O are formed. Eventually, the two rates will be the same; the reaction will then be in a state of equilibrium. *It is a characteristic of any true equilibrium reaction that the same state of equilibrium will be reached by starting with either the reactants or the products.*

It is very important to note that it is not a requirement of equilibrium that the substances be present in the reaction vessel in the exact ratio in which they react with each other. Thus, if we were to place 2 moles of CO and 5 moles of steam in a 1-liter reaction vessel, 6 moles of CO and 4 moles of steam in a second liter vessel, 8 moles of CO_2 and 3 moles of H_2 in a third vessel, and 3 moles of CO, 2 moles of steam, 4 moles of CO_2 and 7 moles of H_2 in a fourth vessel, a state of equilibrium, represented by the equation, $CO + H_2O \rightleftharpoons CO_2 + H_2$, will be attained in each vessel. However, it will be observed that in each of the 4 vessels, when CO and H_2O react to form CO_2 and H_2, they always do so in the mole ratio represented by the equation, namely, 1 mole of CO with 1 mole of H_2O to form 1 mole of CO_2 and 1 mole of H_2. In any equilibrium reaction the substances will always *react* with each other in the mole ratios represented by the equation, but they can be present in all sorts of ratios.

Suppose we have, in a liter reaction vessel at a given high temperature, an equilibrium mixture consisting of 0.10 mole of CO, 0.80 mole of H_2O (steam), 0.80 mole of CO_2, and 0.50 mole of H_2, represented by the equation

$$CO + H_2O \rightleftharpoons CO_2 + H_2$$
$$0.10 \quad\ 0.80 \qquad 0.80 \quad\ 0.50$$

Now suppose we were to force into the vessel a quantity of hydrogen gas. This, obviously, will increase the concentration of hydrogen. This increase in concentration of the hydrogen will increase the rate at which it will react with CO_2. As a result, the rate of the reaction to the left will be speeded up. This will increase the concentration of the CO and H_2O since they are, at the moment, being formed faster than they are reacting with each other. As time passes and the reaction proceeds the speed to the left gradually decreases as the concentrations of CO_2 and H_2 fall; at the same time the speed to the right gradually increases as the concentrations of CO and H_2O rise. In time the forward and reverse rates will again be equal and equilibrium will be re-established. However, at this new equilibrium the rates in the two directions will be greater than in the previous equilibrium. Also, at this new equilibrium we will find that the concentrations of the CO, H_2O, and H_2 are greater while the concentration of the CO_2 is less than before the H_2 was added. The increase in concentration of the H_2 has *shifted the equilibrium* to the left as evidenced by the fact that the concentrations of the CO and H_2O on the left have gone up, while the concentration of the CO_2 on the right has gone down. We can state that, in effect, the addition of H_2 (on the right) has pushed the equilibrium toward the left. Had we added more CO or H_2O to the reaction vessel the equilibrium would have been shifted to the right. In an equilibrium, if the concentration of a given reactant is increased, the equilibrium will be pushed toward the opposite side of the reaction. If the concentration of a reactant is reduced, by removing some of that reactant, the equilibrium will be *pulled* toward the same side of the reaction.

If we were to examine carefully, at the new equilibrium point, the increase in the *number of moles* of CO and H_2O on the left and the decrease in the *number of moles* of CO_2 on the right resulting from forcing more H_2 into the reaction vessel, we would find that the increase in the number of moles of CO is the same as the increase in the number of moles of H_2O and that the decrease in the number of moles of CO_2 is the same as the increase in the number of moles of CO. This is exactly what we would expect from the equation, $CO + H_2O \rightleftharpoons CO_2 + H_2$, since it tells us that CO_2 and H_2 react in the ratio of 1 mole of CO_2 with 1 mole of H_2 to yield 1 mole of CO and 1 mole of H_2O.

Suppose that we have, in a reaction vessel of fixed volume at a given high temperature, an equilibrium mixture represented by the following equation:

$$4\,NH_3 + 5\,O_2 \rightleftharpoons 6\,H_2O + 4\,NO$$

If we now force into this reaction vessel, at the given temperature, some more NO gas the equilibrium will be shifted to the left. At the new equilibrium the concentrations of the NH_3 and O_2 will be higher than before addition of the extra NO and the concentration of the H_2O will be lower. In this case, however, the increase in the *number of moles* of NH_3 and O_2 and the decrease in the *number of moles* of H_2O will not be the same; we will find, as the equation testifies, that in the shift in equilibrium due to addition of more NO, for every 4 additional moles of NH_3 that are produced there will be 5 additional moles of O_2, and 6 moles of H_2O will be used up. These two examples emphasize the very important fact that, *in every equilibrium shift, the change in the number of moles of the reactants involved is strictly in accord with the mole relationships specified by the equation for the reaction.*

We have noted earlier in this discussion that, for any equilibrium reaction, the reactants can be present in all sorts of ratios. This is illustrated in Table 12-1 which gives the equilibrium concentrations of CO, H_2O, CO_2,

Table 12-1
Relative concentrations of reactants in an equilibrium system at constant temperature

Reaction Vessel	[CO]	[H₂O]	[CO₂]	[H₂]	$\dfrac{[CO_2] \times [H_2]}{[CO] \times [H_2O]}$
1	0.20	0.20	0.50	0.40	5.0
2	0.10	0.18	0.30	0.31	5.2
3	0.10	0.80	0.80	0.50	5.0
4	0.30	0.50	0.90	0.83	5.0
5	0.75	0.20	0.80	0.94	5.0

Note: The notations [CO₂], [H₂O], [CO], and [H₂] mean concentration in moles per liter of the substance within the bracket.

and H_2 for the reaction $CO + H_2O \rightleftharpoons H_2 + CO_2$ for five experiments, all carried out at the same temperature.

An examination of the data in Table 12-1 reveals one very striking fact. The answer obtained when the product of the concentrations of the products, H_2 and CO_2, is divided by the product of the concentrations of the reactants, CO and H_2O, is, within the limits of experimental error, the same for each experiment. That is,

$$\frac{[CO_2] \times [H_2]}{[CO] \times [H_2O]} = \text{a constant}$$

Similar data for the thousands of equilibria that have been studied confirm the fact that *in every reacting system in equilibrium at a given temperature the product of the concentrations of the products divided by the product of the concentrations of the reactants is a constant.* This constant, referred to by the letter, K, is called the *equilibrium constant* for the particular reaction at the particular temperature. For the reaction

$$A + B \rightleftharpoons C + D$$

the formula for the equilibrium constant, K, is,

$$K = \frac{[C] \times [D]}{[A] \times [B]}$$

It is commonly referred to as the *equilibrium formula.*

It is obvious that if,

$$\frac{[C] \times [D]}{[A] \times [B]}$$

is constant, then

$$\frac{[A] \times [B]}{[C] \times [D]}$$

will also be constant; the latter constant will be the reciprocal of the former.

By common agreement among scientists, the product of the concentrations of the products is placed in the numerator.

For the reaction $SO_2 + NO_2 \rightleftharpoons SO_3 + NO$,

$$K = \frac{[SO_3] \times [NO]}{[SO_2] \times [NO_2]}$$

When a reaction involves more than 1 mole of a specific reactant, the concentration of that reactant is raised to a power equal to the number of moles of the reactant in the balanced equation. Thus for the reaction $2 SO_2 + O_2 \rightleftharpoons 2 SO_3$,

$$K = \frac{[SO_3]^2}{[SO_2]^2 \times [O_2]}$$

For the reaction $N_2 + 3 H_2 \rightleftharpoons 2 NH_3$,

$$K = \frac{[NH_3]^2}{[N_2] \times [H_2]^3}$$

For the reaction $H_2 + I_2 \rightleftharpoons 2 HI$,

$$K = \frac{[HI]^2}{[H_2] \times [I_2]}$$

For the reaction $4 NH_3 + 5 O_2 \rightleftharpoons 4 NO + 6 H_2O$,

$$K = \frac{[NO]^4 \times [H_2O]^6}{[NH_3]^4 \times [O_2]^5}$$

And for the general reaction, $aA + bB \rightleftharpoons cC + dD$,

$$K = \frac{[C]^c \times [D]^d}{[A]^a \times [B]^b}$$

The reason for raising the concentration of a reactant to a power equal to the number of moles may be more evident if we show each mole as a separate reactant by writing the equation for the reaction in the form

$$N_2 + H_2 + H_2 + H_2 \rightleftharpoons NH_3 + NH_3$$

The equilibrium constant, K, can then be expressed in the form

$$K = \frac{[NH_3] \times [NH_3]}{[N_2] \times [H_2] \times [H_2] \times [H_2]} = \frac{[NH_3]^2}{[N_2] \times [H_2]^3}$$

The numerical value of the equilibrium constant for a given reaction is obtained by inserting the experimentally determined values of the concentrations in the equilibrium formula for the reaction. Thus, for the reaction,

$$\underset{0.60}{SO_2} + \underset{0.80}{NO_2} \rightleftharpoons \underset{0.90}{SO_3} + \underset{1.1}{NO}$$

the calculation, using the equilibrium concentrations given in moles per liter, becomes

$$K = \frac{[SO_3] \times [NO]}{[SO_2] \times [NO_2]} = \frac{0.90 \times 1.1}{0.60 \times 0.80} = 2.1$$

(In this instance all *units* cancel, so the constant is simply a number, 2.1.) For the reaction,

$$2 SO_2 + O_2 \rightleftarrows 2 SO_3$$
$$0.20 \quad\quad 0.30 \quad\quad 0.60$$

$$K = \frac{[SO_3]^2}{[SO_2]^2 \times [O_2]} = \frac{(0.60)^2}{(0.20)^2 \times (0.30)} = 30 \text{ liters/moles}$$

(In this instance the unit, moles/liters, remains uncancelled in the denominator. Therefore, the constant has the dimension 30/moles/liters or 30 liters/moles.)

It should be noted that a *solid* reactant or product is not included in the equilibrium formula. Thus, for the equilibrium reaction,

$$SiF_4(gas) + 2 H_2O (gas) \rightleftarrows SiO_2 (solid) + 4 HF (gas)$$

$$K = \frac{[HF]^4}{[SiF_4] + [H_2O]^2}$$

and for the reaction,

$$LaCl_3 (solid) + H_2O (gas) \rightleftarrows LaClO (solid) + 2 HCl (gas)$$

$$K = \frac{[HCl]^2}{[H_2O]}$$

The reason why the solid is not included is that the *amount* of excess solid present has no effect whatever on the *equilibrium* point. The same state of equilibrium is attained whether we have a small excess or a large excess. The *rate* at which equilibrium is attained will be affected by the total surface of the solid. However, once equilibrium has been attained, the removal of some of the excess solid (in fact, all of the solid) will have no effect on the equilibrium.

The reason why the amount of solid (or liquid) has no effect on the gas-phase equilibrium is this: Just as every liquid has a constant vapor pressure at a given temperature, so every solid has a constant although generally very small vapor pressure. Once a liquid is in equilibrium with its vapor, its vapor presure and, hence, its vapor-phase concentration at constant temperature remain constant regardless of how much or how little liquid is present. Likewise, the concentration of a vapor in equilibrium with its solid remains constant and this concentration is independent of the amount of solid.

In a reaction such as the one given above the equilibrium formula applies to the gaseous reactants in the homogeneous gas phase, that is, to the H_2O gas and $LaCl_3$ *gas* on the left and the HCl gas and the LaClO *gas* on the right. If we write the equilibrium formula to include all gaseous species it becomes

$$K_x = \frac{[LaClO] \times [HCl]^2}{[LaCl_3] \times [H_2O]}$$

But [LaClO] and [LaCl$_3$] are both constant as long as some of each solid is present. Therefore, since they are constant, they can be combined with K_x to give a new constant, K_y, whose value is [HCl]^2/[H$_2$O]. The net effect is that solid (and liquid) reactants do not appear in the equilibrium formula.

It should be noted that the equilibrium formula as presented in the preceding pages assumes that the kinetic behavior of every reacting molecule is completely unaffected by the other molecules that are present in the system; in other words, it assumes that the system is *ideal*. In such an ideal system the *effective concentration* of each species is, in fact, its molar concentration, since each molecule is 100% free to do as it pleases; hence, the equilibrium constant is a precise function of the molar concentrations, as the equilibrium formula testifies.

Most gaseous systems are not ideal; the kinetic behavior of each molecule is affected, to a small degree, by its neighbors; the neighboring molecules impede the freedom of a molecule and, as a result, cause it to waste some of its energy. As a consequence, the *effective concentration* of a given species, its so-called *activity*, is slightly less than its *molar concentration*; it is equal to the product of its molar concentration and its *activity coefficient* in the particular system. It is this *activity* or *effective concentration*, rather than the molar concentration, which should appear in the equilibrium formulas for nonideal systems.

We will assume, in the problems given in this book, that all systems are ideal. In such systems the activity coefficient of each species has a value of 1 and the molar concentration is, in fact, the effective concentration. Making this assumption will in no way detract from the value of a problem as a vehicle for developing logical thinking and reasoning.

PROBLEMS

12.1 At equilibrium at a given temperature and in a liter reaction vessel HI is 20 mole percent dissociated into H$_2$ and I$_2$ according to the equation, 2 HI $\rightleftharpoons$ H$_2$ + I$_2$. If 1 mole of pure HI is introduced into a liter reaction vessel at the given temperature, how many moles of each component will be present when equilibrium is established?

Solution: 20% of 1 mole = 0.2 mole = the number of moles of HI that dissociate. 1 − 0.2 = 0.8 mole of HI that is not dissociated. The equation, 2 HI = H$_2$ + I$_2$, tells us that 2 moles of HI yield 1 mole of H$_2$ and 1 mole of I$_2$. Therefore 0.2 mole of HI yields 0.1 mole of H$_2$ and 0.1 mole of I$_2$.

12.2 PCl$_5$ is 20 mole percent dissociated into PCl$_3$ and Cl$_2$ at equilibrium at a given temperature and in a liter vessel in accordance with the equation, $PCl_5 \rightleftharpoons PCl_3 + Cl_2$. One mole of pure PCl_5 was introduced into a liter reaction vessel at the given temperature. How many moles of each component were present at equilibrium?

12.3 A reaction vessel in which the following reaction had reached a state of equilibrium, $CO + Cl_2 \rightleftharpoons COCl_2$, was found, on analysis, to contain 0.30 mole of CO, 0.20 mole of Cl_2, and 0.80 mole of $COCl_2$, in a liter of mixture. Calculate the equilibrium constant for the reaction.

Solution: First, write the equilibrium formula.

$$K = \frac{[COCl_2]}{[CO] \times [Cl_2]}$$

The notation $[COCl_2]$, by definition, means concentration of $COCl_2$ in moles of $COCl_2$ per liter. Likewise, $[CO]$ and $[Cl_2]$ mean, respectively, moles of CO per liter and moles of Cl_2 per liter. In all problems involving chemical equilibrium the concentrations will be expressed in moles per liter. Substituting in the above equation,

$$K = \frac{0.80}{0.30 \times 0.20} = 13 \text{ liters/mole}$$

12.4 A reaction vessel with a capacity of 1 liter in which the following reaction had reached a state of equilibrium, $2 SO_2 + O_2 \rightleftharpoons 2 SO_3$, was found to contain 0.6 mole of SO_3, 0.2 mole of SO_2, and 0.3 mole of O_2. Calculate the equilibrium constant.

Solution: First, write the equilibrium formula.

$$K = \frac{[SO_3]^2}{[SO_2]^2 \times [O_2]}$$

The total concentrations of O_2, SO_3, and SO_2 are substituted in this formula to give

$$\frac{(0.6)^2}{(0.2)^2 \times (0.3)} = \frac{0.36}{0.012} = 3 \times 10^1 \text{ liters/mole}$$

The reason why the total concentrations of SO_2 and SO_3 and not half of the total concentrations of each are used is that the 2 molecules of SO_2 that react with O_2 can each be picked from the total supply of SO_2 molecules available. The same is true of the 2 SO_3 molecules.

12.5 A quantity of PCl_5 was heated in a liter vessel at 250°C. At equilibrium the concentrations of the gases in the vessel were as follows:

$$PCl_5 = 7.05 \text{ moles/liter}; \quad PCl_3 = 0.54 \text{ mole/liter};$$

$$Cl_2 = 0.54 \text{ mole/liter}$$

Calculate the equilibrium constant, K, for the dissociation of PCl_5 at 250°C.

12.6   An equilibrium mixture of N_2, H_2, and NH_3, which reacts according to the equation, $N_2 + 3\,H_2 \rightleftarrows 2\,NH_3$, was found to consist of 0.800 mole of NH_3, 0.300 mole of N_2 and 0.200 mole of H_2 in a liter. Calculate the equilibrium constant.

12.7   An equilibrium mixture, $CO + Cl_2 \rightleftarrows COCl_2$, contained 1.50 moles of CO, 1.00 mole of Cl_2, and 4.00 moles of $COCl_2$ in a 5-liter reaction vessel at a specific temperature. Calculate the equilibrium constant for the reaction at this temperature.

Solution: In calculating the equilibrium constant, concentration must be expressed in moles *per liter*.

12.8   An equilibrium mixture, $2\,SO_2 + O_2 \rightleftarrows 2\,SO_3$, contained in a 2.0-liter reaction vessel at a specific temperature was found to contain 96 g of SO_3, 25.6 g of SO_2 and 19.2 g of O_2. Calculate the equilibrium constant for the reaction at this temperature.

Solution: Concentrations must be expressed in *moles* per *liter*.

12.9   The equilibrium constant for the reaction, $2\,SO_2 + O_2 \rightleftarrows 2\,SO_3$, is 4.5 liters/mole at 600°C. A quantity of SO_3 gas was placed in a liter reaction vessel at 600°C. When the system reached a state of equilibrium the vessel was found to contain 2.0 moles of O_2 gas. How many moles of SO_3 gas were originally placed in the reaction vessel?

Solution: Note that, when SO_3 decomposes to yield SO_2 and O_2, the products are formed in the ratio of 2 moles of SO_2 and 1 mole of O_2 for every 2 moles of SO_3 decomposed. Since, in this problem, there are 2 moles of O_2 in the vessel there must also be 4 moles of SO_2 present and 4 moles of SO_3 must have decomposed to yield 4 moles of SO_2 and 2 moles of O_2. That means that, at equilibrium, the reaction vessel must contain 4 less moles of SO_3 than were originally introduced. With these facts, and knowing that the equilibrium constant is 4.5 liters/mole, the number of moles of SO_3 originally added can be calculated.

12.10   The equilibrium constant for the reaction, $N_2 + 3\,H_2 \rightleftarrows 2\,NH_3$, is 2.00 liter² mole⁻² at 300°C. A quantity of NH_3 gas was introduced into a liter reaction vessel at 300°C. When equilibrium was established the vessel was found to contain 2.00 moles of N_2. How many moles of NH_3 were originally introduced into the vessel?

12.11   A liter reaction vessel in which the reaction,

$$C\,(solid) + H_2O\,(gas) \rightleftarrows CO\,(gas) + H_2\,(gas),$$

has reached a state of equilibrium contains 0.16 mole of C, 0.58 mole of H_2O, 0.15 mole of CO, and 0.15 mole of H_2. Calculate the equilibrium constant for the reaction.

Solution: Since C is a solid it is not included in the equilibrium formula

$$K = \frac{[CO] \times [H_2]}{[H_2O]} = \frac{[0.15] \times [0.15]}{[0.58]} = 3.9 \times 10^{-3} \text{ mole} \times \text{liter}^{-1}$$

12.12 The equilibrium mixture, $SO_2 + NO_2 \rightleftharpoons SO_3 + NO$, in a liter vessel, was found to contain 0.600 mole of SO_3, 0.400 mole of NO, 0.100 mole of NO_2, and 0.800 mole of SO_2. How many moles of NO would have to be forced into the reaction vessel, volume and temperature being kept constant, in order to increase the amount of NO_2 to 0.300 mole?

Solution: First calculate the equilibrium constant.

$$K = \frac{[SO_3] \times [NO]}{[SO_2] \times [NO_2]} = \frac{0.6 \times 0.4}{0.8 \times 0.1} = 3.00$$

Let X = moles of NO that must be added. We can see from the equation, $SO_2 + NO_2 \rightleftharpoons SO_3 + NO$, that in order to produce 0.2 more mole of NO_2 we must also produce 0.2 more mole of SO_2 and we must use up 0.2 mole of SO_3 and 0.2 mole of NO. Therefore, when we have 0.3 mole of NO_2 we will have 1 mole (0.8 + 0.2) of SO_2, 0.4 mole (0.6 − 0.2) of SO_3 and $X + 0.2$ mole (0.4 + X − 0.2) of NO. Substituting the values in the equilibrium formula, we have

$$\frac{0.4 \times (X + 0.2)}{1 \times 0.3} = 3.00 \qquad (X = 2.05 \text{ moles})$$

12.13 An equilibrium mixture, $CO + H_2O \rightleftharpoons CO_2 + H_2$, contains 0.2 mole of H_2, 0.80 mole of CO_2, 0.10 mole of CO, and 0.40 mole of H_2O in a liter. How many moles of CO_2 would have to be added at constant temperature and volume to increase the amount of CO to 0.20 mole?

12.14 A reaction system in equilibrium according to the equation, $2 SO_2 + O_2 \rightleftharpoons 2 SO_3$, in a liter reaction vessel at a given temperature was found to contain 0.11 mole of SO_2, 0.12 mole of SO_3, and 0.050 mole of O_2. Another liter reaction vessel contains 64 g of SO_2 at the above temperature. How many grams of O_2 must be added to this vessel in order that, at equilibrium, half of the SO_2 is oxidized to SO_3?

Solution: Calculate K from data in first sentence. In second situation 1 mole (64 g) of SO_2 is initially present. Let X = g of O_2 added. Then $X/32$ = moles of O_2 added. At equilibrium $[SO_2] = 0.5$, $[SO_3] = 0.5$, $[O_2] = X/32 − 0.25$. The value of X can then be calculated.

➡ **12.15** The equilibrium constant for the reaction, $PCl_5 \rightleftharpoons PCl_3 + Cl_2$, at 250°C is 0.041 mole × liter^{-1}. Set up, but do not solve, an algebraic equation in one unknown, X, which, if solved for X, will give the number of *grams* of Cl_2 that will be present at equilibrium when 0.3 mole of PCl_5 is heated in a liter vessel at 250°C.

➡ **12.16**   A mixture of 2 moles of CH_4 gas and 1 mole of H_2S gas was placed in an evacuated container which was then heated to and maintained at a temperature of 727°C. When equilibrium was established in the gaseous reaction, $CH_4 + 2 H_2S \rightleftharpoons CS_2 + 4 H_2$, the total pressure in the container was 0.92 atm and the partial pressure of the hydrogen gas was 0.20 atm. What was the volume of the container?

Solution: To calculate V when T is constant we must know P and n. Before any reaction has occurred we know the number of moles, 3, but we do not know the pressure. At equilibrium, we know P, 0.92 mm, but we do not know the number of moles. Therefore, to calculate V we must calculate either the number of moles at equilibrium or the pressure at the start.

To calculate moles at equilibrium, let $X =$ moles of CS_2. Then $4X =$ moles of H_2, $2 - X =$ moles of CH_4, and $1 - 2X =$ moles of H_2S. Total moles at equilibrium = the sum of the above quantities = $3 + 2X$. In a system at constant V and T the number of moles is directly proportional to the pressure. Therefore,

$$\frac{\text{total moles}}{\text{moles of } H_2} = \frac{\text{total pressure}}{\text{partial pressure of } H_2}$$

$$\frac{3 + 2X}{4X} = \frac{0.92}{0.20}$$

Solving, $X = 0.183$ mole,
total moles $= 3.366$, and $V = 300$ liters.

To calculate the total pressure of the 3 moles at the start let us assume that we take the system at equilibrium and, somehow, force the reaction back to the left so that all of the CS_2 and H_2 interact to form CH_4 and H_2S; we would then be back at the start. Since $P_{H_2} = 0.20$ atm, P_{CS_2} must be $\frac{1}{4}$ of that amount or 0.05 atm; the sum of the two partial pressures will be 0.25 atm. Since, when the reaction goes from right to left, the number of moles *decreases* from 5 to 3, that is by a factor of $\frac{2}{5}$, the pressure will decrease by a factor of $\frac{2}{5}$. $\frac{2}{5} \times 0.25$ atm $= 0.10$ atm. Since the equilibrium pressure was 0.92 atm and the calculated decrease is 0.10 atm, the original pressure must have been 0.82 atm. Substituting this value in $PV = nRT$ when $n = 3$ moles and $T = 1000°$ gives a volume of 300 liters.

➡ **12.17**   To determine the equilibrium constant at a given temperature for the gas-phase reaction, $N_2 + 3 H_2 \rightleftharpoons 2 NH_3$, 0.326 mole of H_2 and 0.439 mole of N_2 were mixed in a 1.00-liter vessel. At equilibrium the system was found to contain a total of 0.657 mole.

(a) Calculate the equilibrium constant for the reaction as written above and state the units in which this value is expressed.

(b) Call the constant calculated in part (a) K_1. Call the constant for the following reaction K_2; $NH_3 \rightleftharpoons \frac{1}{2} N_2 + \frac{3}{2} H_2$. State the units in which K_2 would be expressed. State the algebraic relation between K_1 and K_2.

➡ **12.18** When N_2O_5 gas is heated it dissociates into N_2O_3 gas and O_2 gas according to the reaction, $N_2O_5 \rightleftharpoons N_2O_3 + O_2$. K_1 for this reaction at a specific temperature, t°C, is 7.75 moles/liter. The N_2O_3 dissociates to give N_2O gas and O_2 gas according to the reaction, $N_2O_3 \rightleftharpoons N_2O + O_2$. K_2 for this reaction at the same specific temperature, t°C, is 4.00 moles/liter.

When 4.00 moles of N_2O_5 are heated in a 1.00-liter reaction vessel at t°C the concentration of O_2 at equilibrium is 4.50 moles/liter. Calculate the concentrations in moles per liter of all other species in the equilibrium system.

Solution: Let X equal the moles per liter of O_2 derived from the N_2O_5 and Y equal the moles per liter of O_2 derived from the N_2O_3. Then:

$$[O_2] = X + Y \qquad [N_2O_5] = 4.00 - X$$
$$[N_2O_3] = X - Y \qquad [N_2O] = Y$$

Since three equations involving X and Y are available, the value of X and of Y, and hence the concentrations of all species, can be calculated.

➡ **12.19** At 1227°C the equilibrium constant for the reaction, $CaCO_3$ (s) $\rightleftharpoons$ CaO (s) $+ CO_2$ (g), is 0.50 moles/liter. Also, at 1227°C CO_2 decomposes according to the reaction, CO_2 (g) $\rightleftharpoons$ CO (g) $+ \frac{1}{2} O_2$ (g).

One mole of solid $CaCO_3$ is placed in an evacuated 1-liter container and heated to 1227°C. When equilibrium is established the mole fraction of O_2 in the gaseous mixture in the container is 0.15.

How many moles of CaO are there in the container at equilibrium?

Solution hint: Moles of CaO = moles of CO_2 + moles of CO.

➡ **12.20** At 1227°C the partial pressure of the CO_2 gas which is in equilibrium with solid CaO and solid $CaCO_3$ in a one-liter reaction vessel according to the equation, $CaCO_3$ (solid) $\rightleftharpoons$ CaO (solid) $+ CO_2$ (gas), is 61.5 atmospheres. Also, at 1227°C, CO_2 gas decomposes according to the following equilibrium reaction: CO_2 (gas) $\rightleftharpoons$ CO (gas) $+ \frac{1}{2} O_2$ (gas).

One mole of solid $CaCO_3$ is placed in an evacuated one-liter container and heated to 1227°C. When equilibrium is established the mole fraction of the CO_2 in the gaseous mixture of CO_2, CO, and O_2 is 0.55. How many moles of solid $CaCO_3$ are present at equilibrium?

Equilibrium constants in units of pressure

When the gas law equation, $PV = nRT$, is written in the form, $n/V = P/RT$, the term, n/V, represents the concentration of the gas in moles per liter. The equation, $n/V = P/RT$, tells us that, at constant temperature, the concentration of a gas in moles per liter is directly proportional to its partial pressure. It follows, therefore, that for a gaseous equilibrium reaction, an equilibrium constant, K_p, can be written in terms of the partial pressures of the reacting gases. The forms of the K_p expressions for some typical equilibria are given below:

Reaction	Equilibrium Expression
(1) $CO + H_2O \rightleftharpoons CO_2 + H_2$	$K_p = \dfrac{P_{CO_2} \times P_{H_2}}{P_{CO} \times P_{H_2O}}$
(2) $COCl_2 \rightleftharpoons CO + Cl_2$	$K_p = \dfrac{P_{CO} \times P_{Cl_2}}{P_{COCl_2}}$
(3) $2\,NH_3 \rightleftharpoons N_2 + 3\,H_2$	$K_p = \dfrac{P_{N_2} \times (P_{H_2})^3}{(P_{NH_3})^2}$
(4) $4\,H_2 + CS_2 \rightleftharpoons CH_4 + 2\,H_2S$	$K_p = \dfrac{P_{CH_4} \times (P_{H_2S})^2}{(P_{H_2})^4 \times P_{CS_2}}$

Note that the form of the K_p expression is the same as that for the K_c expression (concentration expressed in mole units) except that P_a is substituted for [a]. K_p is expressed in pressure units of atmospheres or millimeters.

Relation between K_p and K_c for a specific equilibrium

Since, as has been noted above, $n/V = P/RT$, and since n/V represents concentration in moles per liter, it follows that, at constant temperature, P/RT can be substituted for the concentration term, [ ], in the K_c expression. Thus for the reaction, $CO + H_2O \rightleftharpoons CO_2 + H_2$,

$$K_c = \frac{[CO_2] \times [H_2]}{[CO] \times [H_2O]} = \frac{\dfrac{P_{CO_2}}{RT} \times \dfrac{P_{H_2}}{RT}}{\dfrac{P_{CO}}{RT} \times \dfrac{P_{H_2O}}{RT}}$$

Since the temperature is constant all of the RT terms will cancel, leaving

$$K_c = \frac{P_{CO_2} \times P_{H_2}}{P_{CO} \times P_{H_2O}}$$

But we have learned that

$$\frac{P_{CO_2} \times P_{H_2}}{P_{CO} \times P_{H_2O}} = K_p$$

Therefore, for this particular reaction, K_c is numerically equal to K_p. K_c will be equal to K_p for all equilibria in which the number of moles of gaseous reactants equals the number of moles of gaseous products.

For the equilibrium, $COCl_2 \rightleftarrows CO + Cl_2$,

$$K_c = \frac{[CO] \times [Cl_2]}{[COCl_2]} = \frac{\dfrac{P_{CO}}{RT} \times \dfrac{P_{Cl_2}}{RT}}{\dfrac{P_{COCl_2}}{RT}} = \frac{P_{CO} \times P_{Cl_2}}{P_{COCl_2}} \times \frac{1}{RT} = K_p \times \frac{1}{RT}$$

For the equilibrium, $2\,NH_3 \rightleftarrows N_2 + 3\,H_2$,

$$K_c = \frac{[N_2] \times [H_2]^3}{[NH_3]^2} = \frac{\dfrac{P_{N_2}}{RT} \times \left(\dfrac{P_{H_2}}{RT}\right)^3}{\left(\dfrac{P_{NH_3}}{RT}\right)^2}$$

$$= \frac{P_{N_2} \times (P_{H_2})^3}{(P_{NH_4})^2} \times \left(\frac{1}{RT}\right)^2 = K_p \times \left(\frac{1}{RT}\right)^2$$

For the equilibrium, $4\,H_2 + CS_2 \rightleftarrows CH_4 + 2\,H_2S$,

$$K_c = \frac{[CH_4] \times [H_2S]^2}{[H_2]^4 \times [CS_2]} = \frac{\dfrac{P_{CH_4}}{RT} \times \left(\dfrac{P_{H_2S}}{RT}\right)^2}{\left(\dfrac{P_{H_2}}{RT}\right)^4 \times \dfrac{P_{CS_2}}{RT}}$$

$$= \frac{P_{CH_4} \times (P_{H_2S})^2}{(P_{H_2})^4 \times P_{CS_2}} \times \left(\frac{1}{RT}\right)^{-2} = K_p \times \left(\frac{1}{RT}\right)^{-2}$$

Examining the four examples given above we conclude that for any gas-phase equilibrium

$$K_c = K_p \times \left(\frac{1}{RT}\right)^{\Delta n}$$

where Δn is the change in the number of moles of gas when the reaction goes from left to right.

PROBLEMS

12.21 In an equilibrium mixture of N_2, H_2, and NH_3, contained in a 5.00-liter reaction vessel at 450°C and a total pressure of 332 atm the partial pressures of the gases were: $N_2 = 47.55$ atm, $H_2 = 142.25$ atm, $NH_3 = 142.25$ atm. Calculate the equilibrium constant for the reaction.

Solution: To calculate the equilibrium constant as a function of concentrations we must know the concentration of each reactant in *moles* per *liter*. To this end we will first calculate the moles of each substance per liter by use of the formula, $PV = nRT$.

12.22 The equilibrium constant for the reaction, $CO + H_2O \rightleftarrows CO_2 + H_2$, is 4.0 at a given temperature. An equilibrium mixture of the above substances at the given temperature was found to contain 0.60 atm of

CO, 0.20 atm of steam, and 0.50 atm of CO_2 in a liter. How many atm of H_2 were there in the mixture?

12.23   Exactly 1 atm of NH_3 was introduced into a liter reaction vessel at a certain high temperature. When the reaction, $2\,NH_3 \rightleftharpoons N_2 + 3\,H_2$, had reached a state of equilibrium 0.6 atm of H_2 was found to be present. Calculate the equilibrium constant for the reaction.

Solution: Note that, when NH_3 dissociates to give N_2 and H_2, the products are formed in the ratio of 1 atm of N_2 and 3 atm of H_2 for every 2 atm of NH_3 which dissociate. To yield 0.6 atm of H_2, 0.4 atm of NH_3 must have dissociated. This leaves 0.6 atm of undissociated NH_3. The 0.4 atm of NH_3 which dissociates will yield 0.2 atm of N_2 and 0.6 atm of H_2. The equilibrium mixture will contain 0.6 atm of NH_3, 0.2 atm of N_2, and 0.6 atm of H_2.

12.24   One atm of SO_3 was placed in a reaction vessel at a certain temperature. When equilibrium was established in the reaction, $2\,SO_3 \rightleftharpoons 2\,SO_2 + O_2$, the vessel was found to contain 0.60 atm of SO_2. Calculate the equilibrium constant for the reaction.

12.25   An equilibrium mixture, $CO_2 + H_2 \rightleftharpoons CO + H_2O$, was found to contain 0.6 atm of CO_2, 0.2 atm of H_2, 0.8 atm of CO, and 0.3 atm of H_2O. How many atm of CO_2 would have to be removed from the system at constant volume and temperature in order to reduce the amount of CO to 0.6 atm?

➡ **12.26**   In an equilibrium mixture, $CO_2 + H_2 \rightleftharpoons CO + H_2O$, contained in a 6.0-liter reaction vessel at 1007°C the partial pressures of the reactants are: $CO_2 = 63.1$ atm, $H_2 = 21.1$ atm, $CO = 84.2$ atm, $H_2O = 31.6$ atm. Enough CO_2 was then removed from the vessel to reduce the partial pressure of the CO to 63.0 atm, temperature being kept constant.

(a) Calculate the partial pressure of the CO_2 in the new equilibrium system.

(b) For the above reaction how does the numerical value of K_c, in which concentration is expressed in moles per liter, compare with the numerical value of K_p, in which concentrations are expressed in atmospheres?

(c) Suppose the volume of the new equilibrium system was reduced to 3 liters by depressing a piston, what would the partial pressure of the CO_2 be?

Solution:

(a)

$$\underset{63.1\text{ atm}}{CO_2} + \underset{21.1\text{ atm}}{H_2} \rightleftharpoons \underset{84.2\text{ atm}}{CO} + \underset{31.6\text{ atm}}{H_2O}$$

$$K_p = \frac{P_{CO} \times P_{H_2O}}{P_{CO_2} \times P_{H_2}} = \frac{84.2 \text{ atm} \times 31.6 \text{ atm}}{63.1 \text{ atm} \times 21.1 \text{ atm}} = 2.0$$

Let $X =$ the partial pressure of the CO_2 in the system after removal of CO_2. Since the partial pressure of the CO is reduced to 63.0 atm, a quantity of CO with a partial pressure of 21.2 atm must have reacted with H_2O to form CO_2 and H_2. Therefore, 21.2 atm worth of H_2O must have reacted and 21.2 atm worth of both CO_2 and H_2 must have been produced. The partial pressures of each reactant in the new equilibrium system will then be

$$\underset{X\text{ atm}}{CO_2} + \underset{42.3\text{ atm}}{H_2} \rightleftarrows \underset{63.0\text{ atm}}{CO} + \underset{10.4\text{ atm}}{H_2O}$$

$$K_p = \frac{63.0 \text{ atm} \times 10.4 \text{ atm}}{X \text{ atm} \times 42.3 \text{ atm}} = 2.0$$

Solving, $X = 7.8$ atm

(b) Since there is no change in the number of moles, $K_c = K_p$.

(c) Since, in the gaseous equilibrium represented by the reaction, $CO_2 + H_2 \rightleftarrows CO + H_2O$, there is no change in the number of moles, increase in pressure by reducing the volume to one half its original value will not shift the equilibrium. All that will happen will be that the partial pressure of each reactant will be doubled. Therefore, the partial pressure of the CO_2 will be 2×7.8 atm or 15.6 atm.

➡ **12.27** An equilibrium mixture, $H_2 + I_2 \rightleftarrows 2 HI$, contains 3 moles of H_2, 2 moles of I_2, and 2 moles of HI in a liter. How many moles of I_2 must be added at constant temperature to have half the added I_2 react to form HI? Let $X =$ moles of I_2 added. Set up the equilibrium expression but do not solve.

➡ **12.28** To the system $LaCl_3$ (solid) $+ H_2O$ (gas) $+$ heat $\rightleftarrows LaClO$ (solid) $+ 2 HCl$ (gas) already at equilibrium we add more water vapor without changing either the temperature or the volume of the system. When equilibrium is re-established the pressure of water vapor is found to have been doubled. Hence, the pressure of HCl present in the system has been multiplied by what factor?

Solution:

$$K = \frac{[HCl]^2}{[H_2O]}$$

Assume that, in the first equilibrium, $[HCl] = a$ and $[H_2O] = b$. Then $K = a^2/b$. Since T is constant K will have the value a^2/b in the second equilibrium.

Let $X =$ the factor by which the concentration of HCl is multiplied in the second equilibrium; since, at constant volume and temperature the concentration of a gas is directly proportional to its partial pressure, this will be the factor by which its partial pressure is multiplied. Then, at the second equilibrium, $[HCl] = Xa$ and $[H_2O] = 2b$

$$K = \frac{[Xa]^2}{2b} = \frac{X^2a^2}{2b} \qquad \text{But } K = a^2/b$$

Therefore, $\frac{X^2a^2}{2b} = \frac{a^2}{b}$. Solving, $X = \sqrt{2} = 1.41$.

➡ **12.29** The equilibrium mixture, $SO_2 + NO_2 \rightleftharpoons SO_3 + NO$, was found to contain 0.60 mole of SO_3, 0.40 mole of NO, 0.80 mole of SO_2 and 0.10 mole of NO_2 per liter. One mole of NO was then forced into the reaction vessel, temperature and volume being kept constant. Calculate the number of moles of each gas in the new equilibrium mixture.

Solution: First calculate the equilibrium constant. Its value is 3.0. We can see from the equation, $SO_2 + NO_2 \rightleftharpoons SO_3 + NO$, that if the concentration of NO is increased the equilibrium will be shifted to the left. Let X be the number of additional moles of SO_2 formed as a result of this shift. There will then be $0.8 + X$ moles of SO_2. But when X new moles of SO_2 are formed, X new moles of NO_2 will also be formed; X moles of SO_3 and X moles of NO will be used up. There will, therefore, be present in the new mixture $0.8 + X$ moles of SO_2, $0.1 + X$ moles of NO_2, $0.6 - X$ moles of SO_3, and $1 + 0.4 - X$ or $1.4 - X$ moles of NO. If we insert these values in the equilibrium formula, for which the constant 3 has been calculated, we have

$$\frac{[SO_3] \times [NO]}{[SO_2] \times [NO_2]} = \frac{(0.6 - X) \times (1.4 - X)}{(0.8 + X) \times (0.1 + X)} = 3.0$$

Solving, $X = 0.12$ mole. Substituting this value of X we find the concentrations of the 4 reactants to be $SO_3 = 0.48$ mole, NO = 1.3 moles, $SO_2 = 0.92$ mole, and $NO_2 = 0.22$ mole.

➡ **12.30** At a given temperature a liter reaction vessel contained 0.60 atm of $COCl_2$, 0.30 atm of CO, and 0.10 atm of Cl_2. $CO + Cl_2 \rightleftharpoons COCl_2$. An amount of 0.40 atm of Cl_2 was added to the vessel at constant temperature and volume. Calculate the number of moles of CO, Cl_2, and $COCl_2$, in the new equilibrium system.

➡ **12.31** The equilibrium constant for the reaction,

$$CO + H_2O \rightleftharpoons CO_2 + H_2$$

is 4.00 at a given temperature. A combination of 0.400 atm of CO and 0.600 atm of steam was brought together at this temperature. How many atm of CO_2 were present when the system reached a state of equilibrium?

➡ **12.32** A reaction vessel, at 27°C, contains a mixture of SO_2 and O_2 in which the partial pressures of SO_2 and O_2 are 3.00 atm and 1.00 atm, respectively. When a catalyst is added, the reaction, $2 SO_2 + O_2 \rightleftharpoons 2 SO_3$, occurs. At equilibrium, at 27°C, the total pressure is 3.75 atm. Calculate K_p and K_c.

➡ **12.33** Pure water vapor is present at a pressure of 1.5 atm in a reaction vessel. To the vessel we add, without change of volume or temperature, excess solid $LaCl_3$. When equilibrium is established the total pressure in the vessel is found to be 2.0 atm. What is the equilibrium constant, in terms of atmospheres, for the reaction?

$$LaCl_3 \text{ (s)} + H_2O \text{ (g)} \rightleftharpoons LaClO \text{ (s)} + 2 HCl \text{ (g)}$$

➡ **12.34** A sample of gas that was initially pure NO_2 was heated to a temperature of 337°C. The NO_2 partially dissociates according to the equation $2 NO_2 \rightleftharpoons 2 NO + O_2$. At equilibrium, the observed density of the gas mixture at 0.750 atm pressure is 0.520 g per liter. Calculate K_c and K_p for this reaction.

➡ **12.35** A 1-liter reaction vessel in which the reaction, $A \text{ (g)} + B \text{ (g)} \rightleftharpoons AB \text{ (g)}$, has reached a state of equilibrium at 727°C contains 0.0200 mole of solid B. The partial pressures of the gaseous reactants in the equilibrium system are: $A = 8.20$ atm; $B = 4.92$ atm; $AB = 11.48$ atm. Calculate the minimum number of moles of A that must be added to the above equilibrium system at 727°C in order that no solid B shall be present at equilibrium.

➡ **12.36** A reaction vessel at 850°C contains $SrCO_3$ (s), SrO (s), and C (s) in equilibrium with CO_2 (g) and CO (g). The total pressure of the CO_2 and CO is 169 mm. K_p for the reaction, $SrCO_3 \text{ (s)} \rightleftharpoons SrO \text{ (s)} + CO_2 \text{ (g)}$, is 2.45 mm at 850°C. Calculate K_p for the reaction, $C \text{ (s)} + CO_2 \text{ (g)} \rightleftharpoons 2 CO \text{ (g)}$, at 850°C.

➡ **12.37** Pure PCl_5 gas is introduced into an evacuated reaction vessel and comes to equilibrium at 250°C, the reaction being $PCl_5 \rightleftharpoons PCl_3 + Cl_2$, all substances being gases. The total pressure is 2.00 atm and the mole fraction of the Cl_2 is 0.407.

(a) What are the partial pressures of PCl_3 and PCl_5?

(b) Calculate K_p for the reaction at 250°C.

Free energy and *K*

In Chapter 9 we defined a thermodynamic function, *G*, the free energy and saw that the sign of ΔG, the difference in free energy between a final and an initial state could be used as a criterion to predict the direction of spontaneous change. The magnitude of ΔG for a process can be used to evaluate the extent to which a spontaneous change occurs. In order to do this most conveniently we again employ the definition of a standard state outlined in Chapter 9, and tabulate experimentally determined values for the standard free energy of formation $\Delta G_f°$ of various substances. In a manner analogous to that used for $\Delta H°$, it is possible to calculate $\Delta G°$ for a reaction.

$$\Delta G° = \sum \Delta G_f° \text{ (products)} - \sum \Delta G_f° \text{ (reactants)}$$

where $\Delta G°$ represents the free energy change of a process in which all the components of a system are at a pressure of 1 atm. The value of $\Delta G°$ is related to the value of K, the equilibrium constant for the process by the extremely important relationship

$$\Delta G° = -2.3RT \log K$$

PROBLEMS

12.38 The standard free energy of formation $(\Delta G_f°)$ of NO_2 is 12.4 kcal/mole. Calculate the value of K at 25°C for the reaction

$$2 O_{2(g)} + N_{2(g)} \rightleftarrows 2 NO_{2(g)}$$

Solution: Since this reaction is the reaction for the formation of 2 moles of $NO_{2(g)}$ from its elements in their standard states, $\Delta G° = 2 (\Delta G_f°) = 24.8$ kcal $= 24,800$ cal. Substitution into $\Delta G° = -2.3RT \log K$ gives

$$\log K = -\frac{24800}{(1.987)(298)} = -41.9, \text{ using } R = 1.987 \text{ cal/deg-mole}$$

$$K = 1.26 \times 10^{-42}$$

12.39 The equilibrium constant K for the reaction $CO_{2(g)} + H_{2(g)} \rightleftarrows CO_{(g)} + H_2O_{(g)}$ is 2.0 at $T = 1007°K$. Calculate the value of $\Delta G°$.

Solution: Substituting into $\Delta G° = -2.3RT \log K$

$$\Delta G° = (-2.3)(1.987)(1007 + 273) \log 2$$

$$= -1.79 \times 10^3 \text{ cal}$$

12.40 The reaction $N_{2(g)} + 3 H_{2(g)} \rightarrow 2 NH_{3(g)}$ is found to have $\Delta G° = -7.94$ kcal. Calculate K at 300°K and 600°K.

12.41 For the reaction $H_{2(g)} + Cl_{2(g)} \rightleftarrows 2 HCl$, $\Delta G° = -45.5$ kcal. Calculate K at 300°K.

12.42 For the reaction $N_{2(g)} + \frac{1}{2} O_{2(g)} \rightleftarrows N_2O_{(g)}$, $\Delta G° = 24.9$ kcal. Calculate K at 323°K.

12.43 For the reaction $Zn_{(s)} + Cu^{++} \rightleftarrows Cu_{(s)} + Zn^{++}$, $K = 2 \times 10^{37}$ at 25°C. Calculate $\Delta G°$.

12.44 For the reaction $AgCl_{(s)} \rightleftharpoons Ag^+ + Cl^-$, $K = 2.12 \times 10^{-10}$ at 27°C. Calculate $\Delta G°$.

➡ **12.45** Calculate what the value of $\Delta G°$ for a reaction must be in order for K for the reaction to increase by a factor of 2 in raising the temperature from 300°K to 400°K.

13

Ionic equilibria.
Ionization constants.
Ionization equilibrium of water.
pH. Formality. Hydrolysis.
Neutralization. Equivalents.
Buffers.
Main reaction approximation.

All weak electrolytes are incompletely ionized in water solution, the ionization reaching a state of equilibrium as represented by an equation such as:

(1) $$HC_2H_3O_2 \rightleftharpoons H^+ + C_2H_3O_2^-$$

Most of the weak electrolytes we shall deal with are either weak acids or weak bases. An acid can be defined as a proton [H$^+$] donor and a base as a proton acceptor. This is called the Brønsted-Lowry definition. With this in mind we can recognize that equation (1) is a convenient simplification of a reaction between an acid and a base. The acid is $HC_2H_3O_2$, acetic acid, and the base is H_2O. The reaction could also be written as

(2) $$HC_2H_3O_2 + H_2O \rightleftharpoons H_3O^+ + C_2H_3O_2^-$$

to emphasize this point. In practice it is common to write the form of equation (1) for the reaction of a weak acid with water.

Since the ionization reaches a state of equilibrium it can, as with all reactions that reach a state of equilibrium, be represented by an equilibrium constant, K, called in this instance an ionization constant.

(3) $$K_i = \frac{[H^+] \times [C_2H_3O_2^-]}{[HC_2H_3O_2]}$$

This equilibrium constant is also referred to as K_A. The bracketed formulas [H$^+$], [C$_2$H$_3$O$_2^-$], and [HC$_2$H$_3$O$_2$] represent concentrations in moles/liter.

The numerical value of the ionization constant for $HC_2H_3O_2$ at 25°C has been determined to be 1.8×10^{-5} moles/liter, commonly expressed as $1.8 \times 10^{-5}M$, where M represents moles/liter.

Weak bases also undergo partial ionization in aqueous solution due to an acid-base reaction with water acting as an acid or proton donor and the weak base acting as a proton acceptor. The most common weak base is ammonia, NH_3. Its ionization is best represented by the equation

(4) $$NH_3 + H_2O \rightleftharpoons NH_4^+ + OH^-$$

and the equilibrium constant for this reaction is

(5) $$K_i = \frac{[NH_4^+][OH^-]}{[NH_3]}$$

This equilibrium constant is also referred to as K_B. It should be noted that $[H_2O]$ does not appear in the equilibrium constant expression since its concentration does not undergo any appreciable changes as the result of any of these equilibria. The anion of an acid is a Brønsted-Lowry base. Since the strength of a base is determined by its attraction for protons it follows that *the weaker the acid the greater the basic strength of its anion.* Therefore, the OH^- ion is the strongest base among the anions of the acids listed in Table 2, page 275. The order of strength of the five strongest bases in this table is:

$$OH^- > S^{--} > AsO_4^{---} > PO_4^{---} > CO_3^{--}$$

For polyprotic acids such as H_3PO_4, H_2S, H_2CO_3, etc., the ionization takes place in steps and each step has its own ionization constant. For H_3PO_4 the three steps, with their ionization constants, are:

(6) $H_3PO_4 \rightleftharpoons H^+ + H_2PO_4^-$ $\quad K_1 = \dfrac{[H^+] \times [H_2PO_4^-]}{[H_3PO_4]} = 7.5 \times 10^{-3}M$

(7) $H_2PO_4^- \rightleftharpoons H^+ + HPO_4^{--}$ $\quad K_2 = \dfrac{[H^+] \times [HPO_4^{--}]}{[H_2PO_4^-]} = 6.2 \times 10^{-8}M$

(8) $HPO_4^{--} \rightleftharpoons H^+ + PO_4^{---}$ $\quad K_3 = \dfrac{[H^+] \times [PO_4^{---}]}{[HPO_4^{--}]} = 1.0 \times 10^{-12}M$

The overall ionization constant for a polyprotic acid is the product of the constants for the separate steps. Thus for H_2S the total ionization is

$$H_2S \rightleftharpoons 2H^+ + S^{--}$$

and the two steps are

$$H_2S \rightleftharpoons H^+ + HS^- \qquad K_1 = 1.0 \times 10^{-7}M$$
$$HS^- \rightleftharpoons H^+ + S^{--} \qquad K_2 = 1.3 \times 10^{-13}M$$

and

$$K_i = K_1 \times K_2 = \frac{[H^+] \times [HS^-]}{[H_2S]} \times \frac{[H^+] \times [S^{--}]}{[HS^-]} = \frac{[H^+]^2 \times [S^{--}]}{[H_2S]}$$

$$= 1.0 \times 10^{-7} \times 1.3 \times 10^{-13} = 1.3 \times 10^{-20}M^2$$

The numerical values of the ionization constants for a number of weak electrolytes are given in Table 2. The smaller the ionization constant the weaker the electrolyte.

Pure water itself is also slightly ionized as follows:

$$H_2O \rightleftharpoons H^+ + OH^-$$

The concentration of H^+ ions has been shown experimentally to be 1×10^{-7} mole/liter at 25°C. The concentration of OH^- ions is the same as the concentration of H^+ ions, 1×10^{-7} mole/liter; for this reason water is neutral.

The ionization constant for water is expressed by the familiar equation:

(9) $$K_w = [H^+] \times [OH^-]$$

At 25°C $K_w = 1 \times 10^{-14} M^2$, which means that at this temperature for any aqueous solution at equilibrium the product of $[H^+]$ and $[OH^-]$ *must* be 1×10^{-14}.

If the H^+ concentration is greater than 10^{-7} mole/liter (e.g., 10^{-6}) the solution will be acidic, while if the hydrogen-ion concentration is less than 10^{-7} mole/liter (e.g., 10^{-8}) the solution will be alkaline.

It is common in equilibrium calculations to deal with small numbers written in exponential form, which are often inconvenient to handle. Often such numbers are expressed as a negative logarithm instead. This is denoted by the symbol p preceding the quantity of interest. For example pH is the negative logarithm of $[H^+]$; pOH, the negative logarithm of $[OH^-]$; and pK, the negative logarithm of the equilibrium constant.

To convert an exponential to its negative logarithm we use the relationship $-\log(a \times 10^{-b}) = -(\log a + \log 10^{-b}) = b - \log a$. For example if $[H^+] = 2.0 \times 10^{-4}$, pH $= -\log[H^+] = -\log(2.0 \times 10^{-4}) = 4 - \log 2 = 3.7$.

A solution whose pH is 7 is neutral. A solution whose pH is greater than 7 is alkaline while one whose pH is less than 7 is acidic. It is important to remember that an increase in pH represents a decrease in hydrogen-ion concentration, that is, a decrease in acidity.

The concept of formality and a second definition of molarity

We learned in Chapter 10 that $1.6M$ K_2CO_3 would be prepared by dissolving 1.6 moles of K_2CO_3 in enough water to give 1 liter of solution. But we know that when K_2CO_3 is dissolved in water it immediately dissociates completely into K^+ and CO_3^{--} ions; furthermore, the CO_3^{--} ions react to a limited extent with water to form HCO_3^-, H_2CO_3, H^+, and OH^-. Accordingly, a solution formed by dissolving 1.6 moles of K_2CO_3 in enough water to form a liter of solution will in fact contain several species none of which has a concentration of precisely 1.6 moles per liter. To provide for this situation the concepts of *formality* and *formal solutions* have been introduced and are now widely used. According to these concepts the term *formality* (abbreviated F) is used to designate *the number of moles of solute that were used in preparing 1 liter of solution;* the term *molarity* is reserved for designating *the actual concentration, in moles per liter of solution, of a particular species that is present in the solution.* Thus, the above solution would be 1.6 F in K_2CO_3, but it would be 3.2 M in K^+ and slightly less than 1.6 M in CO_3^{--}. A solution prepared by dissolving 0.10 mole of $NaC_2H_3O_2$ in enough water to give 1 liter of solution would be labeled 0.10 F $NaC_2H_3O_2$; it would be found to be 0.10 M in Na^+, 7.5×10^{-6} M in OH^-, 1.3×10^{-9} M in H^+, 7.5×10^{-6} M in $HC_2H_3O_2$, and $(0.10 - 7.5 \times 10^{-6})$ M in $C_2H_3O_2^-$.

It is important to note that *formality*, as defined above, is synonymous with *molarity* as that term was defined and used in Chapter 10.

From this point on in this book we will use the concept of formality and with it the new definition of molarity. *Formality*, abbreviated F, will always designate *the number of moles of solute used in preparing 1 liter of solution;* there should never be any ambiguity about the meaning of the notations, 0.25 F Na_2CO_3 and 0.10 F $HC_2H_3O_2$. The concentration of a particular species in a solution will always be expressed in terms of *molarity*, denoted by the abbreviation M; *molarity* in this usage *represents the number of moles of the particular species present in 1 liter of solution.*

It should be noted that the formula for the ionization constant as given above assumes that the system to which it is applied is *ideal*; that is, it assumes that the behavior of each species in the system is completely unaffected by the presence of other species. In such an ideal system the *molar concentration* is, in fact, the *effective concentration*. For nonideal systems the effective concentration, or *activity*, of a given ion or molecule is the product of its molar concentration and its *activity coefficient* in the particular system. It is this *activity* which should appear in the ionization constant formula for a nonideal system. As with the gas-phase equilibria discussed in Chapter 12, we will assume that all solutions encountered in the problems in this book are ideal. Accordingly, the activity coefficient of each species will have a value of 1, and the ionization equilibria can all be expressed in terms of the *molar* concentrations of each species.

PROBLEMS

(See Table 2 for ionization constants.)

13.1 A 0.010 F solution of $HC_2H_3O_2$ is 4.17% ionized. Calculate the ionization constant of $HC_2H_3O_2$.

Solution:

$$HC_2H_3O_2 \rightleftharpoons H^+ + C_2H_3O_2^-$$

$$K = \frac{[H^+] \times [C_2H_3O_2^-]}{[HC_2H_3O_2]}$$

4.17% expressed in decimal form, is 0.0417.

0.0417×0.010 moles $= 0.00042$ moles of $HC_2H_3O_2$ ionized. Since 1 mole of $HC_2H_3O_2$ yields 1 mole of H^+ and 1 mole of $C_2H_3O_2^-$, the 4.2×10^{-4} mole of $HC_2H_3O_2$ will yield 4.2×10^{-4} mole each of H^+ and $C_2H_3O_2^-$; $[H^+]$ and $[C_2H_3O_2^-]$ will each be 4.2×10^{-4} mole/liter. The concentration of un-ionized $HC_2H_3O_2$ molecules will then be $0.010 - 0.00042$ or 0.0096 mole/liter. Substituting these values in the equilibrium formula,

$$K = \frac{(4.2 \times 10^{-4}) \times (4.2 \times 10^{-4})}{(9.6 \times 10^{-3})}$$

$$= 1.8 \times 10^{-5} \text{ mole/liter} = 1.8 \times 10^{-5} M$$

When calculating the ionization constants of acids and bases in water solution, the small concentrations of H^+ or OH^- ions due to the ionization of water are generally ignored.

13.2 From the facts given calculate the ionization constant of each substance.

(a) A 0.10 F solution of NH_3 is 1.3% ionized.

(b) A 0.0010 F solution of $HC_2H_3O_2$ is 12.6% ionized.

(c) A 0.01 F solution of HCN is 0.02% ionized.

13.3 A 0.100 F aqueous solution of the weak acid, HY, freezes at $-0.240°C$. The freezing point constant for water is 1.86. Calculate the ionization constant for HY.

Solution: $HY \rightleftharpoons H^+ + Y^-$

Let $X = [H^+] = [Y^-]$.

$0.100 - X = [HY]$

Total moles of solute $= [H^+] + [Y^-] + [HY] = 0.100 + X$

Moles of solute $= \dfrac{0.240°}{1.86°} = 0.100 + X$

Solving, $X = 0.029$ $M = [H^+] = [Y^-]$.

$[HY] = 0.100 - X = 0.071$ M

$$K = \frac{[H^+] \times [Y^-]}{[HY]} = \frac{0.029 \times 0.029}{0.071} = 1.18 \times 10^{-2} M$$

13.4 The ionization constant for HCN is 4×10^{-10} M at 25°C. Calculate the formality of and the H^+ ion concentration of a solution of HCN which is 0.010% ionized.

Solution:

$$HCN \rightleftharpoons H^+ + CN^-$$

(1)
$$K = \frac{[H^+] \times [CN^-]}{[HCN]} = 4 \times 10^{-10} \, M$$

Let X = formality. Since the solution is 0.010% ionized, and since 0.010%, expressed as decimal, is 0.00010, the concentration of H^+ will be 0.00010 X. The concentration of CN^- will also be 0.00010 X. The concentration of un-ionized HCN will be $X - 0.00010 \, X$. Substituting these values in the above equilibrium formula:

(2)
$$\frac{0.00010 \, X \times 0.00010 \, X}{X - 0.00010 \, X} = 4 \times 10^{-10} \, M$$

Solving, $X = 4 \times 10^{-2}$ = the formality

$$[H^+] = 0.00010 \, X = 1 \times 10^{-4} \times 4 \times 10^{-2}$$
$$= 4 \times 10^{-6} \, M$$

13.5 The ionization constant for NH_3 is $1.8 \times 10^{-5} \, M$. Calculate the formality and OH^- concentration of a solution in which the NH_3 is 1.3% ionized.

13.6 The ionization constant for $HC_2H_3O_2$ is $1.8 \times 10^{-5} \, M$. Calculate the hydrogen-ion concentration of 0.01 $F \, HC_2H_3O_2$.

Solution:

$$HC_2H_3O_2 \rightleftharpoons H^+ + C_2H_3O_2^-$$

Let $X = [H^+] = [C_2H_3O_2^-]$.

$$0.01 - X = [HC_2H_3O_2]$$

$$K = \frac{[H^+] \times [C_2H_3O_2^-]}{[HC_2H_3O_2]} = 1.8 \times 10^{-5} \, M$$

Substituting the values of $[H^+]$, $[C_2H_3O_2^-]$, and $[HC_2H_3O_2]$ in the equilibrium formula:

(1)
$$\frac{X^2}{0.01 - X} = 1.8 \times 10^{-5} \, M$$

The term, X, can be dropped from the expression, $0.01 - X$, in the denominator of Equation (1) if the value of X is so small that, within the limits imposed by the number of allowable significant figures, $0.01 - X = 0.01$. We can estimate the value of X as follows: Drop X from the expression, $0.01 - X$. As a result, $X^2 = 1.8 \times 10^{-7}$, and X is about 4×10^{-4} or 0.0004. $0.01 - 0.0004 = 0.0096$. When rounded off to one significant figure 0.0096 becomes 0.01. Therefore, X can be dropped.

(2)
$$\frac{X^2}{0.01} = 1.8 \times 10^{-5} \, M$$

$$X = 4 \times 10^{-4} \, M = [\text{H}^+]$$

This example illustrates the rule that a term, n, *which is added to or subtracted from* a term, m, in an expression, $m + n$ or $m - n$, can be dropped if it is so small that $m - n$ (or $m + n$), when rounded off to the permissible significant figures, is equal to m.

It should be emphasized that a small term can be dropped only in expressions involving its *addition to* or *subtraction from* a large term, never in an expression involving its multiplication or division by a large term.

13.7   The ionization constant for NH_3 is $1.8 \times 10^{-5} \, M$. Calculate the hydroxide-ion concentration of $0.10 \, F \, \text{NH}_3$.

13.8   Referring to Table 2, page 275, for the ionization constants, calculate the concentrations of HCO_3^- and CO_3^{--} ions in a $0.034 \, F$ solution of CO_2 in water.

Solution: The reactions that occur when CO_2 is dissolved in water are:

(1) $\text{CO}_2 + \text{H}_2\text{O} = \text{H}_2\text{CO}_3$

(2) $\text{H}_2\text{CO}_3 \rightleftharpoons \text{H}^+ + \text{HCO}_3^-$

(3) $\text{HCO}_3^- \rightleftharpoons \text{H}^+ + \text{CO}_3^{--}$

The ionization constant for (2) is

$$\frac{[\text{H}^+][\text{HCO}_3^-]}{[\text{H}_2\text{CO}_3]} = 4.2 \times 10^{-7} \, M = K_1$$

and for (3) is

$$\frac{[\text{H}^+][\text{CO}_3^{--}]}{[\text{HCO}_3^-]} = 4.8 \times 10^{-11} \, M = K_2$$

Since the two equilibria represented by Equations (2) and (3) occur in the same solution, the $[\text{H}^+]$ that appears in the formula for K_1 must be the same $[\text{H}^+]$ that appears in the formula for K_2 and this $[\text{H}^+]$ must be the sum of the hydrogen ions provided by reactions (2) and (3). However, since the equilibrium constant for reaction (3) is so much smaller than that for reaction (2), the amount of H^+ provided by reaction (3) is so very small compared with that provided by reaction (2) that, within the limitations imposed by the number of allowable significant figures, it can be neglected. Accordingly, we will proceed as follows:

Referring to reaction (2) and its ionization constant, K_1, let

$$X = \text{conc of } \text{HCO}_3^- \qquad X = \text{conc of } \text{H}^+$$

$$0.034 - X = \text{conc of } \text{H}_2\text{CO}_3$$

Substituting these values in the equation for K_1

$$\frac{X^2}{0.034 - X} = 4.2 \times 10^{-7} \, M$$

X is so small by comparison with 0.034 that it can be dropped from the expression, $0.034 - X$.

$$X^2 = 1.43 \times 10^{-8} \, M$$

$$X = 1.2 \times 10^{-4} \, M = [HCO_3^-] = [H^+]$$

Since K_2 is so much smaller than K_1, the value of $[H^+]$ calculated from K_1 will be the $[H^+]$ of the solution and can be inserted in the formula for K_2.

Letting $Y = [CO_3^{--}]$ and $1.2 \times 10^{-4} = [H^+]$

$$\frac{1.2 \times 10^{-4} \, Y}{1.2 \times 10^{-4} - Y} = 4.8 \times 10^{-11} \, M$$

Since K_2 is very small, Y will be so small by comparison with 1.2×10^{-4} that it can be dropped from the term, $1.2 \times 10^{-4} - Y$. Therefore,

$$\frac{1.2 \times 10^{-4} \, Y}{1.2 \times 10^{-4}} = 4.8 \times 10^{-11} \, M$$

$$Y = 4.8 \times 10^{-11} \, M = [CO_3^{--}]$$

Note: The equation for the overall ionization is

$$H_2CO_3 \rightleftharpoons 2 \, H^+ + CO_3^{--}$$

The overall ionization constant, K_i, is

$$K_i = \frac{[H^+]^2 \times [CO_3^{--}]}{[H_2CO_3]} = K_1 \times K_2 = 2.0 \times 10^{-17}$$

It should be emphasized that this overall ionization equation and constant cannot be used to solve for $[CO_3^{--}]$, $[HCO_3^-]$, and $[H^+]$ in a solution of CO_2 in pure water. Using the overall equation assumes that $[H^+] = 2 \times [CO_3^{--}]$. This is not a valid assumption. Suppose we solve for $[H^+]$ and $[CO_3^{--}]$ by using K_i.

Let $X = [CO_3^{--}]$; $2 \, X = [H^+]$; $0.034 - X = [H_2CO_3]$.

$$\frac{(2 \, X)^2 X}{0.034 - X} = 2.0 \times 10^{-17}$$

$$4 \, X^3 = 6.8 \times 10^{-19}$$

$$X = 5.5 \times 10^{-7} \, M = [CO_3^{--}]$$

$$2 \, X = 1.1 \times 10^{-6} \, M = [H^+]$$

These answers are quite different from those obtained by the correct method, in which K_1 and K_2 were used separately. That these new answers are not correct can be seen by using them to calculate $[HCO_3^-]$.

$$[HCO_3^-] = K_1 \frac{[H_2CO_3]}{[H^+]} \quad \text{and} \quad [HCO_3^-] = \frac{[H^+][CO_3^{--}]}{K_2}$$

The value of $[HCO_3]$ thus obtained, $1.3 \times 10^{-2} M$, is obviously incorrect, for this high a concentration of HCO_3^- could not be present without an equally high $[H^+]$.

It is worth noting, however, that although the overall ionization constant *cannot* be used to calculate correctly $[H^+]$ and $[CO_3^{--}]$, the value of this overall ionization constant *must be satisfied.* Substituting the correct values, $[H^+] = 1.2 \times 10^{-4} M$, $[CO_3^{--}] = 4.8 \times 10^{-11} M$, and $[H_2CO_3] = 3.4 \times 10^{-2} M$, into the equation for the overall ionization, it is seen that this is indeed the case.

If the *H^+ concentration is fixed* by the addition of a strong acid to a solution of CO_2 in water K_t can then be used in solving for $[CO_3^{--}]$. This case is presented in Problem 13.29

13.9   Calculate the concentrations of $H_2PO_4^-$, HPO_4^{--}, and PO_4^{---} ions in $0.10 F H_3PO_4$.

Solution: The three equilibria and their constants are

(1) $H_3PO_4 \rightleftharpoons H^+ + H_2PO_4^-$ $K_1 = \dfrac{[H^+] \times [H_2PO_4^-]}{[H_3PO_4]}$

$$= 7.5 \times 10^{-3} M$$

(2) $H_2PO_4^- \rightleftharpoons H^+ + HPO_4^{--}$ $K_2 = \dfrac{[H^+] \times [HPO_4^{--}]}{[H_2PO_4^-]}$

$$= 6.2 \times 10^{-8} M$$

(3) $HPO_4^{--} \rightleftharpoons H^+ + PO_4^{---}$ $K_3 = \dfrac{[H^+] \times [PO_4^{---}]}{[HPO_4^{--}]}$

$$= 1.0 \times 10^{-12} M$$

Since all three equilibria occur in the same solution, the value of $[H^+]$ must be the same in each. Since K_1 is so much larger than K_2 and K_3, it will determine the value of $[H^+]$. Likewise, K_1 will determine the value of $[H_2PO_4^-]$. Accordingly, we will first calculate $[H^+]$ and $[H_2PO_4^-]$ from K_1.

Let $X = [H^+]$. Then $X = [H_2PO_4^-]$ and $0.10 - X = [H_3PO_4]$.

$$\frac{X^2}{0.10 - X} = 7.5 \times 10^{-3} M$$

It is obvious that X will be too large in comparison with 0.10 to allow it to be dropped.

Solving this quadratic, $X = 2.4 \times 10^{-2} M = [H^+] = [H_2PO_4^-]$

We will next substitute these values in K_2, letting $Y = [HPO_4^{--}]$

$$K_2 = \frac{(2.4 \times 10^{-2})(Y)}{(2.4 \times 10^{-2} - Y)} = 6.2 \times 10^{-8} M$$

Since K_2 is very small, Y will be so small by comparison with 2.4×10^{-2} that it can be dropped from the term, $2.4 \times 10^{-2} - Y$. This leaves

$$\frac{2.4 \times 10^{-2} \; Y}{2.4 \times 10^{-2}} = 6.2 \times 10^{-8} \; M$$

$$Y = 6.2 \times 10^{-8} \; M = [HPO_4^{--}]$$

We will then substitute the calculated values of $[H^+]$ and $[HPO_4^{--}]$ in K_3, letting $Z = [PO_4^{---}]$.

$$K_3 = \frac{(2.4 \times 10^{-2})(Z)}{(6.2 \times 10^{-8} - Z)} = 1.0 \times 10^{-12} \; M$$

Since K_3 is extremely small, Z will be so small by comparison with 6.2×10^{-8} that it can be dropped from the term, $6.2 \times 10^{-8} - Z$. That leaves

$$\frac{2.4 \times 10^{-2} Z}{6.2 \times 10^{-8}} = 1.0 \times 10^{-12} \; M$$

$$Z = 2.5 \times 10^{-18} \; M = [PO_4^{---}]$$

13.10   Calculate the concentration of S^{--} in $0.10 \; F \; H_2S$.

13.11   Calculate the concentration of CrO_4^{--} in $0.10 \; F \; H_2CrO_4$.

The common ion effect

We have noted in Chapter 12 that if after a chemical reaction of the general type, $A + B \rightleftharpoons C + D$, has reached a state of equilibrium, more C is added to the reaction vessel, the equilibrium is shifted to the left. In exactly the same manner the equilibrium in an ionic equilibrium such as, $HC_2H_3O_2 \rightleftharpoons H^+ + C_2H_3O_2^-$, will be shifted to the left if acetate ions are added to the system. This shift in an ionization equilibrium by increasing the concentration of one of the ions involved is called *the common ion effect*.

PROBLEMS

(See Table 2 for ionization constants.)

13.12   Calculate the OH^- concentration in moles of OH^- per liter of a solution which contains 1×10^{-2} mole of H^+ per liter. Will the solution be neutral, acidic, or alkaline?

Solution: In any water solution the product of the concentration of H^+ and the concentration of OH^-, when these concentrations are expressed in moles per liter, is always equal to 1×10^{-14}. That is

$$[H^+] \times [OH^-] = 1 \times 10^{-14} \, M^2$$

$$[OH^-] = \frac{1 \times 10^{-14}}{[H^+]} = \frac{1 \times 10^{-14}}{1 \times 10^{-2} \text{ mole } H^+ \text{ per liter}}$$

$$= 1 \times 10^{-12} \text{ mole } OH^- \text{ per liter}$$

If the concentration of H^+ is 1×10^{-7} mole of H^+ per liter, the concentration of OH^- will also be 1×10^{-7} mole of OH^- per liter, and the solution will be neutral. If the concentration of H^+ is greater than 1×10^{-7} mole/liter, the solution will be acidic; if less, it will be alkaline. Since 1×10^{-2} mole of H^+ per liter is a higher concentration than 1×10^{-7} mole of H^+ per liter, the solution will be acidic.

13.13 Calculate the OH^- concentration, in g of OH^- per liter, of a solution containing 1×10^{-10} mole of H^+ per liter.

Solution:

$$[H^+] \times [OH^-] = 1 \times 10^{-14} \, M^2$$

$$[OH^-] = \frac{1 \times 10^{-14}}{[H^+]} = \frac{1 \times 10^{-14}}{1 \times 10^{-10} \text{ mole } H^+ \text{ per liter}}$$

$$= 1 \times 10^{-4} \text{ mole of } OH^- \text{ per liter}$$

$$1 \times 10^{-4} \text{ mole of } OH^-/\text{liter} \times \frac{17 \text{ g of } OH^-}{1 \text{ mole of } OH^-}$$

$$= 1.7 \times 10^{-3} \text{ g of } OH^-/\text{liter}$$

13.14 Calculate the OH^- concentration, in grams of OH^- per liter, of a solution whose H^+ concentration is:

(a) 1.0×10^{-6} mole of H^+ per liter

(b) 3.0×10^{-4} g of H^+ per liter

13.15 Calculate the H^+ concentration, in grams of H^+ per liter, of a solution whose OH^- concentration is:

(a) 2.0×10^{-5} mole of OH^- per liter

(b) 3.4×10^{-2} g of OH^- per liter

13.16 Calculate the pH of a solution which contains 1×10^{-5} mole of H^+ per liter.

Solution: By definition, pH is the negative logarithm of the hydrogen-ion concentration when this concentration is expressed in moles of H^+ per liter.

$$pH = -\log [H^+] = -\log 10^{-5} = 5$$

13.17 Calculate the pH of a solution which contains 3×10^{-4} mole of H^+ per liter.

Solution:

$$pH = \log \frac{1}{[H^+]} = \log \frac{1}{3 \times 10^{-4}} = \log \frac{10^4}{3}$$

$$pH = \log \frac{10^4}{3} = \log 10^4 - \log 3$$

$$\log 10^4 = 4 \qquad \log 3 = 0.477$$

$$pH = 4 - 0.477 = 3.523$$

13.18 Calculate the pH of a solution which contains:

(a) 1×10^{-8} mole of H^+/liter

(b) 0.0020 g of H^+/liter

(c) 0.0030 mole of H^+/liter

(d) 0.00017 g of OH^-/100 cc

(e) 2.0×10^{-3} mole of OH^-/liter

(f) 0.010 F $HC_2H_3O_2$ which is 4.17% ionized

(g) 0.10 F NH_4OH which is 4.10% ionized

(h) 0.010 F KOH

(i) 1.00×10^{-8} F HCl

13.19 Calculate the H^+ concentration in moles of H^+ per liter, of a solution whose pH is 5.

Solution: pH is, by definition, the negative of the logarithm of the H^+ concentration. Since the pH is 5, $[H^+]$ must be 10^{-5} mole/liter.

13.20 Calculate the H^+ concentration in moles per liter of a solution whose pH is 4.8.

Solution: Since pH is 4.8, $[H^+]$ must be $10^{-4.8}$ mole/liter. But $10^{-4.8}$ $= 10^{-5} \times 10^{0.2}$

$$10^{0.2} = 1.59 \qquad (\log 10^{0.2} = 0.2 \text{ and antilog of } 0.2 = 1.59)$$

Therefore, $10^{-4.8} = 1.59 \times 10^{-5}$

$$[H^+] = 1.59 \times 10^{-5} \text{ mole/liter}$$

13.21 Calculate the H^+ concentration in moles of H^+ per liter of a solution whose:

(a) pH is 1.5

(b) pH is 13.6

13.22 Calculate the OH^- concentration in moles of OH^- per liter of a solution whose:

(a) pH is 3.6

(b) pH is 6.2

13.23 Which is more strongly acid:

(a) a solution with a pH of 2?

(b) a solution containing 0.020 g of H^+ per liter?

13.24 A 0.0010 F solution of HF has a pH of 4. Calculate the percent ionization of the HF.

Hydrolysis

Since a base can be defined as a proton acceptor, it follows that the anion derived from an acid is a base. This point is underscored by the common practice of referring to the species remaining after an acid donates a proton as the conjugate base of the acid. For example the conjugate base of acetic acid $HC_2H_3O_2$ is the acetate ion $C_2H_3O_2^-$. When the conjugate base of an acid is in aqueous solution an equilibrium will be established in which water acts as an acid and donates a proton to the conjugate base of an acid.

$$(1) \qquad C_2H_3O_2^- + H_2O \rightleftharpoons HC_2H_3O_2 + OH^-$$

This type of process is called hydrolysis.

An analogous situation arises with the species formed when a base accepts a proton. For example when NH_3 is dissolved in water some NH_4^+ is formed. This species is called the conjugate acid of the base, since it is a proton donor. When NH_4^+ is in aqueous solution, hydrolysis can also occur.

$$(2) \qquad NH_4^+ + H_2O \rightleftharpoons NH_3 + H_3O^+$$

or written more simply

$$(3) \qquad NH_4^+ \rightleftharpoons NH_3 + H^+$$

The equilibrium constant for this type of process is called the hydrolysis constant K_h and does not include $[H_2O]$. For (1)

$$(4) \qquad K_h = \frac{[HC_2H_3O_2] \times [OH^-]}{[C_2H_3O_2^-]}$$

and for (3)

$$(5) \qquad K_h = \frac{[NH_3][H^+]}{[NH_4^+]}$$

The numerical value of the hydrolysis constant for a particular ion can be calculated from the ion product constant for water and the ionization constant for the weak acid or weak base formed during hydrolysis in the following way:

In Equation (4) we will multiply both numerator and denominator by $[H^+]$, giving thereby,

$$(6) \qquad K_h = \frac{[HC_2H_3O_2]}{[H^+] \times [C_2H_3O_2^-]} \times [OH^-] \times [H^+]$$

The first term to the right of the $=$ sign is the reciprocal of the ionization constant, K_i, for $HC_2H_3O_2$ and the rest of the expression is the ion product constant, K_w, for water.

If we substitute the numerical values of K_w and K_i we obtain the numerical value of K_h,

(7) $$K_h = \frac{K_w}{K_i} = \frac{1 \times 10^{-14} \; M^2}{1.8 \times 10^{-5} \; M} = 5.6 \times 10^{-10} \; M$$

It is obvious from these calculations that *the weaker the acid* (or base) *the greater the percent hydrolysis of its anion* (or cation).

If the salt is derived from a weak acid and a weak base both the cation and the anion undergo hydrolysis. Thus, for the salt NH_4F, the reactions are:

(8) $$NH_4^+ + H_2O \rightleftharpoons NH_3 + H^+ + H_2O$$

(9) $$F^- + H_2O \rightleftharpoons HF + OH^-$$

The H^+ and OH^- formed in Equations (8) and (9) will react

(10) $$H^+ + OH^- = H_2O$$

Equations (8), (9), and (10), when totaled, give the net equation for the hydrolysis,

(11) $$NH_4^+ + F^- + H_2O \rightleftharpoons NH_3 + HF + H_2O$$

The equilibrium constant for this reaction is

(12) $$K = \frac{[NH_3] \times [HF] \times [H_2O]}{[NH_4^+] \times [F^-] \times [H_2O]}$$

As in the previous hydrolysis, $[H_2O]$ is constant, and can be omitted to give the hydrolysis constant K_h.

(13) $$K_h = \frac{[NH_3] \times [HF]}{[NH_4^+] \times [F^-]}$$

By multiplying both numerator and denominator by $[H^+] \times [OH^-]$, Equation (13) is resolved into three separate equilibria.

(14) $$K_h = \frac{[NH_3]}{[NH_4^+] \times [OH^-]} \times \frac{[HF]}{[H^+] \times [F^-]} \times [H^+] \times [OH^-]$$

(15) $$K_h = \frac{1}{K_{NH_4OH}} \times \frac{1}{K_{HF}} \times K_w = \frac{K_w}{K_{NH_4OH} \times K_{HF}}$$

By substituting, in Equation (14), the numerical values of the three constants, the numerical value of K_h can be calculated.

(16) $$K_h = \frac{1.0 \times 10^{-14}}{1.8 \times 10^{-5} \times 6.9 \times 10^{-4}} = 8.0 \times 10^{-7}$$

Other equilibria involving weak electrolytes

The technique of resolving a given equilibrium constant into its component constants, which was used in calculating hydrolysis constants, can be applied to other systems. To illustrate, when sodium formate ($NaCHO_2$) is added to a solution of acetic acid, the following equilibrium is set up.

$$CHO_2^- + HC_2H_3O_2 \rightleftharpoons HCHO_2 + C_2H_3O_2^-$$

The equilibrium constant for this reaction is

$$K = \frac{[HCHO_2] \times [C_2H_3O_2^-]}{[CHO_2^-] \times [HC_2H_3O_2]}$$

By multiplying both numerator and denominator by $[H^+]$ we obtain

$$K = \frac{[HCHO_2]}{[H^+] \times [CHO_2^-]} \times \frac{[H^+] \times [C_2H_3O_2^-]}{[HC_2H_3O_2]}$$

$$K = \frac{1}{K_{HCHO_2}} \times K_{HC_2H_3O_2} = \frac{1.8 \times 10^{-5}}{2 \times 10^{-4}} = 9 \times 10^{-2}$$

The fact that the equilibrium constant has the small value, 9×10^{-2}, indicates that, at equilibrium, the concentrations of the species on the left are somewhat larger than the concentrations of those on the right. This is another way of stating that the equilibrium lies quite far to the left. This situation is consistent with the facts, testified by the values of the equilibrium constants, that $HCHO_2$ is a stronger acid than $HC_2H_3O_2$ and $C_2H_3O_2^-$ is a stronger base than CHO_2^-. Since both of the stronger electrolytes lie on the right it is logical to expect that the equilibrium should lie far to the left.

PROBLEMS

(See Table 2 for ionization constants.)

13.25 The ionization constant for $HC_2H_3O_2$ is 1.8×10^{-5} M. How many moles of hydrogen ions will there be in a liter of 0.10 F $HC_2H_3O_2$ containing 0.20 mole of $NaC_2H_3O_2$?

Solution:

$$HC_2H_3O_2 \rightleftharpoons H^+ + C_2H_3O_2^-; \qquad NaC_2H_3O_2 = Na^+ + C_2H_3O_2^-$$

Let X = concentration of H^+ at equilibrium
$0.20 + X$ = concentration of $C_2H_3O_2^-$ at equilibrium
$0.10 - X$ = concentration of $HC_2H_3O_2$ at equilibrium

$$K = \frac{[H^+] \times [C_2H_3O_2^-]}{[HC_2H_3O_2]} = 1.8 \times 10^{-5} \ M$$

Substituting the values of $[H^+]$, $[C_2H_3O_2^-]$, and $[HC_2H_3O_2]$ in the above equation,

$$\frac{X(0.20 + X)}{0.10 - X} = 1.8 \times 10^{-5} \, M$$

In solving for X in the above equation we will assume that, since acetic acid is weak, the value of X (the concentration of H^+ ions) will be very much less than 0.10. If this assumption is true X can then be dropped from the terms, $0.20 + X$ and $0.10 - X$. That leaves the expression

$$\frac{0.20 \, X}{0.10} = 1.8 \times 10^{-5} \, M \qquad (X = 9.0 \times 10^{-6} \text{ mole of } H^+ \text{ per liter})$$

The fact that the value of X turns out to be very much less than 0.10 means that the assumption made above is justified.

13.26 The ionization constant for NH_3 is $1.8 \times 10^{-5} \, M$. How many moles of OH^- are there in a liter of $0.10 \, F \, NH_3$ which contains 0.10 mole of NH_4Cl?

13.27 The weak base, NH_3, has an ionization constant of $1.8 \times 10^{-5} \, M$. What is the OH^- concentration of a solution prepared by dissolving 0.25 mole of NH_3 and 0.75 mole of NH_4Cl in enough water to make a liter of solution?

13.28 How many moles of NaCN must be dissolved in a liter of $0.2 \, F$ HCN to yield a solution with a hydrogen-ion concentration of 1×10^{-6} mole per liter?

13.29 The hydrogen ion concentration of a $0.034 \, F$ solution of CO_2 in dilute HCl is $0.10 \, M$. Calculate the molar concentration of CO_3^{--}.

Solution: This problem differs from 13.8 in that H^+ has been added (in the form of HCl) to give a total $[H^+]$ of $0.10 \, M$. Using the first ionization constant for H_2CO_3 we will first solve for $[HCO_3^-]$.

$H_2CO_3 \rightleftharpoons H^+ + HCO_3^-$

Let $X = [HCO_3^-]$.

$0.10 = [H^+]$

$0.034 - X = [H_2CO_3]$

$$K_1 = \frac{[H^+][HCO_3^-]}{[H_2CO_3]} = 4.2 \times 10^{-7} \, M = \frac{0.10 \, X}{0.034 - X}$$

Since K_1 is very small and $[H^+]$ is high $(0.10 \, M)$, $[HCO_3^-]$ will be very small in comparison with $[H^+]$ and $[H_2CO_3]$. Therefore we can drop the X in the term, $0.034 - X$. That leaves

$$\frac{0.10 \, X}{0.034} = 4.2 \times 10^{-7} \, M$$

$0.10 \, X = 1.4 \times 10^{-8} \, M$

$X = 1.4 \times 10^{-7} \, M = [HCO_3^-]$

We can now substitute this value of $[HCO_3^-]$ and the value of 0.10 M for $[H^+]$ in the second ionization constant to give

$$K_2 = \frac{[H^+][CO_3^{--}]}{[HCO_3^-]} = 4.8 \times 10^{-11} \, M = \frac{0.10 \times Y}{1.4 \times 10^{-7} - Y}$$

In this formula $Y = [CO_3^{--}]$.

Since Y will be very small by comparison with 1.4×10^{-7}, it can be dropped from the term, $1.4 \times 10^{-7} - Y$. This leaves

$$\frac{0.10 \, Y}{1.4 \times 10^{-7}} = 4.8 \times 10^{-11} \, M$$

$$0.10 \, Y = 6.7 \times 10^{-18} \, M \qquad Y = 6.7 \times 10^{-17} \, M = [CO_3^{--}]$$

Note: It should be pointed out that, when the hydrogen ion concentration of a solution of a weak polybasic acid such as H_2CO_3 (H_2S, H_3PO_4, etc.) is fixed by the addition of a strong acid, the overall ionization constant, K_i, can be used in solving for $[CO_3^{--}]$. Thus

$$K_i = K_1 \times K_2 = \frac{[H^+]^2 \times [CO_3^{--}]}{[H_2CO_3]} = 2.0 \times 10^{-17}$$

If we substitute the values of $[H^+]$ and $[H_2CO_3]$ in this equation:

$$\frac{(0.10)^2[CO_3^{--}]}{0.034} = 2.0 \times 10^{-17}$$

$$[CO_3^{--}] = 6.8 \times 10^{-17}$$

This, it will be noted, is practically the same value for $[CO_3^{--}]$ that was obtained when the calculation was made via K_1 and K_2.

13.30   Calculate the sulfide ion concentration of a 0.10 F solution of H_2S in 0.10 F HCl.

13.31   Calculate the molar concentrations of $H_2AsO_4^-$, $HAsO_4^{--}$, and AsO_4^{---} in 0.20 F H_3AsO_4 which is 0.10 F in HCl.

13.32   Calculate the OH^- ion concentration of 0.2 F KCN.

Solution: The net equation for the hydrolysis is

$$CN^- + H_2O \rightleftharpoons HCN + OH^-$$

$$K_h = \frac{[HCN] \times [OH^-]}{[CN^-]} = \frac{[HCN]}{[H^+] \times [CN^-]} \times [H^+] \times [OH^-]$$

$$= \frac{K_w}{K_{HCN}} = \frac{1.0 \times 10^{-14}}{4.0 \times 10^{-10}} = 2.5 \times 10^{-5} \, M$$

Let $X = [OH^-] = [HCN]$

$$0.2 - X = [CN^-]$$

$$\frac{X^2}{0.2 - X} = 2.5 \times 10^{-5} \, M$$

$$X^2 = 5 \times 10^{-6} \, M^2$$

$$X = 2 \times 10^{-3} \, M = [OH^-]$$

13.33   Calculate the pH of 0.10 F $KCHO_2$ (potassium formate).

13.34   Calculate the H^+ ion concentration of 0.10 F NH_4Cl.

13.35   Calculate the concentration of HCN and of OH^- in 0.20 F NH_4CN.

13.36   In a 0.5 F solution of KClO, the OH^- ion concentration is 3×10^{-4} M. What is the ionization constant for HClO?

13.37   A liter of solution prepared by dissolving H_2SO_4 in pure water has a pH of 3.00.

   (a) How many moles of H_2SO_4 were dissolved? The first ionization of H_2SO_4 is complete. The ionization constant for HSO_4^- is 1.20×10^{-2} M.

   (b) Calculate the molarity of each species in solution.

Solution: Let X equal the moles of H_2SO_4 dissolved. Let Y equal the moles of SO_4^{--}. Then, since the first ionization of H_2SO_4 is complete,

$$\overset{X}{} \quad \overset{X}{} \quad \overset{X-Y}{}$$
(a) $H_2SO_4 = H^+ + HSO_4^-$

$$\overset{X-Y}{} \quad \overset{Y}{} \quad \overset{Y}{}$$
(b) $HSO_4^- \rightleftharpoons H^+ + SO_4^{--}$

(c) $X + Y = 1.00 \times 10^{-3} = [H^+]$

(d) $Y/(X - Y) = [SO_4^{--}]/[HSO_4^-] = K_2/[H^+] = 12.0$

From (c) and (d) we find that
$X = 5.20 \times 10^{-4}$ moles = moles of H_2SO_4 dissolved
$Y = 4.80 \times 10^{-4}$ moles/liter = $[SO_4^{--}]$
$X - Y = 4.0 \times 10^{-5}$ moles/liter = $[HSO_4^-]$
$1.00 \times 10^{-14} \div 1.00 \times 10^{-3} = 1.00 \times 10^{-11} = [OH^-]$

Neutralization and equivalents

A very common procedure in chemistry involves the reaction of stoichiometric quantities of an acid and a base usually to produce water and a salt. This procedure can provide a great deal of useful information about the nature and quantity of an acid or base. The net equation for a neutralization depends on the type of acid and base involved in the reaction and may be one of four types.

   (a) strong acid and strong base:

(1) $H^+ + OH^- \rightleftharpoons H_2O$ $K = \dfrac{1}{K_w} = 10^{14}$

   (b) strong acid and weak base:

(2) $H^+ + NH_3 \rightleftharpoons NH_4^+$ $K = \dfrac{K_B}{K_w}$

(c) weak acid and strong base:

(3)
$$HCN + OH^- \rightleftharpoons H_2O + CN^- \quad K = \frac{K_A}{K_w}$$

(d) weak acid and weak base:

(4)
$$HCN + NH_3 \rightleftharpoons NH_4^+ + CN^- \quad K = \frac{K_A K_B}{K_w}$$

Inspection of these reactions reveals that they all have the same stoichiometric relationship: one mole of H^+ is transferred from an acid to a base. Thus in any neutralization procedure it is possible to determine the number of moles of H^+ transferred if it is known how many moles of acid or base are involved in the neutralization and how many moles of H^+ each mole of acid supplies or each mole of base can accept. For example, if we find that one mole of a base which accepts one mole of H^+ per mole of base is required to neutralize a quantity of HCl, we know that quantity of HCl is one mole. If we find that one mole of this base is required to neutralize a quantity of H_2SO_4, we know that quantity of H_2SO_4 is $\frac{1}{2}$ mole; and if the acid being neutralized is H_3PO_4 then the quantity of H_3PO_4 is $\frac{1}{3}$ mole. The same reasoning applies to bases. For example one mole of an acid which can donate one mole of H^+ per mole of acid is required to neutralize one mole of NaOH; two moles of this acid are required to neutralize one mole of $Ba(OH)_2$; and three moles of this acid are required to neutralize one mole of $Al(OH)_3$.

A convenient approach to calculations involving the stoichiometry of neutralization involves the use of a quantity called an equivalent, which is closely related to a mole. One equivalent of an acid is one mole of acid divided by the number of moles of H^+ available per mole of acid. One equivalent of a base is one mole of base divided by the number of moles of H^+ which can be accepted by one mole of base, or in the case of hydroxides, the number of OH^- groups per mole of base. For example 1 mole of H_2SO_4 weighs 98 g, 1 equivalent weighs $\frac{98}{2} = 49$ g; 1 mole of $Al(OH)_3$ weighs 78 g, 1 equivalent weighs $\frac{78}{3} = 26$ g. Just as the mass of 1 mole is called the molecular weight, the mass of 1 equivalent is called the equivalent weight. By analogy with molarity and formality, a unit of concentration called normality (N) is defined as the number of equivalents of solute per liter of solution. From the definition of an equivalent it follows that *one equivalent of an acid reacts with one equivalent of a base.*

(5)
$$\text{equivalents}_a = \text{equivalents}_b$$

It is thus possible to write a number of relationships which can be used to handle the stoichiometry of neutralization. Since equivalents = masses of substance/equivalent weight,

(6)
$$\frac{\text{wt}_a}{\text{eq} \cdot \text{wt}_a} = \frac{\text{wt}_b}{\text{eq} \cdot \text{wt}_b}$$

Also, equivalents = normality × volume of solution so that for a solution of acid reacting with a solution of base

(7) $$N_a V_a = N_b V_b$$

Equations (5), (6), and (7) or appropriate combinations of them can be used to simplify calculations involving neutralization.

Another feature of interest in neutralization is related to the question of determining experimentally when neutralization has occurred. For example suppose that a quantity of base is slowly being added to a quantity of acid and that the addition of base must be stopped at the point when the required number of equivalents have been added, how can one determine when this point called the equivalence point is reached? This is usually done by monitoring the pH of the solution and stopping the addition when the pH of the solution is that of the equivalence point. The pH at the equivalence point depends on the type of acid and base involved in the neutralization. This can be seen by inspection of equations (1)–(4). For equation (1), a strong acid and a strong base, the pH at the equivalence point will be 7. However in equations (2)–(4) this will usually not be the case, since the salt formed as a result of neutralization undergoes hydrolysis, which can be thought of as the reverse of neutralization. For example the pH at the equivalence point for equation (2) is less than 7 and for equation (3) is greater than 7.

One way of monitoring the pH is by use of a substance called an indicator, which is a dye whose color depends on pH. Many indicators are weak acids, whose dissociation can be represented as $HIn \rightleftharpoons H^+ + In^-$. HIn and In^- are different colors and therefore the color of the solution will depend on the ratio $[In^-]/[HIn]$, which depends on $[H^+]$ since

(8) $$K_{In} = \frac{[In^-][H^+]}{[HIn]} \quad \text{or} \quad \frac{[In^-]}{[HIn]} = \frac{K_{In}}{[H^+]}$$

At some pH of the solution the indicator will be changing from the color of HIn to the color of In^-. At this point $[In^-] = [HIn]$ and $[In^-]/[HIn] = 1$ and $[H^+] = K_{In}$ or $pH = pK_{In}$. When monitoring a neutralization with an indicator it is desirable to choose an indicator whose pK_{In} is close to the pH at the equivalence point.

PROBLEMS

13.38 If 12 g of NaOH were required to neutralize 400 ml of a solution of HCl in water calculate the concentration of the HCl solution in equivalents of HCl per liter.

Solution: equivalents of NaOH = equivalents of HCl. The number of equivalents of NaOH in 12 g of NaOH equals the number of equivalents of HCl in 400 ml of solution. The equivalent weight of NaOH is equal to its formula weight, 40. Therefore, 12 g of NaOH is $\frac{12}{40}$ or 0.30 equivalent of NaOH, and 0.30 equivalent of NaOH will neutralize 0.30 equivalent of HCl. That means that there is 0.30 equivalent of HCl in 400 ml of solution. The number of equivalents per liter will then be

$$1000 \text{ ml} \times \frac{0.30 \text{ equivalent}}{400 \text{ ml}} = 0.75 \text{ equivalent}$$

13.39 How many grams of KOH will be required for the preparation of 500 ml of 0.400 *N* KOH for use in neutralization reactions?

Solution: 0.400 *N* KOH contains 0.400 equivalent of KOH per liter (1000 ml). Therefore, 500 ml will contain 0.200 equivalent of KOH. The equivalent weight of KOH is its formula weight, 56.1.

$$0.200 \text{ equivalent of KOH} \times \frac{56.1 \text{ g}}{\text{equivalent}} = 11.2 \text{ g of KOH}$$

The problem can be solved in one operation.

$$500 \text{ ml} \times \frac{0.400 \text{ equivalent}}{1000 \text{ ml}} \times \frac{56.1 \text{ g of KOH}}{1 \text{ equivalent}} = 11.2 \text{ g of KOH}$$

13.40 What is the normality of a solution of NaOH which contains 8 g of NaOH per 400 ml of solution?

Solution: The solution, 8 g of NaOH in 400 ml, is the same concentration as 20 g of NaOH in 1000 ml of solution. One equivalent weight of NaOH is 40 g; 20 g of NaOH is 0.5 of an equivalent weight. Since 0.5 equivalent of NaOH is present in a liter of solution the normality is 0.5 *N*. The problem can be solved in one operation.

$$1000 \text{ ml} \times \frac{8 \text{ g of NaOH}}{400 \text{ ml}} \times \frac{1 \text{ equivalent of NaOH}}{40 \text{ g of NaOH}}$$
$$= 0.5 \text{ equivalent of NaOH}$$

13.41 A sample of 50 ml of hydrochloric acid was required to react with 0.40 g of NaOH. Calculate the normality of the hydrochloric acid.

Solution: To find the normality we must find the number of equivalents of HCl in a liter of acid.

$$\text{equivalents of HCl} = \text{equivalents of NaOH}$$

The equivalent weight of NaOH is 40. Therefore 0.40 g of NaOH is 0.010 equivalent of NaOH. Since there is 0.010 equivalent of NaOH there must be 0.010 equivalent of HCl in the 50 ml of hydrochloric acid that were required. To find the number of equivalents per 1000 ml, which is the normality,

$$1000 \text{ ml} \times \frac{0.010 \text{ equivalent}}{50 \text{ ml}} = 0.20 \text{ equivalent}$$

Therefore, the solution is 0.20 *N*.

The above calculation can be carried out in one operation:

$$\frac{0.40 \text{ g of NaOH} \times \dfrac{1 \text{ equivalent}}{40 \text{ g of NaOH}}}{50 \text{ ml of HCl}}$$

$$\times \; 1000 \text{ ml of HCl} = 0.20 \text{ equivalent}$$

13.42 How many grams of KOH will be required to react with 100 ml of 0.80 *N* HCl?

Solution: equivalents of KOH = equivalents of HCl.

$$\text{equivalents of KOH} = \frac{\text{g of KOH}}{56.1 \text{ g of KOH per equivalent}}$$

$$\text{equivalents of HCl} = 100 \text{ ml of HCl} \times \frac{0.80 \text{ equivalent of HCl}}{1000 \text{ ml of HCl}}$$

$$\frac{\text{g of KOH}}{56.1 \text{ g of KOH per equivalent}}$$

$$= 100 \text{ ml of HCl} \times \frac{0.80 \text{ equivalent of HCl}}{1000 \text{ ml of HCl}}$$

$$\text{g of KOH} = \frac{56.1 \times 100 \times 0.80}{1000} = 4.5 \text{ g}$$

Since normality is defined as equivalents of solute per liter of solution, since equivalents of KOH equals equivalents of HCl, and since 100 ml is 0.10 liter, the entire calculation can take the simple form,

0.10 liter × 0.80 equivalent/liter × 56.1 g/equivalent = 4.5 g

13.43 How many milliliters of 0.30 *N* HNO_3 will be required to react with 24 ml of 0.25 *N* KOH?

Solution: equivalents of HNO_3 = equivalents of KOH.

$$\frac{\text{ml of } HNO_3 \times 0.30 \text{ equivalent}}{1000 \text{ ml}} = \frac{24 \text{ ml of KOH} \times 0.25 \text{ equivalent}}{1000 \text{ ml}}$$

$$\text{milliliters of } HNO_3 = \frac{24 \text{ ml} \times 0.25 \text{ equivalent}}{0.30 \text{ equivalent}} 20 \text{ ml}$$

Since, in the above calculation, equivalents/1000 ml represents normality, ml of HNO_3 × normality of HNO_3 = ml of KOH × normality of KOH.

13.44 If 18 ml of 0.1 *F* H_2SO_4 were required to liberate the CO_2 from 82 ml of sodium carbonate solution, calculate the normality of the sodium carbonate solution.

Solution: ml of H_2SO_4 × normality of H_2SO_4 = ml of Na_2CO_3 × normality of Na_2CO_3.

13.45 How many milliliters of 0.250 *F* HCl will be required to neutralize 500 ml of solution containing 8.00 g of NaOH?

13.46   Calculate the normality of a H_3PO_4 solution, 40 ml of which neutralized 120 ml of 0.53 N NaOH.

13.47   How many milliliters of 0.25 N NaOH will be required to neutralize 116 ml of 0.0625 N H_2SO_4?

13.48   It took 40 ml of 0.10 F H_2SO_4 to precipitate completely the Ba^{++} ion (as $BaSO_4$) from a $BaCl_2$ solution. Calculate the number of grams of $BaCl_2$ that were originally present in the $BaCl_2$ solution.

13.49   If 0.664 g of phthalic acid, $H_6C_9O_4$, was required to neutralize 20.0 ml of 0.400 N NaOH, calculate the equivalent weight of the acid.

13.50   It is found that 30.0 ml of 0.20 F H_2SO_4 are required to neutralize 2.61 g of an unknown base. Calculate the equivalent weight of the base.

13.51   How many milliliters of 0.1 F H_3PO_4 solution are required to neutralize 1.35 g of a base whose equivalent weight is 120 g?

13.52   A sample of 200 ml of 1.000 N H_2SO_4 was treated with an excess of Na_2CO_3. How many liters of dry CO_2 gas, measured at STP, were given off?

13.53   A volume of 600 ml of HCl of a certain normality was mixed with 400 ml of NaOH of the same normality. The resulting solution had a pH of 1. Calculate the normality of the HCl and NaOH.

13.54   Calculate the pH at the equivalence point when 50 ml of 0.1 F $HC_2H_3O_2$ are neutralized by 50 ml of 0.1 F NaOH.

Solution: Where the two solutions are mixed the volume of the solution is 100 ml. Assuming the neutralization equilibrium of the weak acid-strong base lies completely to the right,

$$HC_2H_3O_2 + OH^- \rightleftarrows H_2O + C_2H_3O_2^-$$

The moles of $C_2H_3O_2^-$ produced equal moles of $HC_2H_3O_2$ = molarity $\times$ volume = (0.1)(.05) and $[C_2H_3O_2^-]$ = (0.1)(.05)/(0.05 + 0.05). Now the problem is treated as if it read: calculate the pH of 0.05 M $C_2H_3O_2^-$ solution as the result of hydrolysis of this ion.

13.55   Calculate the pH at the equivalence point when 40 ml of 0.2 F NH_3 are neutralized by 80 ml of 0.1 F HCl.

13.56   Calculate the pH at the equivalence point when 100 ml of 0.220 F HCN are neutralized by 0.125 F NaOH solution.

➔ **13.57**   When a solution of a polyprotic acid is neutralized by slow addition of a solution of base, the protons are neutralized one at a time. For H_2CO_3 we can write

$$H_2CO_3 + OH^- \rightleftarrows HCO_3^- + H_2O$$
$$HCO_3^- + OH^- \rightleftarrows CO_3^{--} + H_2O$$

Calculate the *difference* in pH between the equivalence point for the neutralization of the first proton and the second when a 0.2 F solution of NaOH is added to a 0.2 F solution of H_2CO_3.

13.58 What should be the pK of an indicator used to monitor the neutralization of (a) 0.1 M HCl by 0.1 M NaOH; (b) 0.1 M HCl by 0.1 M NH$_3$; (c) 0.1 M HCN by 0.1 M NaOH?

➡ **13.59** A common experimental procedure for carrying out a neutralization is titration. Suppose that we place 100 ml of 0.10 M NaOH in a flask and slowly add 0.10 M HCl solution from a buret. Calculate the pH of the original NaOH solution, and after the addition of 50 ml of HCl, 90 ml of HCl, 99 ml of HCl, 99.9 ml of HCl, and 100 ml of HCl. Show from these calculations that a suitable indicator can have a pK in a range around the equivalence point.

Solution hint: Calculate the number of moles of OH$^-$ and the total volume of the solution after each of the specified volumes has been added.

➡ **13.60** Perform the same calculations as Problem 13.59 for the addition of 0.1 M HCl solution to 100 ml of 0.1 M NaCN.

Solution hint: The reaction of interest for the titration is $H^+ + CN^- \rightleftharpoons HCN$ which allows calculation of [CN$^-$] and [HCN] after each addition.

Buffers

Buffers are solutions of a weak acid and the salt of its conjugate base or of a weak base and the salt of its conjugate acid whose pH will stay relatively constant when relatively small amounts of a strong acid or base are added. How much strong acid or base the pH of a buffer can resist is called the capacity of the buffer. Buffers are also solutions whose pH does not change when they are diluted or concentrated. The pH of the buffer solution just depends on the ratio of the two species composing the buffer. For example a buffer made from 1 mole of hydrocyanic acid and 1 mole of sodium cyanide has [H$^+$] $= 4 \times 10^{-10}$ no matter what the volume of the solution, since

$$[H^+] = K_A \frac{[HCN]}{[CN^-]} \quad \text{and} \quad K_A = 4 \times 10^{-10}$$

Similarly a buffer of 1 mole of ammonia and 1 mole of ammonium chloride has [OH$^-$] $= 1.8 \times 10^{-5}$ since

$$[OH^-] = K_B \frac{[NH_3]}{[NH_4^+]} \quad \text{and} \quad K_B = 1.8 \times 10^{-5}$$

Buffers are best when prepared from roughly comparable molar quantities of the two required species. Thus, in preparing a buffer solution of a desired pH, one uses a weak acid-salt combination whose pK_A is close to the desired pH.

PROBLEMS

(See Table 2 for ionization constants.)

13.61 Calculate the pH of a liter of solution which is 0.1 F in $NaC_2H_3O_2$ and 0.001 F in HCl. K for $HC_2H_3O_2 = 1.8 \times 10^{-5}$ M.

Solution: 0.1 F $NaC_2H_3O_2$ will yield 0.1 M Na^+ and 0.1 M $C_2H_3O_2^-$.
0.001 F HCl will yield 0.001 M H^+ and 0.001 M Cl^-. H^+ is a very strong acid; $C_2H_3O_2^-$ is a moderately strong base.
The 0.001 mole of H^+ will combine with 0.001 mole of $C_2H_3O_2^-$ to form 0.001 mole of $HC_2H_3O_2$.
That will leave $0.1 - 0.001$ or 0.1 mole of $C_2H_3O_2^-$.
Therefore, $[C_2H_3O_2^-] = 0.1$ M and $[HC_2H_3O_2] = 0.001$ M.

$$K = \frac{[H^+] \times [C_2H_3O_2^-]}{[HC_2H_3O_2]} = 1.8 \times 10^{-5} \ M$$

Substituting in the above formula:

$$\frac{[H^+] \times 0.1 \ M}{0.001 \ M} = 1.8 \times 10^{-5} \ M$$

$$[H^+] = 2 \times 10^{-7} \ M$$

13.62 Calculate the pH of a solution which is 0.2 F in NaF and 0.002 F in HCl. K for HF $= 6.9 \times 10^{-4}$.

13.63 What concentrations of $NaC_2H_3O_2$ and $HC_2H_3O_2$ must be used in preparing a solution with a pH of 6.0?

13.64 Which acid and its salt listed in Table 2 would be suitable for preparing a buffer of pH $= 3.2$?

13.65 What is the pH of a buffer prepared from 40.0 g of NaH_2PO_4 and 47.3 g of Na_2HPO_4 in 3 liters of water?

Solution hint: Use K_2 for H_3PO_4.

13.66 Calculate the number of moles of Na_3PO_4 which should be added to 1 liter of 0.2 M Na_2HPO_4 solution to produce a buffer whose pH is 11.7.

➤ **13.67** How many moles of HCl gas must be added to 1 liter of a buffer solution which is 2 M in acetic acid and 2 M in sodium acetate to produce a solution whose pH $= 4.0$? What would be the pH if this quantity of HCl were dissolved in 1 liter of pure water?

➡ **13.68** In a solution prepared by dissolving $NaC_2H_3O_2$ and $HC_2H_3O_2$ in pure water, the sum of the formalities of the 2 solutes is 1.0. The pH of the solution is 5.0.

(a) Calculate the formality of each of the 2 solutes.

(b) Calculate the molarity of each of the species in solution. K_i for $HC_2H_3O_2$ is 1.8×10^{-5} M.

Solution: Since $[H^+]$ is 1.0×10^{-5} and $K = 1.8 \times 10^{-5}$ M, the ratio, $[C_2H_3O_2^-]/[HC_2H_3O_2]$ will be 1.8. Since we know that $[C_2H_3O_2^-] + [HC_2H_3O_2] = 1.0$, the values of $[C_2H_3O_2^-]$ and $[HC_2H_3O_2]$ can be calculated to be 0.64 M and 0.36 M, respectively. The fact that $[H^+]$ is 1×10^{-5} means that a negligible amount of $HC_2H_3O_2$ ionizes and a negligible amount of $C_2H_3O_2^-$ is produced by this ionization. Therefore, the molarity of the $C_2H_3O_2^-$ equals the formality of the $NaC_2H_3O_2$ and the molarity of the $HC_2H_3O_2$ equals its formality.

Note: The solution of this problem illustrates the following fact which can be very useful in solving problems: When the pH or $[H^+]$ or $[OH^-]$ of a solution is known the *ratio* of anion to its acid can be calculated provided the ionization constant of the acid is known. This *ratio* can often be used in solving the problem. If the ratio of two quantities and the sum of the two quantities are known the value of each quantity can be calculated.

➡ **13.69** How many milliliters of 0.200 F NaOH must be added to 100 ml of 0.150 F $HC_2H_3O_2$ to give a solution with a pH of 4.046?

➡ **13.70** How many milliliters each of 1.00 F solutions of NaOH and $HC_2H_3O_2$ must be mixed to give 1 liter of solution having a pH of 4.00?

➡ **13.71** How many moles of solid NaOH must be added to a liter of 0.10 F H_2CO_3 to produce a solution whose hydrogen ion concentration is 3.2×10^{-11} moles per liter? There is no measurable change in volume when the NaOH is dissolved.

Solution: The situation in this problem is complicated by the fact that two reactions are possible, namely

$$H_2CO_3 + OH^- = HCO_3^- + H_2O$$
$$H_2CO_3 + 2\,OH^- = CO_3^{--} + 2\,H_2O$$

Since $[H^+]$ is 3.2×10^{-11} M we can calculate, from K_1 and K_2 for H_2CO_3, that

$$\frac{[CO_3^{--}]}{[HCO_3^-]} = \frac{4.8 \times 10^{-11}}{3.2 \times 10^{-11}} = 1.5 \quad \text{and} \quad \frac{[HCO_3^-]}{[H_2CO_3]} = \frac{4.2 \times 10^{-7}}{3.2 \times 10^{-11}}$$

$$= 1.3 \times 10^4$$

This means that all of the H_2CO_3 is consumed.

Let $X =$ moles of HCO_3^- formed; $1.5\,X =$ moles of CO_3^{--} formed.
$2.5\,X =$ total moles of carbonate $= 0.10$ mole.
Solving, $X = 0.040$ mole of HCO_3^- and $1.5\,X = 0.060$ mole of CO_3^{--}.
To form 0.040 mole of HCO_3^- requires 0.040 mole of OH^-.
To form 0.060 mole of CO_3^{--} requires 0.120 mole of OH^-.
Total moles of OH^- consumed $= 0.16$ mole.
Since $[H^+]$ is 3.2×10^{-11} M, the number of moles of excess OH^- added
was $1.0 \times 10^{-14} \div 3.2 \times 10^{-11} = 3.1 \times 10^{-4}$ or 0.00031 mole. There-
fore, total moles of NaOH added $= 0.16 + 0.00031 = 0.16$.

The main reaction approximation

Many equilibrium states involve a number of simultaneous equilibria which
can make the quantitative treatment of such systems complex. However very
often it is found that one of these simultaneous equilibria will have a K much
larger than any of the others. This equilibrium is called the main reaction and
when its K is at least 100 times greater than any other, problem solving can be
greatly simplified by the use of the main reaction approximation. This
approximation states that only the main reaction need be considered in
calculation of the equilibrium concentrations of all the species which are
involved in it and that the concentrations of other species in the system can
be calculated from those in the main reaction by using the principle that at
equilibrium the values of K for all equilibria must be satisfied.

A problem will specify that certain components are present in solution
or are being mixed. All of these components should be listed in their correct
states before any equilibria are established or at the instant of mixing according to the following rules:

(a) Insoluble or slightly soluble substances are written out of solution
as pure, undissolved components with the appropriate subscript to indicate
their state. Thus, we write $AgCl_{(s)}$ or $H_2S_{(g)}$ or $Br_{2(l)}$ or $Ca(OH)_{2(s)}$.

(b) Soluble substances fall into various categories:

(i) Salts are written as dissociated ions. Thus, calcium nitrate,
$Ca(NO_3)_2$ is written as $Ca^{++} + NO_3^-$ since these are the species
present. Sodium cyanide is $Na^+ + CN^-$.

(ii) Strong acids are written as dissociated. Thus, hydrochloric acid
is $H^+ + Cl^-$, sulfuric acid is $H^+ + HSO_4^-$.

(iii) Strong bases (which are soluble) are written as dissociated.
Sodium hydroxide is $Na^+ + OH^-$.

(iv) Weak acids and weak bases are written as being undissociated.
Thus, acetic acid is HOAc, ammonia is NH_3, phosphoric acid is
H_3PO_4.

(v) water is written undissociated.

Example: Calculate the pH of 0.1 M NaHCO$_3$ solution. Given $K_1 = 4.2 \times 10^{-7}$, $K_2 = 4.8 \times 10^{-11}$ for H$_2$CO$_3$.

The species present under the conditions specified are: Na$^+$, HCO$_3^-$, and H$_2$O. No others may be considered.

Step 1. List all the possible reactions which may occur between the species present. These reactions will fall primarily into one of several categories:

  (i) reactions between an acid and a base. This will often be the most important category.

  (ii) insoluble materials going into solution

  (iii) formation of precipitates from species in solution

  (iv) formation or dissociation of complex ions

  (v) evolution of gases

  (vi) oxidation-reductions

  (vii) combinations of the above

In the problem that we are considering there are only three species to consider: Na$^+$, HCO$_3^-$, and H$_2$O. In general, Na$^+$ (or K$^+$ or other group I cations) never enter into the kinds of reactions we are discussing and can be ignored. Thus, we need only consider HCO$_3^-$ and H$_2$O. The only type of reactions possible in this system are acid–base reactions. In order to list them we must identify what is an acid and what is a base. As we have already seen, water is both an acid and a base. Similarly, HCO$_3^-$ is both an acid and a base—an acid because it can donate a proton, and a base because it can accept one. The reactions which can occur are:

  (1) $\quad$ H$_2$O $\rightleftharpoons$ H$^+$ + OH$^-$ (water with itself)

  (2) $\quad$ H$_2$O + HCO$_3^-$ $\rightleftharpoons$ H$_3$O$^+$ + CO$_3^{--}$ or
      HCO$_3^-$ $\rightleftharpoons$ H$^+$ + CO$_3^{--}$ (water as a base, bicarbonate as an acid)

  (3) $\quad$ H$_2$O + HCO$_3^-$ $\rightleftharpoons$ H$_2$CO$_3$ + OH$^-$ (water as an acid, bicarbonate as a base)

  (4) $\quad$ HCO$_3^-$ + HCO$_3^-$ $\rightleftharpoons$ H$_2$CO$_3$ + CO$_3^{--}$ (bicarbonate with itself, as an acid and a base)

Step 2. Evaluate the value of K for each reaction from the information given. Often this will require combination of various given K's and their associated reactions. Remember chemical equations can be added, in which case their K's are multiplied. Equations can be subtracted, in which case their K's are divided. Equations can be multiplied by a numerical factor, in which case their K's are raised to the power of the numerical factor. Evaluating K's:

K for (1) is $K_w = 10^{-14}$

K for (2) is K_2 which is given as 4.8×10^{-11}.

Reaction (2) is simply the second dissociation of H_2CO_3. K for (3) can be evaluated by the following procedure: Equation (3) has HCO_3^- on the left and H_2CO_3 on the right. Are we given data for an equilibrium which relates HCO_3^- and H_2CO_3? Yes, K_1 refers to

(a) $H_2CO_3 \rightleftharpoons H^+ + HCO_3^-$

which has H_2CO_3 on the left and HCO_3^- on the right. Since we want to reverse this equation it will have to be subtracted from some other equation. Equation (3) also has OH^- on the right. Do we have the K of a reaction with OH^- on the right? Yes, K_w refers to

(b) $H_2O \rightleftharpoons H^+ + OH^-$.

If we now take (b) — (a) we get Equation (3). Note that subtraction of an equation is like reversing this equation and then adding. If we get an equation from (b) — (a), then its equilibrium constant is K_b/K_a or in this case,

$$K_w/K_1 = \frac{1 \times 10^{-14}}{4.2 \times 10^{-7}} = 2.4 \times 10^{-8}.$$

Notice that this is simply the hydrolysis of the salt of a weak acid which is K_w/K_a.

K for (4) must also be evaluated in this way. An equation relating HCO_3^- and CO_3^{--} is

(a) $HCO_3^- \rightleftharpoons H^+ + CO_3^{--} \quad K_2$

and one relating HCO_3^- and H_2CO_3 is

(b) $H_2CO_3 \rightleftharpoons H^+ + HCO_3^- \quad K_1$

which will be subtracted

(a) — (b) = $2HCO_3^- \rightleftharpoons H_2CO_3 + CO_3^{--}$

$$K = K_2/K_1 = \frac{4.8 \times 10^{-11}}{4.2 \times 10^{-7}} = 1.1 \times 10^{-4}.$$

Step 3.   Examine all the calculated K's. If one of them is at least 10^2 greater than the others the main reaction approximation will work. This largest K and its associated reaction is selected as the main reaction and all the other ones are temporarily ignored. Here the main reaction is (4).

Step 4. Calculate the equilibrium concentrations of everything involved in the main reaction:

$$2\ HCO_3^- \rightleftharpoons H_2CO_3 + CO_3^{--}$$

start	0.1	0	0
equil.	$0.1 - 2x$	x	x

$$K = \frac{[H_2CO_3][CO_3^{--}]}{[HCO_3^-]^2} = \frac{(x)(x)}{(0.1 - 2x)^2} = 1.1 \times 10^{-4}$$

$x = 1.1 \times 10^{-3} = [CO_3^{--}] = [H_2CO_3]$ (assuming $0.1 - 2x = 0.1$)
$[HCO_3^-] = 0.98(0.100 - 2(0.001)) = 9.8 \times 10^{-2}$

Step 5. To calculate the concentration of species not involved in the main reaction, choose one of the secondary reactions which involves the species of interest and has all the other species in it involved in the main reaction. Plug into the expression for K.

To get $[H^+]$ we can use $HCO_3^- \rightleftharpoons H^+ + CO_3^{--}$

$$K_2 = \frac{[H^+][CO_3^{--}]}{[HCO_3^-]} = \frac{[H^+](1.1 \times 10^{-3})}{(9.8 \times 10^{-2})} = 4.8 \times 10^{-11}$$

$[H^+] = 4.3 \times 10^{-9}$ $pH = 9 - \log 4.3 = 8.37$

To get OH^- use $[H^+][OH^-] = 10^{-14} = (4.3 \times 10^{-9})[OH^-]$

$$[OH^-] = 2.3 \times 10^{-6}$$

$$pOH = 14 - pH = 5.63$$

Example: Given for H_3PO_4 that $K_1 = 7.5 \times 10^{-3}$, $K_2 = 6.2 \times 10^{-8}$, $K_3 = 1.0 \times 10^{-12}$; calculate the pH of 0.1 M NaH_2PO_4 and the concentration of all species at equilibrium.

Species present: Na^+, $H_2PO_4^-$, H_2O
Reactions: (ignore Na^+)

(1) $H_2O \rightleftharpoons H^+ + OH^-$ $K_w = 10^{-14}$

(2) $H_2PO_4^- \rightleftharpoons H^+ + HPO_4^{--}$ $K_2 = 6.2 \times 10^{-8}$

(3) $H_2PO_4^- \rightleftharpoons 2H^+ = PO_4^{---}$ $K = K_2K_3 = 6.2 \times 10^{-20}$

(4) $H_2PO_4^- + H_2O \rightleftharpoons H_3PO_4 + OH^-$ $K = K_w/K_1 = \dfrac{1 \times 10^{-14}}{7.5 \times 10^{-3}}$
$$= 1.3 \times 10^{-12}$$

(5) $H_2PO_4^- + H_2PO_4^- \rightleftharpoons H_3PO_4 + HPO_4^{--}$ $K = K_2/K_1 = \dfrac{6.2 \times 10^{-8}}{7.5 \times 10^{-3}}$
$$= 8.3 \times 10^{-6}$$

(6) $H_2PO_4^- + 2\ H_2PO_4^- \rightleftharpoons 2\ H_3PO_4 + PO_4^{---}$

This is a combination of:

(a) $H_2PO_4^- \rightleftharpoons H^+ + HPO_4^{--}$ K_2

(b) $HPO_4^- \rightleftharpoons H^+ + PO_4^{---}$ K_3

(c) $2\,H_3PO_4 \rightleftharpoons 2H^+ + 2\,H_2PO_4^-$ K_1^2

$$a + b - c = K_2K_3/K_1^2$$
$$= (6.2 \times 10^{-8})(1.0 \times 10^{-12})/(7.5 \times 10^{-3})^2$$
$$= 1.1 \times 10^{-15}$$

So (5) is the main reaction.

$$2\,H_2PO_4^- \rightleftharpoons H_3PO_4 + HPO_4^{--}$$

start	0.1	0	0
equil.	$0.1 - 2x$	x	x

$$\frac{[H_3PO_4][HPO_4^{--}]}{[H_2PO_4^-]^2} = 8.3 \times 10^{-6} = \frac{(x)(x)}{(0.1 - 2x)^2} \quad \text{Assume } 2x \ll 0.1.$$

$$x = 2.9 \times 10^{-4} = [H_3PO_4] = [HPO_4^{--}]$$

$$[H_2PO_4^-] = 0.1 - 2(0.0003) = 0.099 = 9.9 \times 10^{-2}$$

To get $[H^+]$ we can use:

$$\frac{[H^+][HPO_4^{--}]}{[H_2PO_4^-]} = 6.2 \times 10^{-8} = \frac{[H^+](2.9 \times 10^{-4})}{(9.9 \times 10^{-2})}$$

$$[H^+] = 2.1 \times 10^{-5}$$

$$pH = 5 - \log 2.1 = 4.7$$

Notice that if we check this $[H^+]$ by substituting,

$$\frac{[H^+][H_2PO_4^-]}{[H_3PO_4]} = \frac{(2.1 \times 10^{-5})(9.9 \times 10^{-2})}{(2.9 \times 10^{-4})} = 7.2 \times 10^{-3}$$

which agrees well with the value of K_1 given.

To get $[PO_4^{---}]$ use $K_3 = \dfrac{[PO_4^{---}][H^+]}{[HPO_4^{--}]} = \dfrac{[PO_4^{---}](2.1 \times 10^{-5})}{2.9 \times 10^{-4}}$

To get $[OH^-]$ use $[H^+][OH^-] = 10^{-14} = (2.1 \times 10^{-5})[OH^-]$

$$[OH^-] = 4.8 \times 10^{-10}$$

Thus, all the equilibrium concentrations are:

$$[Na^+] = 0.1;$$
$$[H_2PO_4^-] = 9.9 \times 10^{-2};$$
$$[H_3PO_4] = [HPO_4^{--}] = 2.9 \times 10^{-4};$$
$$[PO_4^{---}] = 1.4 \times 10^{-11};$$
$$[H^+] = 2.1 \times 10^{-5};$$
$$[OH^-] = 4.8 \times 10^{-10}.$$

PROBLEMS

13.72 Calculate the OH^- ion concentration of 0.10 F K_2SO_3. Calculate also the concentrations of H^+, HSO_3^-, SO_3^{--}, H_2SO_3, and K^+.

13.73 Calculate the pH of 0.20 F Na_2CO_3.

13.74 Calculate the formate-ion concentration in moles per liter of a 0.2 F solution of NaF in 0.1 F HCOOH (formic acid).

Solution hint: The main reaction is

$$F^- + HCOOH \rightleftharpoons HF + HCOO^-$$

13.75 Calculate the CN^- ion concentration in moles per liter, of a 0.1 F solution of $NaC_2H_3O_2$ in 0.1 F HCN.

13.76 Calculate the pH of:

(a) 0.10 F $NaHSO_3$.

(b) 0.10 F K_2HPO_4.

(c) 0.10 F KH_2PO_4.

➡ **13.77** Calculate the concentration of NH_3 in a solution which is 0.1 F in $HC_2H_3O_2$ and 0.1 F in NH_4Cl.

Solution: The principal reactions with their equilibrium constants are:

(1) $\quad HC_2H_3O_2 \rightleftharpoons H^+ + C_2H_3O_2^- \qquad K_i = 1.8 \times 10^{-5}$

(2) $\quad NH_4^+ + H_2O \rightleftharpoons H^+ + NH_4OH \qquad K_{hyd} = 5.6 \times 10^{-10}$

Since K_{hyd} is so much smaller than K_i the amount of H^+ derived from the hydrolysis of NH_4^+ is negligible by comparison with the amount derived from the ionization of $HC_2H_3O_2$. Therefore, calculate the $[H^+]$ derived from the ionization of $HC_2H_3O_2$ as in Problem 13.6. The value of $[H^+]$ thus calculated is 1.3×10^{-3} M.

To calculate $[NH_3]$ substitute the above value of $[H^+]$, 1.3×10^{-3} M, for $[H^+]$ and 0.1 M for $[NH_4^+]$ in K_{hyd}.

Note: In the hydrolysis of NH_4^+, as in all equilibrium reactions, the magnitude of the equilibrium constant tells us the extent of completeness of the reaction. In many problems that will be encountered it will be very useful to know this fact.

➡ **13.78** Calculate the molar concentration of each species in a solution which is 0.20 F in NH_3 and 0.20 F in NaCN.

Solution: The principal reactions with their equilibrium constants are:

(1) $\qquad\qquad NH_3 + H_2O \rightleftharpoons NH_4^- + OH^-$

$$K_i = \frac{[NH_4^+] \times [OH^-]}{[NH_3]} = 1.8 \times 10^{-5}$$

(2) $$CN^- + H_2O \rightleftharpoons HCN + OH^-$$

$$K_{hyd} = \frac{[HCN] \times [OH^-]}{[CN^-]} = 2.5 \times 10^{-5}$$

Since the two equilibria occur in the same solution the value of $[OH^-]$ is the same for each. Therefore, K_i and K_{hyd} can be equated to give:

(3) $$\frac{1.8 \times 10^{-5} \times [NH_3]}{[NH_4^+]} = \frac{2.5 \times 10^{-5} \times [CN^-]}{[HCN]}$$

But $[NH_3]$ is 0.20 M and $[CN^-]$ is 0.20 M. Substituting these values in Equation (3)

(4) $$\frac{[HCN]}{[NH_4^+]} = \frac{2.5 \times 10^{-5}}{1.8 \times 10^{-5}}$$

or

(5) $$[HCN] = 1.4 \times [NH_4^+]$$

Let $X = [NH_4^+]$.
Then $1.4\ X = [HCN]$.
Since the OH^- derived in the ionization of NH_3 is, according to Equation (1), equal to $[NH_4^+]$ and since the OH^- derived from the hydrolysis of CN^- is, according to Equation (2), equal to $[HCN]$, the total $[OH^-]$ will be equal to $[NH_4^+] + [HCN]$, or 2.4 X.
Substituting X for $[NH_4^+]$, 2.4 X for $[OH^-]$, and 0.20 M for $[NH_3]$ in the formula for K_i the value of X and, hence, the molar concentration of each species in solution can be calculated.

➡ **13.79** Calculate the molar concentration of each species in a solution which is 0.20 F in NH_3 and 0.20 F in $NaC_2H_3O_2$.

➡ **13.80** A mixture of 500 ml of 1.0 $F\ HNO_3$ and 100 ml of 15 $F\ NH_3$ was diluted with water to 1.0 liter. Calculate the $[H^+]$ of the solution.

➡ **13.81** A volume of 100 ml of a certain solution of NH_3 in water was mixed with 400 ml of 1.00 F HCl. The resulting solution was diluted with water to a volume of one liter; this liter of solution was found to have a hydrogen ion concentration of 2.22×10^{-9} moles per liter.
Calculate the formality of the original solution of NH_3.

➡ **13.82** How many moles of solid KOH must be added to a liter of 0.20 $F\ H_2SO_3$ to yield a solution whose pH is 7.85? There is no change of volume when the KOH is added.

➡ **13.83** A solution which is 0.020 F in oxalate has a pH of 4.0. Calculate the molarity of each species in the solution. For $H_2C_2O_4$, $K_1 = 6.5 \times 10^{-2}$ M and $K_2 = 6.1 \times 10^{-5}$ M.

Solution: Since $[H^+]$ is 1.0×10^{-4}, and since the ionization constants are known, the ratios, $[HC_2O_4^-]/[H_2C_2O_4]$, $[C_2O_4^{--}]/[HC_2O_4^-]$, and

$[C_2O_4^{--}]/[H_2C_2O_4]$ can be calculated. From these ratios, and the fact that $[H_2C_2O_4] + [HC_2O_4^-] + [C_2O_4^{--}] = 0.020$, the molarity of each species in solution can be calculated.

➡ **13.84** A mixture of solid Na_2CO_3 and $NaHCO_3$ weighs 59.2 g. The mixture is dissolved in enough water to give 2.00 liters of solution. The pH of this solution is found to be 10.62. How many grams of Na_2CO_3 were there in the mixture?

➡ **13.85** Calculate the pH of a solution which is 0.10 *F* in HCl and 0.35 *F* in NaCN.

Solution: Since both HCl and NaCN are strong electrolytes and since CN^- is, as the ionization constant of 4.0×10^{-10} attests, a very strong base, the principal reaction is

$$H^+ + CN^- \rightleftharpoons HCN$$

Since an excess of CN^- is present all but a very small amount, X, of the H^+ from the HCl is converted into HCN by reaction with CN^-. Therefore, at equilibrium, $[H^+] = X$, $[HCN] = 0.10 - X$ and $[CN^-] = 0.25 + X$.

$$K = \frac{[H^+] \times [CN^-]}{[HCN]} = \frac{X(0.25 + X)}{0.10 - X} = 4.0 \times 10^{-10}$$

Since X will obviously be very small it can be dropped from the terms, $0.25 + X$ and $0.10 - X$.

$$\frac{0.25\, X}{0.10} = 4.0 \times 10^{-10} \qquad X = 1.6 \times 10^{-10} = [H^+]$$

$$pH = 10 - \log \text{ of } 1.6 = 10 - 0.2 = 9.8$$

It is important to recognize in all problems, as in this problem, that the anion of a weak acid is a strong base and will, therefore, have a strong tendency to combine with the strong acid, H^+.

➡ **13.86** A volume of 4.0 ml of 0.10 *F* NaCN was mixed with 2.0 ml of hydrochloric acid and 4.0 ml of water to give 10.0 ml of solution with a hydrogen ion concentration of 0.10 *M*. What was the formality of the hydrochloric acid?

➡ **13.87**

(a) A water solution is 0.10 *F* in the soluble, strong electrolyte, NaCN. Calculate the pH of this solution.

(b) The above 0.10 *F* solution of NaCN is then made 0.10 *F* in $HC_2H_3O_2$ by adding pure acetic acid; there is no change in the volume of the solution when the pure acetic acid is added.

Calculate the pH of the resulting solution.

➡ **13.88** A mixture of solid KCN and solid $KHSO_4$ totals 0.40 mole. When this mixture is dissolved in enough water to form a liter of solution the pH of this solution is 10. Calculate the number of moles of KCN in the mixture of solids.

Solution hints: What is the net equation for the main reaction? What is the numerical value of the equilibrium constant for this reaction; what does its value tell us about the completeness of the reaction? Which solute is present in excess? How does the number of moles of $KHSO_4$ in the mixture relate to the number of moles of one of the species present at equilibrium?

➡ **13.89**

(a) Calculate the pH of 0.10 F H_2S.

(b) To 1.00 liter of 0.10 F H_2S is added solid KOH until the pH is 7.00. Compute the amount of KOH added.

(c) What is the pH of the solution when 0.090 formula weight of KOH has been added all told?

(d) Calculate how much KOH must be added (in total) to bring the pH to 13.00.

➡ **13.90** A saturated CO_2 solution in pure water is 3.4×10^{-2} molar in CO_2. How many moles of CO_2 will dissolve in 1.00 liter of 0.100 F NaOH?

Solution: Write the net equation for the reaction that takes place when excess CO_2 is added to a solution of a strong base. How many moles of OH^- are present in a liter of 0.100 F NaOH? With how many moles of CO_2 will this OH^- react? What, then, is the total solubility of CO_2 in 0.100 F NaOH?

➡ **13.91** A solution is prepared by dissolving 1.07 moles of NaH_2PO_4 and 3.32 moles of Na_3PO_4 in enough water to make a liter of solution. What is the pH of the solution? What are the molar concentrations of $H_2PO_4^-$, HPO_4^{--}, PO_4^{---}, and H_3PO_4?

Solution: The main equilibrium is

$$PO_4^{---} + H_2PO_4^- \rightleftharpoons 2\,HPO_4^{--}$$

➡ **13.92** To what volume must a liter of a solution of the weak acid, HZ, be diluted with water in order to give a hydrogen ion concentration one-half that of the original solution?

Solution:
$$HZ \rightleftharpoons H^+ + Z^-$$

Let $X = [H^+]$ at original equilibrium.
Then $X = [Z^-]$ at original equilibrium.
Let $Y = [HZ]$ at original equilibrium.

$$K = \frac{X^2}{Y}$$

Since diluting the solution does not affect the value of K,

$$\frac{X^2}{Y} = \frac{\left(\dfrac{X}{2}\right)^2}{\dfrac{Y}{V}} \qquad (V = \text{volume in liters of diluted solution})$$

Solving, $V = 4$ liters.

➡ **13.93** HA is a weak acid with an ionization constant of $1.0 \times 10^{-8}\ M$. HA forms the ion, HA_2^-. The equilibrium constant for the reaction, $HA_2^- \rightleftharpoons HA + A^-$, is $0.25\ M$. Calculate $[H^+]$, $[A^-]$, and $[HA_2^-]$ in $1.0\ F$ HA.

Solution:

$$HA \rightleftharpoons H^+ + A^-$$

$$A^- + HA \rightleftharpoons HA_2^-$$

$$K_i = \frac{[H^+] \times [A^-]}{[HA]} = 1.0 \times 10^{-8}\ M$$

$$K_{inst} = \frac{[HA] \times [A^-]}{[HA_2^-]} = 0.25\ M$$

Since K_i is very small the number of moles of HA that ionize will be so small that $[HA]$ will be $1.0\ M$.
Therefore,

$$[H^+] \times [A^-] = 1 \times 10^{-8}$$

Since $K_{inst} = 0.25\ M$, and $[HA] = 1.0$, $[HA_2^-] = 4 \times [A^-]$.
Also, we see that $[H^+] = [A^-] + [HA_2^-]$.
Therefore, if we let $X = [A^-]$, then $[HA_2^-] = 4\ X$, and $[H^+] = 5X$.
Substituting in K_i:

$$5\ X^2 = 1 \times 10^{-8}\ M^2$$

$$X = 4.5 \times 10^{-5} = [A^-]$$

$$5\ X = 2.2 \times 10^{-4} = [H^+]$$

$$4\ X = 1.8 \times 10^{-4} = [HA_2^-]$$

Solubility products.
Complex ions.

When a saturated solution of sugar is prepared by shaking an excess of sugar with water the following equilibrium is set up:

solid sugar $\rightleftharpoons$ sugar molecules in solution

When a saturated solution of a very slightly soluble salt, such as AgCl, is prepared by shaking excess AgCl with water, a somewhat different type of equilibrium is set up. AgCl is a salt; hence, it is 100% ionized. Therefore, the saturated solution which is in equilibrium with solid AgCl contains silver ions and chloride ions but no un-ionized AgCl molecules. These silver ions and chloride ions are in equilibrium with the excess solid AgCl. We may, therefore, represent the equilibrium which exists in such a saturated solution of AgCl as follows:

$AgCl \text{ (solid)} \rightleftharpoons Ag^+ + Cl^-$

Since this is a true equilibrium it will have an equilibrium constant. As has already been noted in Chapter 12, solid reactants are not involved in the equilibrium equation. Therefore

$$K = [Ag^+] \times [Cl^-]$$

This equation tells us, simply, that the product of the concentrations of the solute ions in a saturated solution of a very slightly soluble electrolyte is constant at a given temperature. This constant, K, is called the *solubility product constant,* or simply the *solubility product,* and is usually designated by the notation K_{sp} or S.P.

If the concentration of silver ions and the concentration of chloride ions, expressed in moles per liter, is such that their product is less than K_{sp} for AgCl, or is just barely equal to K_{sp}, no precipitate of AgCl will form. If, on the other hand, the product of the concentrations of Ag^+ and Cl^- is greater than K_{sp}, silver chloride will precipitate; furthermore, AgCl will keep on precipitating until enough Ag^+ ions and Cl^- ions have been removed to lower the product of their concentrations to the value of K_{sp}.

The solubility product represents a typical ionic equilibrium. As such it behaves exactly like the other ionic equilibria discussed in Chapter 13. In all problems involving solubility products we will assume, as we did in Chapter 13, that all systems are ideal and that, accordingly, the activity coefficient of each ion is 1. Therefore, the effective concentration, or *activity,* of each ion is numerically equal to its molar concentration, and the solubility product can be expressed, correctly, as a function of the molar concentrations of the ions involved. For the majority of monovalent ions the activity coefficients in dilute solution are in fact very nearly equal to 1. However, for polyvalent ions the values of the activity coefficients may be somewhat less than 1. In any case, making the above assumption will not alter the usefulness of the problem as a means of learning how to think.

The solubility product constants used are normally those determined at 20°C. In certain of the problems the calculations are based on data obtained at temperatures other than 20°C. Because the recorded solubility product is a function of the temperature as well as the accuracy of the determination, the value of K_{sp} for a certain substance may not be the same in all problems. Since such lack of agreement will in no way affect the calculation, it need not be the source of any concern.

It should be emphasized that the solubility product concept applies only to *very slightly soluble strong electrolytes.* It does not apply to *highly soluble* strong electrolytes such as NaCl, $MgSO_4$, and KOH, or to *weak* electrolytes, regardless of their solubility.

PROBLEMS

(See Table 2 for values of ionization constants.)

14.1 A solution in equilibrium with a precipitate of AgCl was found, on analysis, to contain 1.0×10^{-4} mole of Ag^+ per liter and 1.7×10^{-6} mole of Cl^- per liter. Calculate the solubility product for AgCl.

Solution: The solubility product is, by definition, the product of the concentrations of the ions in equilibrium with a precipitate of a very sparingly soluble (insoluble) substance.

For AgCl,

$$K_{sp} = [Ag^+] \times [Cl^-]$$
$$= 1.0 \times 10^{-4} \text{ mole } Ag^+ \text{ per liter} \times 1.7 \times 10^{-6} \text{ mole } Cl^- \text{ per liter}$$
$$= 1.7 \times 10^{-10} M^2 \text{ (M means moles per liter.)}$$

Note that, in the equilibrium with which this problem is concerned, the concentration of Ag^+ ions is not the same as the concentration of Cl^- ions. As was pointed out in Chapter 15, the reacting substances in an equilibrium need not be present in the exact ratio called for by the equation. They can be present in an unlimited combination of ratios. However, when they interact they always do so in the mole ratio represented by the equation. Thus, when Ag^+ ions react with Cl^- ions to form AgCl, they always do so in the ratio of 1 mole of Ag^+ to 1 mole of Cl^-. However, the solution which is in equilibrium with the solid AgCl can contain Ag^+ and Cl^- ions in unlimited numbers of ratios. The only requirement is that the product of $[Ag^+]$ and $[Cl^-]$ at the particular temperature must always equal the K_{sp}.

14.2 A solution in equilibrium with a precipitate of Ag_2S was found, on analysis, to contain 6.3×10^{-18} mole of S^{--} per liter and 1.26×10^{-17} mole of Ag^+ per liter. Calculate the solubility product for Ag_2S.

Solution:

$$\text{S.P. for } Ag_2S = [Ag^+]^2 \times [S^{--}]$$
$$= (1.26 \times 10^{-17} \text{ mole } Ag^+ \text{ per liter})^2$$
$$\times (6.3 \times 10^{-18} \text{ mole } S^{--} \text{ per liter})$$
$$= 1.0 \times 10^{-51} M^3$$

14.3 A solution in equilibrium with a precipitate of Ag_3PO_4 was found on analysis to contain 1.6×10^{-5} mole of PO_4^{---} per liter and 4.8×10^{-5} mole of Ag^+ per liter. Calculate the solubility product for Ag_3PO_4.

14.4 A solution in equilibrium with a precipitate of $Pb_3(PO_4)_2$ was found, on analysis, to contain 3.4×10^{-7} mole of PO_4^{---} per liter and 5.1×10^{-7} mole of Pb^{++} per liter. Calculate the solubility product of $Pb_3(PO_4)_2$.

Note: In comparing the solubility products in Problems 14.1, 14.2, 14.3, and 14.4, it will be noted that the units in which K_{sp} is expressed is a function of the number of ions involved in the equilibrium. The unit is "moles per liter" raised to a power equal to the number of ions per mole of solute. In subsequent problems in this chapter the units in which a particular solubility product is expressed will generally not be given.

14.5   A solution in equilibrium with a precipitate of Ag_3PO_4 was found, on analysis, to contain 1.52×10^{-3} g of PO_4^{---} per liter and 5.18×10^{-3} g of Ag^+ per liter. Calculate the solubility product of Ag_3PO_4.

Solution: First find the concentration of each ion in moles per liter. Then solve as in Problem 14.3.

14.6   Exactly 450 ml of 1.00×10^{-4} *F* $BaCl_2$ is placed in a beaker. In order to just start precipitation of $BaSO_4$ it is necessary to add, with constant stirring, exactly 350 ml of 2.00×10^{-4} *F* K_2SO_4. What is the solubility product of $BaSO_4$?

Solution hint: Note that the final volume is 800 ml.
It should be noted in this problem as well as in later problems that when the anion, SO_4^{--}, of a weak acid, HSO_4^-, is dissolved in water hydrolysis occurs according to the equation,

$$SO_4^{--} + H_2O \rightleftarrows HSO_4^- + OH^-$$

Therefore, 2.00×10^{-4} *F* K_2SO_4 will not be *exactly* 2.00×10^{-4} *F* in SO_4^{--}. The value of the hydrolysis constant for SO_4^{--}, calculated as in Problem 13.32, is 8.3×10^{-13}. That means that the amount of hydrolysis is so small that, within the limits of the precision of our measurements, it can be neglected. Only when the anion is a very strong base, such as S^{--}, PO_4^{---}, and CN^-, and the precision is very high must the effects of hydrolysis be considered.

14.7   In each of the following a saturated solution was prepared by shaking the pure solid compound with pure water. The solubilities obtained are given. From these solubilities, calculate the solubility product of each solute. (In each instance ignore the hydrolysis of each ion.)

(a) $AgCl = 1.67 \times 10^{-5}$ mole AgCl per liter
Solution: Since AgCl ionizes, completely, as follows, $AgCl = Ag^+ + Cl^-$, a saturated solution containing 1.67×10^{-5} mole of AgCl per liter will contain 1.67×10^{-5} mole of Ag^+ per liter, and 1.67×10^{-5} mole of Cl^- per liter.

S.P. $= [Ag^+] \times [Cl^-]$

$\qquad = (1.67 \times 10^{-5}$ mole Ag^+ per liter)

$\qquad\qquad\qquad\qquad\qquad\qquad \times (1.67 \times 10^{-5}$ mole Cl^- per liter)

$\qquad = 2.8 \times 10^{-10}$

(b) $AgI = 2.2 \times 10^{-3}$ mg AgI per liter
(c) $AgBr = 5.7 \times 10^{-10}$ equivalent AgBr per milliliter
(d) $BaCrO_4 = 1.4 \times 10^{-5}$ millimole per milliliter
(e) $Ag_2SO_4 = 1.4 \times 10^{-2}$ mole Ag_2SO_4 per liter

Solution: From the equation for the ionization of Ag_2SO_4

$$Ag_2SO_4 = 2\ Ag^+ + SO_4^{--}$$

it is evident that 1 mole of Ag_2SO_4 produces 2 moles of Ag^+ and 1 mole of SO_4^{--}.

Therefore, 1.4×10^{-2} mole of Ag_2SO_4 will produce 2.8×10^{-2} mole of Ag^+ and 1.4×10^{-2} mole of SO_4^{--}.

$$\begin{aligned} S.P. &= [Ag^+]^2 \times [SO_4^{--}] \\ &= (2.8 \times 10^{-2})^2 \times (1.4 \times 10^{-2}) \\ &= 1.1 \times 10^{-5} \end{aligned}$$

(f) $PbI_2 = 1.28 \times 10^{-3}\ M$

14.8 What concentration of Ag^+ in moles per liter must be present to just start precipitation of AgCl from a solution containing 1.0×10^{-4} mole of Cl^- per liter? The solubility product of AgCl is 2.8×10^{-10}.

Solution: A substance will start to precipitate when the product of the concentrations of its ions equals (or just barely exceeds) the solubility product. No precipitate will form until the product of the concentrations of its ions equals the solubility product. In this particular problem precipitation of AgCl will not begin until the product of the mole concentrations of the ions involved equals the solubility product for AgCl.

$$S.P. = [Ag^+] \times [Cl^-] = 2.8 \times 10^{-10}$$

$$[Ag^+] = \frac{2.8 \times 10^{-10}}{[Cl^-]} = \frac{2.8 \times 10^{-10}}{1 \times 10^{-4}\ mole\ Cl^-\ per\ liter}$$

$$= 2.8 \times 10^{-6}\ mole\ Ag^+\ per\ liter$$

14.9 What concentration of OH^-, in moles per liter, is necessary to start precipitation of $Fe(OH)_3$ from a solution containing 2×10^{-6} mole of Fe^{+++} per liter? The solubility product of $Fe(OH)_3$ is 6×10^{-38}.

14.10 What concentration of sulfide ion, expressed in moles per liter, must be present to just start precipitation of the sulfide of the metal from each of the following solutions? The solubility product of each of the sulfides precipitated is given.

(a) $1.0\ F\ CuCl_2$; 4×10^{-36}

(b) $0.2\ F\ FeCl_2$; 4×10^{-17}

(c) $0.0010\ F\ CdCl_2$; 6×10^{-27}

(d) $0.1\ F\ BiCl_3$; 1×10^{-70}

14.11 The solubility of PbI_2 in water is 2×10^{-3} mole per liter. What concentration of lead ion would be required to just start precipitation of PbI_2 from $0.002\ F\ KI$?

14.12 The solubility product of Ag_3PO_4 is 1.8×10^{-18}. Assuming that a precipitate can be seen as soon as it begins to form, what is the minimum concentration of PO_4^{---} in milligrams per liter that can be detected by the addition of Ag^+ until the solution is 0.010 M in silver ions?

14.13 K_{sp} $BaSO_4 = 1.5 \times 10^{-9}$; K_{sp} $Fe(OH)_3 = 6 \times 10^{-38}$.

Amounts of 0.00005 mole of soluble 100% ionized iron (III) sulfate and 0.00001 mole of soluble 100% ionized barium hydroxide are added to enough water to give a liter of solution. Will a precipitate form?

14.14 To a solution containing 0.010 mole of Ag^+, Cl^- was added, the final volume being 1000 cc; 7.0×10^{-3} mole of AgCl precipitated. How much Cl^- remained in solution? S.P. of AgCl is 2.8×10^{-10}.

14.15 A suspension of calcium hydroxide in water was found to have a pH of 12.3. Calculate the solubility product of $Ca(OH)_2$.

Solution hint: How does $[OH^-]$ compare with $[Ca^{++}]$?

14.16 The molar concentration of the Cd^{++} in a solution in equilibrium with a precipitate of CdS was found to be four times as great as the molar concentration of the S^{--}. K_{sp} for CdS is 6×10^{-27}. What was the concentration of the Cd^{++}?

Solution:

$$K_{sp} = [Cd^{++}] \times [S^{--}] = 6 \times 10^{-27}$$

Let X = concentration of Cd^{++}

$$\frac{X}{4} = \text{concentration of } S^{--}$$

Substituting these values in the S.P. equation

$$X \times \frac{X}{4} = 6 \times 10^{-27}$$

$$X^2 = 2.4 \times 10^{-26}$$

$$X = 1.6 \times 10^{-13} \; M$$

14.17 A liter of solution which is in equilibrium with a precipitate of $Cd(OH)_2$ contains four times as many moles of OH^- as Cd^{++}. How many moles of OH^- are present? S.P. of $Cd(OH)_2$ is 1.6×10^{-14}.

➤ **14.18** You are given equal volumes of two lead salt solutions in which the concentration of Pb^{++} is exactly the same. To one is added 4.00×10^{-1} *mole* of KCl and to the other is added 1.00×10^{-3} *mole* of Na_2SO_4. The final volume is in each case one liter. A total of 0.103 *gram* of $PbCl_2$ precipitates from one solution and 0.136 *gram* of $PbSO_4$ precipitates from the other. The solubility product of $PbSO_4$ is 1.10×10^{-8}. Formula weights: $PbCl_2 = 278$; $PbSO_4 = 303$. Calculate the solubility product of $PbCl_2$.

14.19 From the respective solubility products at 20°C, calculate the solubility of each of the following in moles per liter. (By "solubility" is meant the quantity of solute that will go into solution when the pure solid is shaken with pure water, at 20°C, until a saturated solution is obtained.) The solubility product of each solute is given directly after its formula. (Ignore hydrolysis.)

(a) AgSCN; 1×10^{-12}

Solution: When AgSCN dissolves, it is 100% dissociated into Ag^+ and SCN^-. Therefore,

moles of AgSCN dissolved = moles of Ag^+ = moles of SCN^-

Let X = moles of AgSCN dissolved

$$X = \text{moles of } Ag^+ \qquad X = \text{moles of } SCN^-$$

$$K_{sp} = [Ag^+] \times [SCN^-] = X^2 = 1 \times 10^{-12} \, M^2$$

$$X = 1 \times 10^{-6} \, M$$

(b) AgCl; 2.8×10^{-10}

(c) $Mg(OH)_2$; 8.9×10^{-12}

Solution:

$$Mg(OH)_2 = Mg^{++} + 2 \, OH^-$$

Let X = moles of $Mg(OH)_2$ that dissolve

$$X = \text{moles of } Mg^{++} \qquad 2 \, X = \text{moles of } OH^-$$

$$K_{sp} = [Mg^{++}] \times [OH^-]^2 = X \times (2 \, X)^2 = 8.9 \times 10^{-12} \, M^3$$

$$4X^3 = 8.9 \times 10^{-12} \, M^3$$

$$X = 1.3 \times 10^{-4} \, M$$

(d) Ag_2SO_4; 1.1×10^{-5}

(e) $Al(OH)_3$; 5×10^{-33}

14.20 The solubility product of PbI_2 at 30°C is 1×10^{-8}. The solubility product of $BaSO_4$ at 30°C is also 1×10^{-8}. How does the solubility of PbI_2 in moles per liter compare with the solubility of $BaSO_4$ in moles per liter?

14.21 The solubility product of AgCl is 2.8×10^{-10}. How many moles of AgCl will dissolve in a liter of $0.010 \, F$ KCl? The KCl is 100% ionized.

Solution:

$$[Ag^+] \times [Cl^-] = 2.8 \times 10^{-10}$$

$$[Ag^+] = \frac{2.8 \times 10^{-10}}{[Cl^-]} = \frac{2.8 \times 10^{-10}}{1 \times 10^{-2}} = 2.8 \times 10^{-8} \, M$$

To produce this 2.8×10^{-8} mole of Ag^+, 2.8×10^{-8} mole of AgCl must have gone into solution. In making this calculation the Cl^-

derived from the AgCl has been ignored since its concentration is negligible, being about 2.8×10^{-8} *M*.

14.22 The solubility of $BaSO_4$ in water is 1×10^{-5} *M*. What is its solubility in 0.1 *F* K_2SO_4?

14.23 What volume in liters of 0.10 *F* $MgCl_2$ is required to dissolve the same amount of Hg_2Cl_2 that will dissolve in 1.00 liter of pure water? K_{sp} for Hg_2Cl_2 is 4.0×10^{-18}.

Solution: Hg_2Cl_2 ionizes as follows: $Hg_2Cl_2 \rightleftarrows Hg_2^{++} + 2\,Cl^-$ K_{sp} for $Hg_2Cl_2 = [Hg_2^{++}] \times [Cl^-]^2 = 4.0 \times 10^{-18}$.
Its solubility in pure water, calculated as in Problem 14.19, is 1.0×10^{-6} mole per liter.
Its solubility in 0.10 *F* $MgCl_2$, in which $[Cl^-]$ is 0.20 *M*, calculated as in Problem 14.21, is 1.0×10^{-16} mole per liter.

1.0×10^{-6} mole/liter of water $\div$ 1.0×10^{-16} mole/liter of 0.10 *F* $MgCl_2$
$= 1.0 \times 10^{10}$ liters of 0.10 *F* $MgCl_2$ per liter of water.

14.24 When excess solid Ag_2CrO_4 is shaken with a liter of 0.10 *F* K_2CrO_4, 0.723 mg of Ag_2CrO_4 dissolve. Calculate the solubility product of Ag_2CrO_4.

14.25 Silver oxide is in equilibrium with its saturated solution according to the reaction, $Ag_2O + H_2O \rightleftarrows 2\,Ag^+ + 2\,OH^-$. The solubility product for AgOH is $[Ag^+] \times [OH^-] = 2 \times 10^{-8}$. How many moles of Ag_2O will dissolve in a liter of solution whose pH is 11?

➤ **14.26** The first ionization of sulfuric acid, $H_2SO_4 = H^+ + HSO_4^-$, is 100% complete. The ionization constant for the second ionization, $HSO_4^- \rightleftarrows H^+ + SO_4^{--}$, is 1.2×10^{-2}. The solubility product for $BaSO_4$ is 1.0×10^{-10}. Excess solid $BaSO_4$ was shaken with a solution of sulfuric acid until a saturated solution of $BaSO_4$ was obtained. The pH of this saturated solution was 2. How many moles of $BaSO_4$ dissolved per liter of saturated solution?

Solution: K_{sp} for $BaSO_4$ being 1.0×10^{-10}, the maximum number of moles of $BaSO_4$ that can dissolve in a liter of pure water is 1.0×10^{-5}. The number of moles that will dissolve in a solution of H_2SO_4 whose pH is 2 will be much less than 1.0×10^{-5}.
The number of moles of Ba^{++} ions in the final solution will be equal to the number of moles of $BaSO_4$ that dissolve. The number of moles of SO_4^{--} ions in the final solution will equal the number of moles initially present (from the H_2SO_4) plus the number derived from the $BaSO_4$ that dissolves. The number derived from the latter source is so small compared with the number initially present that they can be ignored in the calculation.

Therefore, to solve the problem, first calculate the concentration of SO_4^{--} ions in a solution of H_2SO_4 whose pH is 2 in the manner outlined in Problem 13.37. This value, divided into K_{sp} for $BaSO_4$, gives the concentration of Ba^{++} and, hence, the number of moles of $BaSO_4$ that dissolve. The smallness of the answer justifies the assumption made in solving the problem.

➡ **14.27** A solution in equilibrium with solid CaC_2O_4 has a pH of 4.0. The sum of the $C_2O_4^{--}$, $HC_2O_4^{-}$, and $H_2C_2O_4$ in the solution is 0.20 F. Calculate the concentration of Ca^{++} ions in the solution. K_{sp} for CaC_2O_4 = 1.3×10^{-9}.

Solution hint: See Problem 13.83.

➡ **14.28** A solution contains 0.010 mole Cl^- per liter and 0.0010 mole CrO_4^{--} per liter. The S.P. of $AgCl$ is 1.56×10^{-10}, the S.P. of Ag_2CrO_4 is 9.0×10^{-12}. What will be the concentration of Cl^- in moles per liter when Ag_2CrO_4 just begins to precipitate by the continued addition of Ag^+, the volume of the solution at this point being exactly 1 liter?

Solution: When Ag^+ ions are added to the solution represented by this problem, $AgCl$ will begin to precipitate when the product of $[Ag^+]$ and $[Cl^-]$ equals the solubility product, 1.56×10^{-10}. Since $[Cl^-]$ is 1×10^{-2}, the precipitation of $AgCl$ will begin when $[Ag^+]$ is 1.56×10^{-8}. Since the solubility product of Ag_2CrO_4 is 9.0×10^{-12} and $[CrO_4^{--}]$ is 1×10^{-3}, precipitation of Ag_2CrO_4 will not begin until $[Ag^+]$ is 9.5×10^{-5}; that means that, at the start, only $AgCl$ precipitates. As more Ag^+ ions are added after precipitation of $AgCl$ first begins, more $AgCl$ will precipitate. As more $AgCl$ precipitates the concentration of the Cl^- ions remaining in solution decreases, and as the concentration of Cl^- ions decreases the concentration of Ag^+ ions required to continue precipitation of $AgCl$ increases; during the entire $AgCl$ precipitation process the product of $[Ag^+]$ and $[Cl^-]$ must always be equal to 1.56×10^{-10}. Finally, the concentration of Cl^- ions will be low enough so that a Ag^+ ion concentration of 9.5×10^{-5} will be required to precipitate more $AgCl$. When that happens Ag_2CrO_4 will also begin to precipitate. Since $[Ag^+] \times [Cl^-]$ must always equal 1.56×10^{-10}, when $[Ag^+]$ is 9.5×10^{-5}, $[Cl^-]$ will be $1.56 \times 10^{-10} \div 9.5 \times 10^{-5}$ or 1.6×10^{-6} mole per liter.

➡ **14.29** A solution contains 0.000020 mole of Br^- per liter and 0.010 mole of Cl^- per liter. The S.P. of $AgCl$ is 1.56×10^{-10}, the S.P. of $AgBr$ is 3.25×10^{-13}. Which of these ions will start precipitating first when Ag^+ is added to the above solution? What will be its concentration when the other ion begins to precipitate?

➡ **14.30** The solubility products of $AgIO_3$ and $Ba(IO_3)_2$ are 1.0×10^{-8}

and 6.0×10^{-10}, respectively. A solution is 8.6×10^{-4} molar in Ag^+ and 3.74×10^{-3} molar in Ba^{++}. Iodate ion is added to this solution slowly and with constant stirring.

(a) Which cation precipitates as the iodate salt first? At what IO_3^- ion concentration does this precipitate just start to form?

(b) At what IO_3^- ion concentration does the second cation just start to precipitate as the iodate salt?

(c) What is the concentration of the first cation when the second cation just starts to precipitate?

➡ **14.31** Calculate the concentration of Cl^- in a solution saturated with both $AgCl$ and Ag_2CrO_4 and in which the concentration of chromate ion is 1.0×10^{-3} M. K_{sp} for $Ag_2CrO_4 = 1.7 \times 10^{-12}$, for $AgCl = 1.1 \times 10^{-10}$.

➡ **14.32** If to a certain solution 1.00×10^{-7} F in KI and 0.100 F in NaCl solid $AgNO_3$ is added, how many moles of AgI per liter will be precipitated before the solution is saturated with AgCl? Solubility products: AgI, 1.00×10^{-16}; AgCl, 1.10×10^{-10}.

➡ **14.33** A liter of solution which was in equilibrium with a solid mixture of AgCl and AgI was found to contain 1×10^{-8} mole of Ag^+, 1×10^{-2} mole of Cl^-, and 1×10^{-8} mole of I^-. Enough Ag^+ ions were added, slowly and with constant stirring, to increase the concentration of Ag^+ to 10^{-6} mole per liter; the volume of the solution was kept constant at 1 liter. How many moles of AgCl were precipitated as a result of this addition of Ag^+ ions? How many moles of AgI were precipitated as a result of this addition of Ag^+?

➡ **14.34** One commonly used titrimetric method of determining the chloride ion content of a material utilizes the red color of silver chromate as an end point indicator. A solution of known volume, to which the chromate ion indicator has been added and which contains a weighed sample of the material, is titrated with a standard $AgNO_3$ solution until the red of the Ag_2CrO_4 just appears. From the measured and observed quantities and the appropriate solubility product constants, the amount of Cl^- present can be determined within a small error. Given the following data:

A 1.7750 g sample is dissolved in 203 ml of water.

1 cc of 0.00100 F K_2CrO_4 is added.

46.00 ml of 0.250 N $AgNO_3$ were required to produce the red end point.

K_{sp} of $AgCl = 1.50 \times 10^{-10}$

K_{sp} of $Ag_2CrO_4 = 9.00 \times 10^{-12}$

(a) What is the $[CrO_4^{--}]$ just as the red color appears?

(b) What is the $[Ag^+]$ at this end point?

(c) What is the [Cl⁻] at this end point?

(d) How many moles of Ag^+ were added?

(e) How many moles of AgCl precipitated?

(f) What is the total amount of Cl^- present?

(g) What is the weight percent of Cl^- in the original sample?

➡ **14.35**   A 0.20 F solution of Na_3PO_4 has a pH of 10.5. How many moles of Ag_3PO_4 will dissolve in a liter of this solution? K_{sp} for $Ag_3PO_4 = 1.8 \times 10^{-13}$.

➡ **14.36**   The solubility product of $PbSO_4$ is 1.3×10^{-8}. The solubility product of $Pb(ClO_4)_2$ (lead perchlorate) is 2.4×10^{-15}. Perchloric acid ($HClO_4$) is completely ionized. The first ionization of H_2SO_4 is complete; the ionization constant for the ionization of HSO_4^- is 1.2×10^{-2}.

A solution prepared by dissolving pure $HClO_4$ and pure H_2SO_4 in the same beaker of water has a pH of 1.51. To neutralize 100 ml of this solution requires 90.0 ml of 0.050 F KOH. How many moles of solid $Pb(NO_3)_2$ must be dissolved in a liter of this solution before a precipitate begins to form? Give the formula of the solid that begins to precipitate.

➡ **14.37**   Calculate the concentration of I^- in a solution obtained by shaking 0.100 F KI with an excess of AgCl. K_{sp} of AgCl $= 1.1 \times 10^{-10}$, K_{sp} of AgI $= 1.0 \times 10^{-16}$.

Solution: The two equilibria involved are:

(1) $$AgCl \, (s) \rightleftharpoons Ag^+ + Cl^-$$

(2) $$Ag^+ + I^- \rightleftharpoons AgI \, (s)$$

The net equation for the principal reaction is

(3) $$AgCl \, (s) + I^- \rightleftharpoons AgI \, (s) + Cl^-$$

The equilibrium constant for reaction (3) is

(4) $$K = \frac{[Cl^-]}{[I^-]} = \frac{[Cl^-] \times [Ag^+]}{[I^-] \times [Ag^+]} = \frac{1.1 \times 10^{-10}}{1.0 \times 10^{-16}} = 1.1 \times 10^6$$

Let $X = [I^-]$

$$0.100 - X = [Cl^-]$$

Substituting these values in Equation (4) gives us

$$\frac{0.100 - X}{X} = 1.1 \times 10^6$$

Since $[I^-]$ is only 10^{-6} as large as $[Cl^-]$, and since the maximum value of $[Cl^-]$ is 0.100 M, it is obvious that the value of X in the expression, $0.100 - X$, is so small that it can be dropped.

$$\frac{0.100}{X} = 1.1 \times 10^6$$

$$X = 9.1 \times 10^{-8} \, M = [I^-]$$

Note: Since 0.100 M I^- is added to solid AgCl, a natural inclination, when substituting in Equation (4), is to let $X = [Cl^-]$ and $0.100 - X = [I^-]$. If this is done, X, being very large (about 0.100), cannot be dropped from the expression, $0.100 - X$. The equation will then be,

$$\frac{X}{0.100 - X} = 1.1 \times 10^6$$

Solving,

$$X = 1.1 \times 10^5 - 1.1 \times 10^6 \, X$$

$$1.1 \times 10^6 \, X + X = 1.1 \times 10^5$$

If X is dropped from the expression, $1.1 \times 10^6 \, X + X$, the calculated value of X is 0.100. The value of $[I^-]$, since it is $0.100 - X$, will then be $0.100 - 0.100$, or zero. Obviously, X cannot be dropped in this instance.

The correct procedure in situations of this type is to *let X equal that quantity which we know is very small*; X, being very small, can then be dropped from expressions in which it is subtracted from or added to a number which is very large by comparison with X.

Note that Equation (4) above tells us that in a saturated solution in equilibrium with the two solids, AgCl and AgI, containing the common ion, Ag^+, the concentrations of the two dissimilar ions, Cl^- and I^-, are to each other as the solubility products of the parent species. That is

$$\frac{[Cl^-]}{[I^-]} = \frac{K_{sp} \text{ for AgCl}}{K_{sp} \text{ for AgI}}$$

This relationship can also be derived as follows:

For AgCl, $K_{sp} = [Ag^+] \times [Cl^-] = 1.1 \times 10^{-10}$ and $[Ag^+]$

$$= \frac{1.1 \times 10^{-10}}{[Cl^-]}$$

For AgI, $K_{sp} = [Ag^+] \times [I^-] = 1.0 \times 10^{-16}$ and $[Ag^+] = \dfrac{1.0 \times 10^{-16}}{[I^-]}$

Since $[Ag^+]$ is the same for both equilibria

$$\frac{1.0 \times 10^{-16}}{[I^-]} = \frac{1.1 \times 10^{-10}}{[Cl^-]} \quad \text{and} \quad \frac{[Cl^-]}{[I^-]} = \frac{1.1 \times 10^{-10}}{1.0 \times 10^{-16}}$$

$$= \frac{K_{sp} \text{ for AgCl}}{K_{sp} \text{ for AgI}}$$

This relationship will be an important one to keep in mind in certain problems.

➡ **14.38** Calculate the concentration of Ag^+ in a solution prepared by mixing 100 ml of a solution 0.200 F in both NaCl and KI with 100 ml of 0.100 F AgNO$_3$. K_{sp} for AgCl $= 1.1 \times 10^{-10}$, for AgI $= 1.0 \times 10^{-16}$.

➡ **14.39**   A solution is 0.10 M in Cl^-, in Br^-, and in I^-. To 1 liter of this solution is added 0.15 mole of $AgNO_3$. What are the final concentrations of Cl^-, Br^-, and I^- in the solution?

$$K_{AgCl} = 1.5 \times 10^{-10} \quad K_{AgBr} = 5.0 \times 10^{-13} \quad K_{AgI} = 8.3 \times 10^{-17}$$

➡ **14.40**   A liter of solution known to contain Zn^{++} and Ni^{++} in equal molar concentrations was kept saturated with H_2S. When precipitation was complete it was found that

(a) the volume of the solution in equilibrium with the precipitate was 1 liter

(b) the pH of the solution was 4.000

(c) 99.000% of the Ni^{++} originally present was precipitated as NiS

What percent of the Zn^{++} originally present was precipitated as ZnS? Solubility products: $ZnS = 1.3 \times 10^{-20}$, $NiS = 1.3 \times 10^{-22}$.

➡ **14.41**   The solubility of $CaCO_3$ in water at 25°C is 1.3×10^{-4} moles per liter. The solubility product, calculated from this solubility, is $4.8 \times 10^{-9}\ M^2$, not 1.7×10^{-8}. Explain.

Solution: Hydrolysis of CO_3^{--} is not ignored.

➡ **14.42**   To 1 liter of a 1 F solution of Na_2CO_3 is added 10^{-7} moles of $MgCl_2$. Will a precipitate form; and, if so, what is the precipitate?

$$K_{sp} \text{ for } MgCO_3 = 4.0 \times 10^{-5} \quad K_{sp} \text{ for } Mg(OH)_2 = 1.3 \times 10^{-11}$$

Solution: Calculate, from the hydrolysis constant for CO_3^{--}, the concentrations of CO_3^{--} and OH^- in 1 F Na_2CO_3.

➡ **14.43**   A certain trivalent metal ion forms an insoluble hydroxide and an insoluble carbonate; the solubility products are 1.0×10^{-20} and 5.5×10^{-25}, respectively. If a mixture of a very small amount of these 2 solids is placed in 0.20 F Na_2CO_3, will the conversion hydroxide $\rightarrow$ carbonate occur, or will the reverse change take place?

Solubility products and the hydrogen sulfide equilibrium

A saturated solution of hydrogen sulfide in water at 18°C and standard barometric pressure is approximately 0.10 M in H_2S. Hydrogen sulfide is a very weak acid; hence, its percent ionization is small. Although H_2S ionizes in two stages,

$$H_2S \rightleftharpoons H^+ + HS^- \quad K_1 = 1.0 \times 10^{-7}\ M$$
$$HS^- \rightleftharpoons H^+ + S^{--} \quad K_2 = 1.3 \times 10^{-13}\ M$$

the overall ionization can be represented by one equation,

$$H_2S \rightleftharpoons 2\,H^+ + S^{--}$$

The ionization constant for this overall reaction is 1.3×10^{-20}. That is,

$$\frac{[H^+]^2 \times [S^{--}]}{[H_2S]} = 1.3 \times 10^{-20} M^2$$

When the concentration of H^+ is fixed, by addition of acid or base to the solution, this overall ionization constant can justifiably be used to calculate $[S^{--}]$. Since the solution is saturated, the concentration of H_2S will be constant, namely $0.10\ M$. We can, therefore, combine this constant value with the ionization constant to get the equation,

$$\frac{[H^+]^2 \times [S^{--}]}{0.10} = 1.3 \times 10^{-20}\ M^2$$

$$[H^+]^2 \times [S^{--}] = 1.3 \times 10^{-21}\ M^3$$

This constant, $1.3 \times 10^{-21}\ M^3$, is the *ion product* for a saturated solution of H_2S. It is a very useful constant in calculations involving reactions in which an acidified or alkalinized saturated solution of hydrogen sulfide is either a reactant or a product.

PROBLEMS

→ **14.44** Given solutions containing 1×10^{-6} mole per liter of Hg^{++}, Cu^{++}, Pb^{++}, Sn^{++}, Ni^{++}, Fe^{++}, and Mn^{++}, respectively. These solutions are saturated with H_2S at 18°C. The ion product, $[H^+]^2 \times [S^{--}]$, for a saturated $(0.10\ M)$ solution of H_2S is 1.3×10^{-21}. In each case what is the greatest H^+ concentration in moles per liter which will just allow precipitation of the sulfide to start? The solubility product for each sulfide is given.

(a) Hg^{++}; 1×10^{-50}

Solution:

$$[Hg^{++}] \times [S^{--}] = 1 \times 10^{-50}$$

$$[S^{--}] = \frac{1 \times 10^{-50}}{1 \times 10^{-6}\ \text{mole } Hg^{++}\ \text{per liter}} = 1 \times 10^{-44}\ M$$

This is the concentration of S^{--} which must be present for precipitation of HgS to start from a solution containing 1×10^{-6} mole of Hg^{++} per liter.

$$[H^+]^2 \times [S^{--}] = 1.3 \times 10^{-21}$$

$$[H^+]^2 = \frac{1.3 \times 10^{-21}}{1 \times 10^{-44}\ \text{mole } S^{--}\ \text{per liter}} = 1.3 \times 10^{23}$$

$[H^+] = 3.6 \times 10^{11}$ *M*. This is the concentration of H^+ which will be in equilibrium with 1×10^{-44} mole S^{--} per liter.

(b) Cu^{++}; 4.0×10^{-36} (e) Ni^{++}; 1×10^{-22}

(c) Pb^{++}; 4×10^{-26} (f) Fe^{++}; 4×10^{-17}

(d) Sn^{++}; 1.0×10^{-24} (g) Mn^{++}; 8×10^{-14}

➤ **14.45** A solution containing 1×10^{-6} mole of Cd^{++} per liter was kept saturated with H_2S until precipitation was complete. The concentration of H^+ was kept at 0.2 *M* during the precipitation. How many grams of CdS were precipitated per liter of solution? S.P. for CdS is 6×10^{-27}.

Solution: $[H^+]^2 \times [S^{--}] = 1.3 \times 10^{-21}$

$$[S^{--}] = \frac{1.3 \times 10^{-21}}{[H^+]^2} = \frac{1.3 \times 10^{-21}}{(0.2)^2} = 3.25 \times 10^{-20} \ M.$$

This is the final concentration of S^{--} in the solution.

$$[Cd^{++}] \times [S^{--}] = 6 \times 10^{-27}$$

$$[Cd^{++}] = \frac{6 \times 10^{-27}}{[S^{--}]} = \frac{6 \times 10^{-27}}{3.25 \times 10^{-20}} = 1.85 \times 10^{-7}$$

This is the concentration of Cd^{++} ions left in the solution.

Moles of CdS precipitated $= 1 \times 10^{-6} - 1.85 \times 10^{-7} = 8.15 \times 10^{-7}$

Grams of CdS precipitated $= 8.15 \times 10^{-7}$ moles $\times$ 144.5 g/mole

$$= 1.18 \times 10^{-4} \text{ g} = 1 \times 10^{-4} \text{ g}$$

➤ **14.46** The solubility product of SnS is 1.1×10^{-24}. One liter of 0.00013 *M* Sn^{++} was kept saturated with H_2S until precipitation was complete, at which time 0.0135 g of SnS precipitated. What was the pH of the solution at the end of the precipitation?

➤ **14.47** How many grams of Ag^+ must be present per liter before Ag_2S will start to precipitate from a saturated solution of H_2S whose pH is 2.0? The solubility product of Ag_2S is 1×10^{-50}.

➤ **14.48** The solubility product of Cu_2S is 4.4×10^{-49}. What must be the pH of a saturated solution of H_2S containing 2.0×10^{-18} mole of Cu^+ per liter if Cu_2S will just barely start precipitating?

➤ **14.49** The solubility product of SnS is 1×10^{-24}. If a 0.5×10^{-11} *M* solution of Sn^{++} whose pH is maintained at 4 is saturated with H_2S, will a precipitate form?

➤ **14.50** A solution is 0.0000010 *M* with respect to Pb^{++} and 0.0050 *M* with respect to Cu^{++}. If the solution is kept saturated with H_2S, what is the hydrogen-ion concentration which will permit the maximum precipitation of CuS but will not allow the precipitation of PbS? K_{sp} for CuS is 3.5×10^{-38} and for PbS is 1.0×10^{-29}.

➤ **14.51** If NiS just begins to precipitate from a 0.0010 *F* $NiCl_2$ solution saturated with H_2S when the pH is 1.0, what is the solubility product for NiS?

Solubility products and the ammonia equilibrium

PROBLEMS

(See Table 2 for values of ionization constants.)

➤ **14.52** When excess solid $Mg(OH)_2$ is shaken with 1 liter of $1.0 F$ NH_4Cl the resulting saturated solution has a pH of 9.0. The net equation for the reaction that occurs is

$$Hg(OH)_2 + 2 NH_4^+ = Mg^{++} + 2 NH_3 + 2 H_2O$$

Calculate the solubility product for $Mg(OH)_2$.

Solution: $NH_3 + H_2O \rightleftharpoons 2 NH_4^+ + OH^-$ $K = \dfrac{[NH_4^+] \times [OH^-]}{[NH_3]}$
$$= 1.8 \times 10^{-5}$$

$$pH = 9.0; [H^+] = 1.0 \times 10^{-9}; [OH^-] = 1.0 \times 10^{-5}$$

$K_{sp} = [Mg^{++}] \times [OH^-]^2$. Since we know $[OH^-]$, all we need is to find $[Mg^{++}]$. The important fact to notice is that the molar concentration of Mg^{++} is $\frac{1}{2}$ the molar concentration of NH_3. The molar concentration of the NH_3 can be calculated as follows:

From the ionization constant formula for NH_3 we can determine that:

$$\frac{[NH_4^+]}{[NH_3]} = \frac{1.8 \times 10^{-5}}{[OH^-]} = \frac{1.8 \times 10^{-5}}{1.0 \times 10^{-5}} = 1.8$$

Since we started with 1 liter of $1.0 F NH_4Cl$, $[NH_4^+] + [NH_3] = 1.0$. Let $X = [NH_3]$; $1.0 - X = [NH_4^+]$.

$$\frac{1.0 - X}{X} = 1.8 \qquad \text{Solving, } X = 0.36 = [NH_3]$$

$$\tfrac{1}{2} X = 0.18 = [Mg^{++}]$$

$$K_{sp} = [Mg^{++}] \times [OH^-]^2 = 0.18 \times (1.0 \times 10^{-5})^2 = 1.8 \times 10^{-11}$$

➤ **14.53** Excess $Mg(OH)_2$ is added to a solution which is $0.20 F$ in NH_4NO_3 and $0.50 F$ in NH_3. Calculate the concentration of Mg^{++} ions at equilibrium. K_{sp} for $Mg(OH)_2 = 1.2 \times 10^{-11}$.

➤ **14.54** The S.P. of $Mg(OH)_2$ is 8.9×10^{-12}. How many grams of NH_4^+ must be present in a liter of $0.10 F NH_3$ containing 0.30 g of Mg^{++}, to prevent $Mg(OH)_2$ from being precipitated?

Solution: First calculate the maximum $[OH^-]$ that will just fail to produce a precipitate of $Mg(OH)_2$ in the solution. Then calculate the amount of NH_4^+ that must be present in $0.10\ F\ NH_4OH$ to give this $[OH^-]$.

➤ 14.55 The S.P. of $Mn(OH)_2$ is 2.0×10^{-13}. How many grams of NH_4Cl must be present in 100 ml of $0.20\ F\ NH_3$ to prevent precipitation of $Mn(OH)_2$ when the solution is added to 100 ml of $0.20\ F\ MnCl_2$?

➤ 14.56 You are given 200 ml of a solution containing Mn^{++} and Mg^{++} each at $0.01\ M$, and 200 ml of a solution $0.40\ F$ in ammonia. How many grams of solid ammonium chloride should be added to the latter so that when the solutions are mixed, the $Mn(OH)_2$ will be precipitated as completely as possible but the $Mg(OH)_2$ will remain unprecipitated? K_{sp} for $Mg(OH)_2 = 1.4 \times 10^{-11}$, for $Mn(OH)_2 = 4.5 \times 10^{-14}$.

Solution hint: Note that NH_4^+ is liberated in the precipitation of $Mn(OH)_2$.

➤ 14.57 The S.P. for $Fe(OH)_3$ is 6×10^{-38}. How much NH_4^+ must be present in order to prevent the precipitation of $Fe(OH)_3$ in a solution $0.1\ F$ in NH_4OH and $0.0010\ M$ in Fe^{+++}? Would it be possible to dissolve that much NH_4^+ in a liter?

Solubility product equilibria involving selected weak acids

PROBLEMS

(See Table 2 for values of ionization constants.)

➤ 14.58 The solubility product of $AgC_2H_3O_2$ is 4.0×10^{-4}. To 1 liter of a solution $1.0\ F$ in $HC_2H_3O_2$ and $0.10\ F$ in HNO_3 is added just enough solid $AgNO_3$ to start precipitation of $AgC_2H_3O_2$. How many moles of $AgNO_3$ are added?

➤ 14.59 A solution was prepared by dissolving 1.80 moles of $NaC_2H_3O_2$ and 1.00 mole of $HC_2H_3O_2$ in enough water to give 1.00 liter of solution. What is the maximum concentration of Fe^{+++} that can exist in this solution without precipitation of $Fe(OH)_3$? Solubility product for $Fe(OH)_3 = 6.0 \times 10^{-38}$.

➤ 14.60 The K_{sp} of $Fe(OH)_3$ is 6.0×10^{-38}.

(a) What is the formal solubility of $Fe(NO_3)_3$ in a solution that is $0.20\ F$ in NH_3 and $0.36\ F$ in NH_4NO_3?

(b) What volume of this NH_3-NH_4NO_3 solution is needed to dissolve the same amount of $Fe(NO_3)_3$ that will dissolve in 1.00 liter of a solution that is $0.10\ F$ in HCOOH and $0.42\ F$ in HCOOK?

➤ **14.61** A saturated solution was prepared by shaking excess solid $CaCO_3$ with water containing a small amount of HCl. When equilibrium was established the pH of the saturated solution was found to be 7.0. How many moles of $CaCO_3$ dissolved per liter of solution? The solubility product of $CaCO_3$ is 7×10^{-9}.

Solution hint: Let $X = [Ca^{++}] =$ moles of $CaCO_3$ dissolved.
X will then equal $[CO_3^{--}] + [HCO_3^-] + [H_2CO_3]$.
Knowing K_1 and K_2 for H_2CO_3 and knowing the pH of the solution the relative amounts of CO_3^{--}, HCO_3^-, and H_2CO_3 are known. The necessary equations for calculating X can then be set up.

➤ **14.62** The solubility of $BaCO_3$ in water saturated with CO_2 at 1 atm is 0.01 mole per liter. The concentration of H_2CO_3 in this solution is 0.04 M. The net equation is $BaCO_3 \text{ (s)} + H_2CO_3 \rightleftharpoons Ba^{++} + 2\,HCO_3^-$. Calculate the equilibrium constant for this reaction and calculate the solubility product of $BaCO_3$.

➤ **14.63** A liter of water in contact with excess solid $BaCO_3$ was kept saturated with CO_2. When equilibrium had been established in the reaction

$$BaCO_3 \text{ (s)} + H_2CO_3 \rightleftharpoons Ba^{++} + 2\,HCO_3^-$$

the concentration of the H_2CO_3 was 0.040 M.
Calculate the concentration of the CO_3^{--} ions in the solution.
The solubility product for $BaCO_3$ is 1.0×10^{-8}.

➤ **14.64** When excess solid $BaSO_3$ is added to a liter of pure dilute HCl, 4.0×10^{-4} moles of $BaSO_3$ are dissolved. No SO_2 gas is evolved in the process, and no complex ions are formed. The pH of the resulting solution is 5.0. Calculate the solubility product for $BaSO_3$.

➤ **14.65** In order to just prevent precipitation of BaF_2 in a solution which is 0.10 F in $BaCl_2$ and 0.10 F in KF, it is necessary to adjust the pH to a value of 1.96 by addition of HCl. Calculate the solubility product of BaF_2.

➤ **14.66** A solution containing 0.01 M Zn^{++}, 0.1 F acetic acid, and 0.05 F $NaC_2H_3O_2$, is saturated with H_2S. What concentration of Zn^{++} remains in solution? K_{sp} for ZnS $= 1.3 \times 10^{-20}$.

Solution hint: Note that 2 moles of H^+ ions are liberated for each mole of ZnS precipitated.

➤ **14.67** A solution contains 0.01 F $Ca(NO_3)_2$, 0.01 F $Sr(NO_3)_2$, and 0.5 F oxalic acid. To what value should the hydrogen ion concentration be adjusted, by addition of HCl or NaOH, in order to precipitate as much as possible of the calcium while leaving all of the strontium in solution? K_{sp} for calcium oxalate $= 2.6 \times 10^{-9}$. K_{sp} for strontium oxalate $= 7.0 \times 10^{-8}$.

Solution hint: Note that 1 mole of oxalate is removed for each mole of CaC_2O_4 precipitated.

Complex ions and solubility products

When a cation (called the *central ion*) combines with one or more anions or neutral molecules (called *ligands*) to form a new ion, this new ion is called a *complex ion*. Complex ions are weak electrolytes and, as such, dissociate incompletely to form the original species. Thus

$$Zn^{++} + 4\,NH_3 \rightleftharpoons Zn(NH_3)_4^{++}$$

and

$$Zn(NH_3)_4^{++} \rightleftharpoons Zn^{++} + 4\,NH_3$$

The equilibrium constants for ionization of complex ions are referred to as *instability constants*. (See Table 3, page 276.) Just as polybasic acids such as H_3PO_4 ionize in stages, each stage having its own ionization constant, k_1, k_2 and k_3, with the overall ionization constant being the product of k_1, k_2, and k_3, so complex ions also dissociate in stages with each stage having its own instability constant.

$$Zn(NH_3)_4^{++} \rightleftharpoons Zn(NH_3)_3^{++} + NH_3$$

$$Zn(NH_3)_3^{++} \rightleftharpoons Zn(NH_3)_2^{++} + NH_3$$

$$Zn(NH_3)_2^{++} \rightleftharpoons Zn(NH_3)_1^{++} + NH_3$$

$$Zn(NH_3)_1^{++} \rightleftharpoons Zn^{++} + NH_3$$

The overall instability constant is the product of the four individual constants and is represented by the formula

$$K_{inst} = \frac{[Zn^{++}] \times [NH_3]^4}{[Zn(NH_3)_4^{++}]}$$

Just as, with polyprotic acids, the overall ionization constant can be used to calculate the concentration of the anion only if excess H^+ or OH^- is present in the system, so, with complex ion equilibria, the overall instability constant can be used only when excess ligand is present. In all problems involving complex ions excess ligand will be present; accordingly, overall instability constants can be used.

PROBLEMS

(See Tables 2 and 3 for values of equilibrium constants.)

➭ **14.68** Calculate the concentration of Cu^{++} in a solution which is 0.10 F in $CuSO_4$ and 1.40 F in NH_3. The instability constant for $Cu(NH_3)_4^{++}$ is 4.7×10^{-15} mole^4/liter^4.

Solution: The very low value of the instability constant means that essentially all the copper ion in solution will be in the form of the ammine complex. Therefore, $[Cu(NH_3)_4^{++}] = 0.10$ mole/liter. Then, $[NH_3] = 1.40 - 4 (0.10) = 1.00$ mole/liter

$$\frac{[Cu^{++}][NH_3]^4}{[Cu(NH_3)_4^{++}]} = 4.7 \times 10^{-15} \text{ mole}^4/\text{liter}^4$$

and

$$[Cu^{++}] = 4.7 \times 10^{-16} \text{ mole/liter}$$

The low $[Cu^{++}]$ thus calculated justifies the assumption that $Cu(NH_3)_4^{++}$ is by far the predominant copper-containing species. It should be noted that, in effect, the NH_3 acts as a *buffer*; it ties up the Cu^{++} in the form of the weak electrolyte, $Cu(NH_3)_4^{++}$.

➡ **14.69** The instability constant for $Ag(NH_3)_2$ is 6.0×10^{-8} mole^2/liter^2.

(a) What is the molar concentration of NH_3 needed to convert exactly 50% of the silver ion to the ammine complex in *YF* $AgNO_3$?

(b) What is the formal concentration of NH_3 in this solution?

Solution hint: At equilibrium $[Ag^+] = Y/2$ and $[Ag(NH_3)_2^+] = Y/2$. Let $X = [NH_3]$ at equilibrium. The formal concentration of NH_3 will be equal to $[NH_3] + 2 \times [Ag(NH_3)_2^+]$.

➡ **14.70** The element Q has stable oxidation states of 2, 3, and 4. A mixture weighing 207 g contains "*a*" moles of QCl_2, "*b*" moles of $Q_2(SO_4)_3$, and "*c*" moles of $Q(CrO_4)_2$.

(1) 1.35 moles of Ba^{++} are needed to precipitate all the sulfate and chromate in a 207 g sample.

(2) A second identical 207 g sample requires 0.180 mole of MnO_4^- to oxidize all the Q to the +4 state (MnO_4^- is reduced to Mn^{++}).

(3) A third identical 207 g sample requires 2.60 moles of ammonia to completely convert Q^{++} and Q^{++++} to their very stable ammine complexes, $Q(NH_3)_4^{++}$, and $Q(NH_3)_6^{++++}$ (Q^{+++} forms neither a stable ammine complex nor an insoluble hydroxide).

(a) How many moles of each compound were in a 207 g sample of this mixture?

(b) Calculate the atomic weight of Q.

➡ **14.71** $Ga(OH)_3$ is practically insoluble in water, its solubility product being about 1×10^{-36}. A beaker containing a liter of a saturated solution of $Ga(OH)_3$ prepared by stirring solid $Ga(OH)_3$ with water contains 4.0×10^{-4} mole of excess solid $Ga(OH)_3$. 1.2×10^{-3} mole of solid KOH is dissolved in the liter of solution. As a result, all of the solid $Ga(OH)_3$ dissolves, a very stable complex hydroxo ion being formed. The resulting solution has a pH of 10.9. Calculate the formula of the complex hydroxo ion.

→ **14.72** Co(OH)$_2$ is practically insoluble in water. It dissolves to some extent in NH$_3$ to form a very stable ammine complex ion. In an effort to determine the composition of the complex ion, a chemist carried out three experiments in which he added *excess* solid Co(OH)$_2$ to three different solutions containing NaOH and NH$_3$. *At equilibrium* he found the *molar* concentrations of NH$_3$, OH$^-$, and complex ion in the solutions to be:

Solutions	Conc. of NH$_3$	Conc. of OH$^-$	Conc. of Complex Ion
First experiment	0.5 *M*	0.1 *M*	3.1 × 10^{-9} *M*
Second experiment	1.0 *M*	0.1 *M*	2.0 × 10^{-7} *M*
Third experiment	2.0 *M*	0.1 *M*	1.3 × 10^{-5} *M*

Calculate the formula of the complex ion.

Solution hints: The formula of the complex ion is Co(NH$_3$)$_x^{++}$.

What will be true of the value of $\dfrac{[\text{Co}^{++}] \times [\text{NH}_3]^x}{[\text{Co(NH}_3)_x^{++}]}$ in each experiment?

Since [OH$^-$] is the same in each experiment what will be true of [Co^{++}] in each experiment?

→ **14.73** The solubility product of AgCl is 1.2 × 10^{-10}. The instability constant for Ag(NH$_3$)$_2^+$ is 6.0 × 10^{-8}. What must be the formal concentration of a solution of NH$_3$ in water if one liter of this solution shall just barely dissolve 0.020 mole of AgCl?

Solution: The net equation for the principal reaction is

$$\text{AgCl (solid)} + 2\,\text{NH}_3 \rightleftharpoons \text{Ag(NH}_3)_2^+ + \text{Cl}^-$$

Since 0.020 mole of AgCl dissolves, [Cl$^-$] = 0.020 *M*.

Since K_{sp} for AgCl = 1.2 × 10^{-10}, $[\text{Ag}^+] = \dfrac{1.2 \times 10^{-10}}{0.20 \times 10^{-1}}$

$$= 6.0 \times 10^{-9}\ M.$$

That means that practically all of the dissolved AgCl is present as Ag(NH$_3$)$_2^+$. Therefore, [Ag(NH$_3$)$_2^+$] = 0.020 *M*.

The equilibrium constant for the principal reaction is:

$$K = \frac{[\text{Ag(NH}_3)_2^+] \times [\text{Cl}^-]}{[\text{NH}_3]^2} = \frac{[\text{Ag(NH}_3)_2^+] \times [\text{Cl}^-] \times [\text{Ag}^+]}{[\text{NH}_3]^2 \times [\text{Ag}^+]}$$

$$= \frac{K_{sp}}{K_{inst}} = \frac{1.2 \times 10^{-10}}{6.0 \times 10^{-8}} = 2.0 \times 10^{-3}$$

Let X = the concentration of NH$_3$ at equilibrium. Substituting the values of the three species in the above equilibrium constant:

$$\frac{(0.020) \times (0.020)}{X^2} = 2.0 \times 10^{-3}.$$ Solving, $X = 0.45\ M = [NH_3]$ at equilibrium. The formation of the 0.020 mole of $Ag(NH_3)_2^+$ required the consumption of 0.040 mole of NH_3. Therefore, the formal concentration of the original solution of NH_3 was $0.45 + 0.040 = 0.49\ F$.

➡ **14.74** Excess solid AgCl is treated with 100 ml of $1\ F\ NH_3$. How many grams of AgCl will dissolve? The solubility product for AgCl is 1.0×10^{-10}. The instability constant for $Ag(NH_3)_2^+$ is 6.0×10^{-8}.

➡ **14.75** The solubility product of $AgIO_3$ is 4.5×10^{-8}. When excess solid $AgIO_3$ is treated with 1 liter of $1\ F\ NH_3$, 85 g of $AgIO_3$ dissolve.

$$AgIO_3 + 2\ NH_3 \rightleftharpoons Ag(NH_3)_2^+ + IO_3^-$$

Calculate the equilibrium constant for the reaction:

$$Ag(NH_3)_2^+ \rightleftharpoons Ag^+ + 2\ NH_3$$

➡ **14.76**

(a) Exactly one millimole of silver chloride is shaken with a liter of water. What is the concentration of silver ion in the saturated solution?

(b) Solid potassium bromide is added to the solution. How many moles of potassium bromide are added to the solution at the point at which the first bit of silver bromide forms?

(c) How many moles of potassium bromide are added at the point at which the last bit of silver chloride disappears?

(d) At the point at which the solid is converted completely to silver bromide, addition of potassium bromide is terminated and ammonia is added until all the silver bromide dissolves. What is the concentration of ammonia in this solution?
K_{sp} for AgCl $= 1.8 \times 10^{-10}$, for AgBr $= 5.2 \times 10^{-13}$. Instability constant for $Ag(NH_3)_2^+ = 6.0 \times 10^{-8}$.

➡ **14.77** The cation M^{++} forms the complex ion MCl_4^{--} whose instability constant is 1.0×10^{-21}. The solubility product of MI_2 is 1.0×10^{-15}. Calculate the number of moles of Cl^- ions that must be present in a liter of aqueous solution in order that 0.010 mole of MI_2 shall dissolve when excess solid MI_2 is added to the liter of solution.

➡ **14.78** K_{sp} for $Zn(OH)_2$ is 1×10^{-16}. When excess $Zn(OH)_2$ is treated with 1 liter of $0.4\ F$ KCN the reaction

$$Zn(OH)_2\ (s) + 4\ CN^- \rightleftharpoons Zn(CN)_4^{--} + 2\ OH^-$$

occurs. When equilibrium is reached, the pH is 13. Calculate the equilibrium constant for the reaction,

$$Zn(CN)_4^{--} \rightleftharpoons Zn^{++} + 4\ CN^-$$

➡ **14.79**   The solubility product of $Cd(OH)_2$ is 2.00×10^{-14}. The instability constant of $Cd(CN)_4^{--}$ is 1.40×10^{-19}. The ionization constant of HCN is 4.00×10^{-10}. Excess solid $Cd(OH)_2$ is added to a liter of KCN solution. When equilibrium is established, the pH of the final solution is 12.4 and the concentration of $Cd(CN)_4^{--}$ in the final solution is $0.0120\ M$.
Calculate the concentration of Cd^{++}, of CN^-, and of HCN in the final solution. Calculate the formal concentration of KCN in the original KCN solution.

➡ **14.80**

(a) Silver cyanide, AgCN, is soluble in water to the extent of 1.34×10^{-4} mg per 100 ml of water. Assuming that when AgCN dissolves the species present are Ag^+ and CN^-, what is the solubility product for AgCN?

(b) Actually a very stable complex is formed; the constant for the reaction, $Ag(CN)_2^- \rightleftharpoons Ag^+ + 2\ CN^-$, is $K = 9.0 \times 10^{-22}$. Considering this fact also, what is the true solubility product for AgCN?

Solution: (a) The solubility product, calculated as in Problem 14.7, is 1.0×10^{-16}.

(b) Assuming the correctness of (a), the equilibrium system would be:

$$AgCN \rightleftharpoons Ag^+ + CN^-;$$

$$[Ag^+]\ \text{and}\ [CN^-]\ \text{would each be}\ 1.0 \times 10^{-8}\ M$$

But the following reaction occurs:

$$Ag^+ + 2\ CN^- \rightleftharpoons Ag(CN)_2^-$$

Since $K_{inst} = 9.0 \times 10^{-22}$ this reaction is nearly complete to the right. At equilibrium, $[CN^-] = X$, $[Ag^+] = 5.0 \times 10^{-9} + 0.50\ X$ and $[Ag(CN)_2^-] = 5.0 \times 10^{-9} - 0.50\ X$.
If we substitute these values for $[CN^-]$, $[Ag^+]$, and $[Ag(CN)_2^-]$ in the formula for the instability constant and solve for X, we will find the correct values of $[Ag^+]$ and $[CN^-]$ and, hence, the correct value of the solubility product.

➡ **14.81**   Calculate the solubility of AgI in $0.1\ F\ Hg(NO_3)_2$. The main reaction is $AgI\ (s) + Hg^{++} \rightleftharpoons HgI^+ + Ag^+$. For the equation $HgI^+ \rightleftharpoons Hg^{++} + I^-$, $K = 10^{-13}$. K_{sp} for AgI is 1×10^{-16}.

➡ **14.82**   The solubility product of AgCN is 2.6×10^{-19}. The instability constant of the dicyanoargentate(I) ion is $9.0 \times 10^{-22}\ M^2$. The ionization constant of HCN is $4.0 \times 10^{-10}\ M$. Calculate the molar concentrations of all ionic and molecular species in $0.100\ F$ HCN to which has been added sufficient solid AgCN to form a saturated solution.

Solution: The net equation for the principal reaction is

$$AgCN\ (s) + HCN \rightleftharpoons Ag(CN)_2{}^- + H^+$$

The equilibrium constant for this reaction is

$$K = \frac{[Ag(CN)_2{}^-] \times [H^+]}{[HCN]}$$

In calculating the numerical value of this equilibrium constant we will, as in previous problems, resolve it into constants whose values we know. In accomplishing this we will start with the most complex equilibrium, that which involves $Ag(CN)_2{}^-$. We will multiply both numerator and denominator by $[Ag^+] \times [CN^-]^2$. Having done this we find that we have also introduced the terms for K_{sp} for AgCN and K for HCN.

$$K = \frac{[Ag(CN)_2{}^-] \times [H^+] \times [Ag^+] \times [CN^-] \times [CN^-]}{[HCN] \times [Ag^+] \times [CN^-]^2}$$

$$= \frac{K_{AgCN} \times K_{HCN}}{K_{inst}} = \frac{2.6 \times 10^{-19} \times 4.0 \times 10^{-10}}{9.0 \times 10^{-22}} = 1.2 \times 10^{-7}$$

With this information we can then solve for $[H^+]$, $[Ag(CN)_2{}^-]$, and $[HCN]$. Then, using the appropriate equilibrium constants, we can calculate the concentrations of the other species in the solution.

➡ **14.83**   To a liter of 0.1 F HCN is added 9.9 g of CuCl. Assuming that there is no change in the volume of the solution when the CuCl is added, calculate the molar concentration of each species in the solution. K_{inst} for $Cu(CN)_2{}^- = 5 \times 10^{-28}\ M^2$; K_{sp} for CuCl $= 3.2 \times 10^{-7}$.

➡ **14.84**   Calculate the pH of a solution prepared by adding excess solid $Cu(OH)_2$ to 1.0 F NH_4Cl. K_{sp} for $Cu(OH)_2 = 1.6 \times 10^{-19}$. K_{inst} for $Cu(NH_3)_4{}^{++} = 4.7 \times 10^{-15}$.

➡ **14.85**   ZnS will precipitate from 0.010 F $Zn(NO_3)_2$ solution on saturation with H_2S only if the pH is greater than 1.00. ZnS will not be precipitated from a solution 0.010 F in $Zn(NO_3)_2$ and 1.00 M in CN^- unless the pH is greater than 9.00. Calculate K for the reaction $Zn(CN)_4{}^{--} \rightleftharpoons Zn^{++} + 4\ CN^-$. A solution saturated with H_2S is 0.10 M in H_2S. The overall ionization constant for H_2S is 1.3×10^{-20}.

➡ **14.86**   At 50°C the concentration of undissociated H_2S in equilibrium with H_2S gas at a partial pressure of 1.00 atm is 0.075 M. In 1.00 liter of solution buffered at pH 4.00, and which is 0.0045 F in $Ni(NO_3)_2$ and 0.500 F in NaCl, the partial pressure of H_2S gas required to just begin precipitation of NiS is 0.0333 atm. Assuming the K's of H_2S and the K_{sp} for NiS given on pages 275–277 are applicable at 50°C, and that $NiCl^+$ is the only chloro complex of Ni^{++} formed, calculate the equilibrium constant for the reaction $NiCl^+ \rightleftharpoons Ni^{++} + Cl^-$.

➡ **14.87** The following solubility equilibria apply to $Zn(OH)_2$:

$$Zn(OH)_2 \text{ (s)} \rightleftharpoons Zn^{++} + 2\,OH^- \qquad K_{sp} = 5.0 \times 10^{-17}\ M^3$$

$$Zn(OH)_2 \text{ (s)} + 2\,OH^- \rightleftharpoons Zn(OH)_4{}^{--} \qquad K = 0.25\ M^{-1}$$

Over what pH range can $Zn(OH)_2$ be quantitatively precipitated in the sense that the total concentration of ions containing zinc in equilibrium with solid $Zn(OH)_2$ is less than $10^{-4}\ M$?

Oxidation-reduction processes.

Balancing redox equations

Many redox equations are so easily balanced that they can be done quickly by *simple inspection*. Thus, for the unbalanced equation, $Al + H^+ = Al^{+++} + H_2$, it is quite apparent that the balanced equation will be $2\,Al + 6\,H^+ = 2\,Al^{+++} + 3\,H_2$. In this equation, as in all balanced equations, *the number of atoms of each element on the left of the equality sign must equal the number of atoms of each element, respectively, on the right and the net sum of the charges on all ions on the left must equal the net sum of the charges on all ions on the right.* Whenever possible equations should be balanced by simple inspection; there is no point in calling upon some relatively cumbersome balancing routine when simple inspection will suffice.

For many redox equations the balancing is too involved to be accomplished, easily, by simple inspection. Such equations can be balanced by the *method of half-reactions* (called also the *ion-electron method*) and by the *method of change in oxidation number* (also called *loss and gain of electrons*).

The method of half-reactions

Every redox reaction is, in reality, the sum of two half-reactions. Thus, the overall reaction represented by the equation, $Zn + Cu^{++} = Zn^{++} + Cu$, is the sum of the two half-reactions, $Zn = Zn^{++} + 2\,e^-$ and $Cu^{++} + 2\,e^- =$

Cu. (It should be noted that each of the above half-reactions, like all half-reactions, is balanced with respect to both number of atoms and net charge; the charge balance is in each instance achieved by the use of electrons. The presence of electrons in a balanced equation means that it represents a half-reaction.)

In the first of the above half-reactions Zn is *oxidized* to Zn^{++}; this *oxidation* is attended by the *loss of 2 electrons* per atom of Zn; the fact that *electrons are lost* is evidence that *oxidation* has occurred; *oxidation may be defined as a reaction in which electrons are lost.* In the second half-reaction Cu^{++} is *reduced* to Cu; this *reduction* is attended by the *gain of 2 electrons* per atom of Cu; the fact that *electrons are gained* is evidence that *reduction* has occurred; *reduction may be defined as a reaction in which electrons are gained.*

When the two half-reactions which constitute the redox equation are added the final balanced equation is obtained.

$$Zn = Zn^{++} + 2\,e^-$$
$$\underline{Cu^{++} + 2\,e^- = Cu}$$
$$Zn + Cu^{++} = Zn^{++} + Cu$$

Note that in this addition process the electrons, being on opposite sides of the equality sign, cancel and hence do not appear in the final balanced equation. The reason they cancel is that the total number of electrons lost in the first half-reaction is equal to the total number of electrons gained in the second half-reaction.

Balancing by the *method of half-reactions* is based upon the concepts and facts that have been cited above, namely, that:

1. Every redox equation is the sum of two balanced equations for two half-reactions.

2. The two balanced half-reaction equations can be added to give the final balanced redox equation provided the total number of electrons lost in the half-reaction in which oxidation occurs is equal to the total number of electrons gained in the half-reaction in which reduction occurs. This is equivalent to stating that, in any balanced redox equation, the total number of electrons lost by the element, or elements, oxidized must equal the total number of electrons gained by the element, or elements, reduced.

In brief, balancing by this method consists of three steps:

First, the unbalanced skeleton equation is broken down into two unbalanced skeleton half-reactions.

Second, each half-reaction is balanced.

Third, the two balanced half-reactions are added together to give the final balanced redox equation; to do this, enough multiples of each balanced half-reaction must, if necessary, be taken so that the total number of electrons lost in one half-reaction equals the total number of electrons gained in the other. As a result, all electrons cancel and do not appear in the balanced equation. (If electrons appear in the final equation it is not balanced.) The following examples will illustrate the use of the *method of half-reactions*.

Example 1. Balance the equation, $H_2SO_3 + MnO_4^- = SO_4^{--} + Mn^{++}$ (In acid solution.).

Step 1. Write the two skeleton half-reactions:

$H_2SO_3 = SO_4^{--}$ (The reducing agent, H_2SO_3, is oxidized to SO_4^{--}.)

$MnO_4^- = Mn^{++}$ (The oxidizing agent, MnO_4^-, is reduced to Mn^{++}.)

Step 2. Balance each half-reaction.

(a) To balance the O, add H_2O.

$$H_2SO_3 + H_2O = SO_4^{--}$$
$$MnO_4^- = Mn^{++} + 4 H_2O$$

(b) To balance the H, add H^+.

$$H_2SO_3 + H_2O = SO_4^{--} + 4 H^+$$
$$MnO_4^- + 8 H^+ = Mn^{++} + 4 H_2O$$

(c) To balance the charge, add electrons.

$$H_2SO_3 + H_2O = SO_4^{--} + 4 H^+ + 2 e^-$$
$$MnO_4^- + 8 H^+ + 5 e^- = Mn^{++} + 4 H_2O$$

Step 3. Add the two half-reactions. Since 2 electrons are lost in the first while 5 are gained in the second, and since the lowest common multiple of 2 and 5 is 10, we will multiply the first by 5 and the second by 2. When the resulting equations are added, the 10 e^- on the two sides of the equations will cancel. Also 16 H^+ and 5 H_2O will cancel from each side.

$$5 H_2SO_3 + 5 H_2O = 5 SO_4^{--} + 20 H^+ + 10 e^-$$
$$2 MnO_4^- + 16 H^+ + 10 e^- = 2 Mn^{++} + 8 H_2O$$

$$\overline{2 MnO_4^- + 5 H_2SO_3 = 2 Mn^{++} + 5 SO_4^{--} + 4 H^+ + 3 H_2O}$$

Example 2. $P_4S_3 + NO_3^- = H_3PO_4 + SO_4^{--} + NO$ (In acid solution.)

Step 1. White the two skeleton half-reactions.

$$NO_3^- = NO$$
$$P_4S_3 = 4 H_3PO_4 + 3 SO_4^{--}$$ (Note that the skeleton half-reaction

contains the minimum number of moles of product per mole of reactant.)

Step 2. *Balance each half-reaction.*

(a) To balance the O, add H_2O.

$$NO_3^- = NO + 2 H_2O$$
$$P_4S_3 + 28 H_2O = 4 H_3PO_4 + 3 SO_4$$

(b) To balance the H, add H^+.

$$NO_3^- + 4 H^+ = NO + 2 H_2O$$
$$P_4S_3 + 28 H_2O = 4 H_3PO_4 + 3 SO_4^{--} + 44 H^+$$

(c) To balance the charge, add e^-.

$$NO_3^- + 4 H^+ + 3 e^- = NO + 2 H_2O$$
$$P_4S_3 + 28 H_2O = 4 H_3PO_4 + 3 SO_4^{--} + 44 H^+ + 38 e^-$$

Step 3. *Add the two half-reactions.* To balance the electrons multiply the first equation by 38 and the second by 3.

$$38 NO_3^- + 152 H^+ + 114 e^- = 38 NO + 76 H_2O$$
$$3 P_4S_3 + 84 H_2O = 12 H_3PO_4 + 9 SO_4^{--} + 132 H^+ + 114 e^-$$

$$3 P_4S_3 + 38 NO_3^- + 20 H^+ + 8 H_2O = 12 H_3PO_4 + 9 SO_4^{--} + 38 NO$$

Example 3. $Cu_2S + SO_4^{--} = Cu^{++} + H_2SO_3$ (In acid solution.)

Step 1. *Write the two skeleton half-reactions.*

$$Cu_2S = 2 Cu^{++} + H_2SO_3$$
$$SO_4^{--} = H_2SO_3$$

(Note that H_2SO_3 is a product in each half-reaction.)

Step 2. *Balance each half-reaction.*

(a) To balance O, add H_2O.

$$Cu_2S + 3 H_2O = 2 Cu^{++} + H_2SO_3$$
$$SO_4^{--} = H_2SO_3 + H_2O$$

(b) To balance H, add H^+

$$Cu_2S + 3 H_2O = 2 Cu^{++} + H_2SO_3 + 4 H^+$$
$$SO_4^{--} + 4 H^+ = H_2SO_3 + H_2O$$

(c) To balance the charge, add e^-.

$$Cu_2S + 3 H_2O = 2 Cu^{++} = H_2SO_3 + 4 H^+ + 8 e^-$$
$$SO_4^{--} + 4 H^+ + 2 e^- = H_2SO_3 + H_2O$$

Step 3. *Add the two half-reactions.* To balance the electrons multiply the second equation by 4.

$$Cu_2S + 3\ H_2O = 2\ Cu^{++} + H_2SO_3 + 4\ H^+ + 8\ e^-$$

$$4\ SO_4^{--} + 16\ H^+ + 8\ e^- = 4\ H_2SO_3 + 4\ H_2O$$

$$\overline{Cu_2S + 4\ SO_4^{--} + 12\ H^+ = 2\ Cu^{++} + 5\ H_2SO_3 + H_2O}$$

Example 4. $HS_2O_4^- + CrO_4^{--} = SO_4^{--} + Cr(OH)_4^-$ (In basic solution.)

Step 1. $$HS_2O_4^- = 2\ SO_4^{--}$$
$$CrO_4^{--} = Cr(OH)_4^-$$

Step 2.

(a) $$HS_2O_4^- + 4\ H_2O = 2\ SO_4^{--}$$
$$CrO_4^{--} = Cr(OH)_4^-\ (O\ is\ balanced)$$

(b) To balance the H add H^+ as in acid solution. However, since the solution is basic, these H^+ ions will react with OH^- ions to form H_2O. Accordingly, the procedure is first to add H^+ to balance H, then have this H^+ react with OH^- to form H_2O, then add the resulting 2 equations.

$$HS_2O_4^- + 4\ H_2O = 2\ SO_4^{--} + 9\ H^+$$
$$9\ H^+ + 9\ OH^- = 9\ H_2O$$
$$\overline{HS_2O_4^- + 9\ OH^- = 2\ SO_4^{--} + 5\ H_2O}$$

$$CrO_4^{--} + 4\ H^+ = Cr(OH)_4^-$$
$$4\ H_2O = 4\ H^+ + 4\ OH^-$$
$$\overline{CrO_4^{--} + 4\ H_2O = Cr(OH)_4^- + 4\ OH^-}$$

(c) $HS_2O_4^- + 9\ OH^- = 2\ SO_4^{--} + 5\ H_2O + 6\ e^-$
    $CrO_4^{--} + 4\ H_2O + 3\ e^- = Cr(OH)_4^- + 4\ OH^-$

Step 3.

$$HS_2O_4^- + 9\ OH^- = 2\ SO_4^{--} + 5\ H_2O + 6\ e^-$$
$$2\ CrO_4^{--} + 8\ H_2O + 6\ e^- = 2\ Cr(OH)_4^- + 8\ OH^-$$
$$\overline{HS_2O_4^- + 2\ CrO_4^{--} + 3\ H_2O + OH^- = 2\ SO_4^{--} + 2\ Cr(OH)_4^-}$$

Example 5. $P_4 = H_2PO_2^- + PH_3$ (In basic solution.)

Step 1. $P_4 = 4\ H_2PO_2^-$ (P_4 is oxidized.)
        $P_4 = 4\ PH_3$ (P_4 is reduced.)
                    (Note that P_4 disproportionates.)

Step 2. (a)
$$8\,H_2O + P_4 = 4\,H_2PO_2^-$$
$$P_4 = 4\,PH_3$$

(b)
$$8\,H_2O + P_4 = 4\,H_2PO_2^- + 8\,H^+$$
$$8\,H^+ + 8\,OH^- = 8\,H_2O$$
$$\overline{P_4 + 8\,OH^- = 4\,H_2PO_2^-}$$

$$P_4 + 12\,H^+ = 4\,PH_3$$
$$12\,H_2O = 12\,H^+ + 12\,OH^-$$
$$\overline{P_4 + 12\,H_2O = 4\,PH_3 + 12\,OH^-}$$

(c) $P_4 + 8\,OH^- \qquad = 4\,H_2PO_2^- + 4\,e^-$
$$P_4 + 12\,H_2O + 12\,e^- = 4\,PH_3 + 12\,OH^-$$

Step 3.
$$3\,P_4 + 24\,OH^- = 12\,H_2PO_2^- + 12\,e^-$$
$$P_4 + 12\,H_2O + 12\,e^- = 4\,PH_3 + 12\,OH^-$$
$$\overline{4\,P_4 + 12\,OH^- + 12\,H_2O = 12\,H_2PO_2^- + 4\,PH_3}$$

or

$$P_4 + 3\,OH^- + 3\,H_2O = 3\,H_2PO_2^- + PH_3$$

Example 6. $HPO_3^- + H_2O_2 = PO_4^{---}$ (In basic solution.)

Step 1. $HPO_3^- = PO_4^{---}$

$H_2O_2 = (H_2O, H^+, \text{ and } OH^- \text{ are generally not included in a skeleton equation.})$

Step 2. (a)
$$H_2O + HPO_3^- = PO_4^{---}$$
$$H_2O_2 = 2\,H_2O$$

(b)
$$H_2O + HPO_3^- = PO_4^{---} + 3\,H^+$$
$$3\,H^+ + 3\,OH^- = 3\,H_2O$$
$$\overline{HPO_3^- + 3\,OH^- = PO_4^{---} + 2\,H_2O}$$

$$H_2O_2 + 2\,H^+ = 2\,H_2O$$
$$2\,H_2O = 2\,H^+ + 2\,OH^-$$
$$\overline{H_2O_2 = 2\,OH^-}$$

(c) $HPO_3^- + 3\,OH^- = PO_4^{---} + 2\,H_2O + 1\,e^-$
$$H_2O_2 + 2\,e^- = 2\,OH^-$$

Step 3.
$$2\,HPO_3^- + 6\,OH^- = 2\,PO_4^{---} + 4\,H_2O + 2\,e^-$$
$$H_2O_2 + 2\,e^- = 2\,OH^-$$
$$\overline{2\,HPO_3^- + H_2O_2 + 4\,OH^- = 2\,PO_4^{---} + 4\,H_2O}$$

The method of change in oxidation number

In any redox reaction the oxidation number of at least one element is increased, and the oxidation number of at least one element is decreased. Thus in the reaction

$$Sn^{++} + 2\,Fe^{+++} = Sn^{++++} + 2\,Fe^{++}$$

tin is *oxidized* from an oxidation number of 2 in Sn^{++} to 4 in Sn^{++++}. At the same time, iron is *reduced* from an oxidation number of 3 in Fe^{+++} to 2 in Fe^{++}. Here Fe^{+++} is said to be the *oxidizing agent*, while Sn^{++} is said to be the *reducing agent*.

In the reaction

$$Cr_2O_7^{--} + 3\,H_2S + 8\,H^+ = 2\,Cr^{+++} + 3\,S + 7\,H_2O$$

chromium is *reduced* from an oxidation number of 6 in $Cr_2O_7^{--}$ to 3 in Cr^{+++}, while sulfur is *oxidized* from an oxidation number of -2 in H_2S to 0 in S. $Cr_2O_7^{--}$ is the *oxidizing agent* or *oxidant*, H_2S is the *reducing agent* or *reductant*.

In determining the oxidation number of a specific element in a given molecule or ion the following rules are applied.

1. In an elemental ion, such as Sn^{++++} or Sn^{--}, the oxidation number is equal to the charge on the ion. Thus, in Sn^{++++} and S^{--} the oxidation numbers of Sn and S are, respectively, $+4$ and -2.

2. The oxidation number of any free element, such as O_2 and S, is zero.

3. In its compounds or ions H has an oxidation number of $+1$. (The exceptions to this rule are the metal hydrides such as CaH_2, where the oxidation number of H is -1.)

4. In its compounds or ions O has an oxidation number of -2. (The exceptions to this rule are OF_2, in which the oxidation number of O is $+2$, and the peroxides, such as H_2O_2 and Na_2O_2, in which its oxidation number is -1.)

5. In any neutral molecule the total positive oxidation numbers equal the total negative oxidation numbers. In an ion the charge on the ion equals the difference between the positive and negative oxidation numbers. In the compound, H_3AsO_4, the total oxidation number for the four atoms of oxygen is -8; since the three hydrogen atoms have a total oxidation number of $+3$, the oxidation number of arsenic is $+5$. In the ion, $Cr_2O_7^{--}$, the total oxidation number for the seven atoms of oxygen is -14; since the charge on the ion is -2, the total positive oxidation number is $+12$, or $+6$ for each of the two atoms of chromium.

If we examine the two equations given above, and any other equation that we wish to select, we will find that, *in every balanced redox equation the total increase in oxidation number of the element (or elements) oxidized equals the total decrease in oxidation number of the element (or elements) reduced.* In the first of the two equations given above one atom of tin has its oxidation number increased from $+2$ to $+4$; the total increase in oxidation number is 2. Two atoms of iron each have their oxidation number reduced from $+3$ to $+2$; the total decrease in oxidation number is 2. In the second balanced equation three atoms of S are each oxidized from an oxidation number of -2 in H_2S to 0 in free S; the total increase in oxidation number is 6. Two atoms of Cr are each reduced from an oxidation number of $+6$ in $Cr_2O_7^-$ to $+3$ in Cr^{+++}; the total decrease in oxidation number is 6.

We will notice also, if we examine the two equations above and any other balanced equation, that *the sum of the charges of the ions on each of the two sides of the balanced equation is the same.* In the first equation the sum of the charges on the one Sn^{++} ion and two Fe^{+++} ions on the left is $+8$; the total net charge on the one Sn^{++++} ion and the two Fe^{++} ions on the right is also $+8$. In the second equation the sum of the charges on the one $Cr_2O_7^{--}$ ion and the eight H^+ ions is $+6$. The total charge on the two Cr^{+++} ions on the right is also $+6$.

The above facts (the equality in the total increase and decrease in oxidation number and the equality of the net charge on the two sides of the equation) are the bases for balancing redox equations by the *method of change in oxidation numbers.* The following examples will illustrate the use of the method.

Example 7. $Fe^{++} + MnO_4^- = Fe^{+++} + Mn^{++}$ (In acid solution.)

Step 1.   Identify the element or elements oxidized and the element or elements reduced. Note the initial and final oxidation number of each of these elements. Note the change in oxidation number of each of these elements.

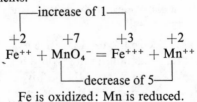

Fe is oxidized: Mn is reduced.

Step 2.   Select a sufficient number of moles of each reactant so that the total increase in oxidation number equals the total decrease.

$$\overbrace{5\ Fe^{++} + MnO_4^- = 5\ Fe^{+++}}^{\text{total increase of 5}} + Mn^{++}$$
$$\underbrace{\qquad\qquad\qquad\qquad}_{\text{total decrease of 5}}$$

Step 3. Balance the charges on each side of the equation by adding the necessary H^+ ions. (If the solution is alkaline, the charges can be balanced by adding OH^- ions. If the solution is neutral, either H^+ or OH^- ions may be added; H_2O will provide these ions.)

As the equation is written in Step 2, the net charge on the left (from the five Fe^{++} ions and the one MnO_4^- ion) is $+9$ and the net charge on the right (from the five Fe^{+++} ions and the one Mn^{++} ion) is $+17$. By adding eight H^+ ions to the left the charge on each side will be $+17$.

$$5 \, Fe^{++} + MnO_4^- + 8 \, H^+ = 5 \, Fe^{+++} + Mn^{++}$$

Step 4. Balance the hydrogen by adding H_2O. If the work has been correct up to this point, balancing the H will also balance the oxygen and, thus, balance the equation.

$$5 \, Fe^{++} + MnO_4^- + 8 \, H^+ = 5 \, Fe^{+++} + Mn^{++} + 4 \, H_2O$$

Example 8. $FeS + NO_3^- = NO + SO_4^{--} + Fe^{+++}$ (In acid solution.)

If more than one element is oxidized (and/or reduced), the total increase (decrease) in oxidation number is the sum of the increases (decreases) for each element. The oxidation of FeS by HNO_3 illustrates such a reaction.

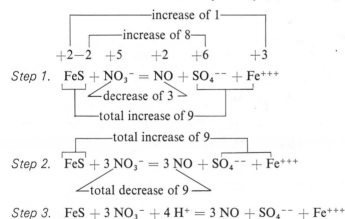

Step 3. $FeS + 3 \, NO_3^- + 4 \, H^+ = 3 \, NO + SO_4^{--} + Fe^{+++}$

Step 4. $FeS + 3 \, NO_3^- + 4 \, H^+ = 3 \, NO + SO_4^{--} + Fe^{+++} + 2 \, H_2O$

Example 9. $As_2S_3 + NO_3^- = NO + SO_4^{--} + H_3AsO_4$ (In acid solution.)

If more than one gram-atom of the element (or elements) oxidized and/or reduced is present in a mole of reactant, the minimum number of moles of product formed per mole of reactant must be given in Step 1.

Thus, when As_2S_3 is oxidized by HNO_3 to yield H_3AsO_4 and SO_4^{--}, one mole of As_2S_3 will yield two moles of H_3AsO_4 and three moles of SO_4^{--}. The successive steps in the balancing process will then be:

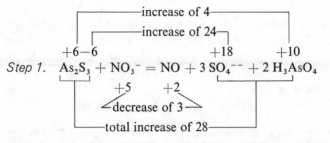

Step 1.

$$As_2S_3 + NO_3^- = NO + 3 SO_4^{--} + 2 H_3AsO_4$$

Step 2. $\quad 3 As_2S_3 + 28 NO_3^- = 28 NO + 9 SO_4^{--} + 6 H_3AsO_4$

Step 3. $\quad 3 As_2S_3 + 28 NO_3^- + 10 H^+ = 28 NO + 9 SO_4^{--} + 6 H_3AsO_4$

Step 4. $\quad 3 As_2S_3 + 28 NO_3^- + 10 H^+ + 4 H_2O = 28 NO + 9 SO_4^{--} + 6 H_3AsO_4$.

Example 10. $\quad Cu_2S + SO_4^{--} = SO_2 + Cu^{++}$ (In acid solution.)

When a given product (SO_2 in the above equation) is derived from two (or more) separate sources, it should appear twice (or more) in the equation.

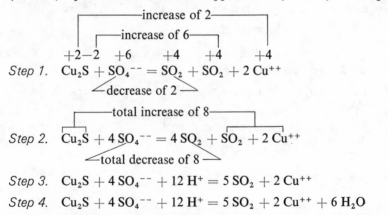

Step 1. $\quad Cu_2S + SO_4^{--} = SO_2 + SO_2 + 2 Cu^{++}$

Step 2. $\quad Cu_2S + 4 SO_4^{--} = 4 SO_2 + SO_2 + 2 Cu^{++}$

Step 3. $\quad Cu_2S + 4 SO_4^{--} + 12 H^+ = 5 SO_2 + 2 Cu^{++}$

Step 4. $\quad Cu_2S + 4 SO_4^{--} + 12 H^+ = 5 SO_2 + 2 Cu^{++} + 6 H_2O$

Example 11. $\quad H_2O_2 + Cr(OH)_4^- = CrO_4^{--}$ (In basic solution.)

When H_2O_2 is an oxidant it must be remembered that, in H_2O_2, the oxidation number of O is -1 whereas in all other oxygen compounds (except OF_2) its value is -2.

Step 1.

$$H_2O_2 + Cr(OH)_4^- = CrO_4^{--}$$

$\qquad\qquad\qquad\qquad\qquad\quad$ (-4 per 2 atoms of O)

Step 2. $3 H_2O_2 + 2 Cr(OH)_4^- = 2 CrO_4^{--}$

┌─increase of 6─┐

└─────decrease of 6──────────┘ $(-12$ for 6 atoms)

Step 3. $3 H_2O_2 + 2 Cr(OH)_4^- + 2 OH^- = 2 CrO_4^{--}$ (In basic solution balance charge by adding OH^-.)

Step 4. $3 H_2O_2 + 2 Cr(OH)_3^- + 2 OH^- = 2 CrO_4^{--} + 8 H_2O.$

For most equations balancing by oxidation number change is generally less time-consuming than by half-reactions. However, in those instances in which the changes in oxidation numbers are not obvious it may be wiser to use the method of half-reactions.

It should be emphasized that *oxidation number* is a *concept* which has been created by scientists for the purpose of expressing, quantitatively, the relative combining capacities of the constituent elements in a molecule or ion. For most binary species and for most ternary species in which oxygen is a constituent the calculation of the oxidation number of each constituent element poses no problem when we, by definition, assign H a value of $+1$ and O a value of -2, and when we recognize that the sum of the oxidation numbers must always equal the net charge on the species. Thus, in MnO_2 the oxidation number of Mn is obviously $+4$, in MnO_4^- it is $+7$, and in H_2MnO_4 it is $+6$.

To determine the oxidation numbers of the two elements in As_2S_3 we can look upon As_2S_3 as having been derived from H_2S, in which the oxidation number of S is -2. Accordingly, we can assume that, in As_2S_3, the oxidation number of S is -2; the oxidation number of As will then be $+3$.

For species such as $CrSCN^{++}$ in the unbalanced equation

$CrSCN^{++} + BrO^-$
$= Br^- + NO_3^- + CO_3^{--} + SO_4^{--} + CrO_4^{--}$ (In basic solution.)

the determination may be less obvious. As a matter of fact it really makes no great difference what oxidation numbers are assigned to Cr, S, C, and N, respectively, *as long as their sum is equal to $+2$, the charge on the ion.* We can, if we choose, arbitrarily assign the values, Cr = $+3$, S = -2, and N = -3. The value for C must then be $+4$. As noted below the total increase in oxidation number per mole of $CrSCN^{++}$ is $+19$.

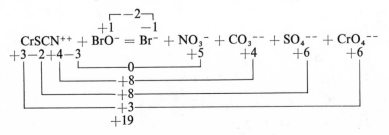

If we assign the values, $Cr = +3$, $N = +5$, $S = -2$, and $C = -4$ the total increase in oxidation number per mole of $CrSCN^{++}$ is again $+19$, as noted below.

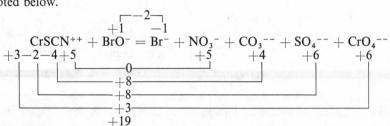

If we assign the values, $Cr = +6$, $N = -4$, $C = -4$, and $S = +4$ the total increase in oxidation number per mole of $CrSCN^{++}$ is again $+19$, as noted below.

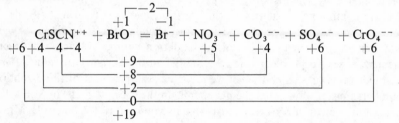

No matter what values we assign to Cr, S, C, and N, as long as the sum equals $+2$, the total change in oxidation number per mole of $CrSCN^{++}$ will always be $+19$ when the indicated products are formed, and the balanced equation, calculated by the 4-step process already outlined, will always turn out to be

$$2\ CrSCN^{++} + 19\ BrO^- + 18\ OH^-$$
$$= 19\ Br^- + 2\ NO_3^- + 2\ CO_3^{--} + 2\ SO_4^{--} + 2\ CrO_4^{--} + 9\ H_2O$$

PROBLEMS

Balance each of the following equations. (H^+, OH^-, and H_2O are not included in the unbalanced equation; addition of these species, where necessary, is a part of the balancing process.)

Group A. *In acidic solution.*

15.1 $Sn^{++} + Ce^{++++} = Ce^{+++} + Sn^{++++}$

15.2 $H_2S + Fe^{+++} = Fe^{++} + S$

15.3 $H_2SO_3 + HNO_2 = NO + SO_4^{--}$

15.4 $Br^- + MnO_4^- = Mn^{++} + Br_2$

15.5 $Sn^{++} + H_2O_2 = Sn^{++++}$

15.6 $I^- + Fe^{+++} = Fe^{++} + I_2$

15.7 $Mn^{++} + HBiO_3 = Bi^{+++} + MnO_4^-$

15.8 $Mn^{++} + MnO_4^- = MnO_2$

15.9 $Sb_2S_3 + NO_3^- = NO_2 + SO_4^{--} + Sb_2O_5$

15.10 $SnS_2O_3 + MnO_4^- = Mn^{++} + SO_4^{--} + Sn^{++++}$

15.11 $FeHPO_3 + Cr_2O_7^{--} = Cr^{+++} + H_3PO_4 + Fe^{+++}$

15.12 $Hg_4Fe(CN)_6 + ClO_3^- = Cl^- + NO + CO_2 + Fe^{+++} + Hg^{++}$

15.13 $Fe_2Fe(CN)_6 + NO_3^- = NO + CO_2 + Fe^{+++}$

15.14 $FeAsS + NO_3^- = NO + SO_4^{--} + H_3AsO_4 + Fe^{+++}$

15.15 $CrSCN^{++} + Cl_2 = Cl^- + NO_3^- + CO_2 + SO_4^{--} + Cr_2O_7^{--}$

15.16 $Sn(S_2O_3)_2^{--} + FeS_2O_8^+ = SO_4^{--} + Sn^{++++} + Fe^{++}$

Group B. *In basic solution.*

15.17 $S^{--} + ClO_3^- = Cl^- + S$

15.18 $CN^- + IO_3^- = I^- + CNO^-$

15.19 $HPO_3^- + OBr^- = Br^- + PO_4^{---}$

15.20 $Fe(OH)_2 + O_2 = Fe(OH)_3$

15.21 $Ni(OH)_2 + OBr^- = Br^- + NiO_2$

15.22 $Co(OH)_2 + H_2O_2 = Co(OH)_3$

15.23 $Bi(OH)_3 + Sn(OH)_4^{--} = Sn(OH)_6^{--} + Bi$

15.24 $SO_3^{--} + Co(OH)_3 = Co(OH)_2 + SO_4^{--}$

15.25 $Sn(OH)_4^{--} + MnO_4^- = MnO_2 + Sn(OH)_6^{--}$

15.26 $HS_2O_4^- + AsO_4^{---} = AsO_2^- + SO_4^{--}$

15.27 $PH_3 + CrO_4^{--} = Cr(OH)_4^- + P$

15.28 $H_2PO_2^- + CNO^- = CN^- + HPO_3^-$

15.29 $OCl^- = Cl^- + ClO_3^-$

15.30 $FeHPO_3 + OCl^- = Cl^- + PO_4^{---} + Fe(OH)_3$

15.31 $Cu_2SnS_2 + S_2O_8^{--} = SO_4^{--} + Sn(OH)_6^{--} + Cu(OH)_2$

15.32 $V = H_2 + HV_6O_{17}^{---}$

Equivalents in redox processes

In the discussion on neutralization we defined an equivalent of acid or base in such a way that one equivalent of acid would react with one equivalent of base. This approach is also useful in handling the stoichiometry of redox reactions in general and electrolysis in particular. In a redox reaction or an electrolysis there is a transfer of a number of electrons from one chemical species to another. The weight of one equivalent of a species involved in such a process is defined as the weight of one mole divided by the number of electrons being transferred per mole of the substance. For example in the process: $Zn_{(s)} + 2 Ag^+ \rightleftharpoons Zn^{++} + 2 Ag_{(s)}$ there is a transfer of two electrons per mole of $Zn_{(s)}$; the equivalent weight of $Zn_{(s)}$ is $65.4/2 = 32.7$ g and 1 mole of $Zn_{(s)}$ represents two equivalents. However there is a transfer of only one electron per mole of Ag^+ and therefore the weight of one equivalent is the same as the weight of one mole: 107.9 g.

Often the number of electrons transferred is not obvious from the inspection of the overall reaction. For example: $2 MnO_4^- + 5 H_2SO_3 \rightarrow 2 Mn^{++} + 5 SO_4^{--} + 4 H^+ + 3 H_2O$. Here the number of electrons transferred can be determined by dividing the overall process into half-reactions:

$$MnO_4^- + 8 H^+ + 5e \rightarrow 4 H_2O + Mn^{++}$$

$$H_2SO_3 + H_2O \rightarrow SO_4^{--} + 4 H^+ + 2e$$

Inspection of the half-reactions reveals that 5 electrons are transferred per mole of MnO_4^-, so that one mole of MnO_4^- is 5 equivalents. Similarly two electrons are transferred per mole of H_2SO_3 so that one mole of H_2SO_3 is 2 equivalents. Notice that in the overall reaction 2 moles or 10 equivalents of MnO_4^- react with 5 moles of 10 equivalents of H_2SO_3.

The relationship between moles and equivalents can also be determined by examination of the oxidation numbers of the atoms in the species involved in the redox process. In any redox reaction the equivalent weight of a substance is its formula weight divided by the change in oxidation number of its component atoms. Thus, when MnO_4^- reacts to form Mn^{++} and H_2O, the oxidation number of Mn decreases from $+7$ in MnO_4^- to $+2$ in Mn^{++}, for a net change of 5. Since there is no change in the oxidation number of O, the equivalent weight of MnO_4^- is one fifth of its formula weight.

PROBLEMS

15.33 How many equivalents of $KMnO_4$ will be required to react with 30 g of $FeSO_4$ in the following reaction?

$$5 Fe^{++} + MnO_4^- + 8 H^+ = 5 Fe^{+++} + Mn^{++} + 4 H_2O$$

Solution: equivalents of $KMnO_4$ = equivalents of $FeSO_4$.

$$\text{equivalent weight of } FeSO_4 = \frac{\text{formula weight of } FeSO_4}{\text{oxidation number change of Fe}}$$

$$= \frac{151.9}{1} = \frac{151.9 \text{ g of } FeSO_4}{\text{equivalent of } FeSO_4}$$

$$\text{equivalents of FeSO}_4 = \frac{30 \text{ g of FeSO}_4}{151.9 \text{ g/equivalent of FeSO}_4}$$

$$= 0.20 \text{ equivalent of FeSO}_4$$

Therefore, 0.20 equivalent of $KMnO_4$ is required.

15.34 What would be the concentration, in grams per liter, of 0.100 N $KMnO_4$ when used in the following reaction?

$$2 \text{ MnO}_4^- + 10 \text{ Cl}^- + 16 \text{ H}^+ = 2 \text{ Mn}^{++} + 5 \text{ Cl}_2 + 8 \text{ H}_2\text{O}$$

Solution: 0.100 N $KMnO_4$ contains 0.100 equivalent (equivalent weights) of $KMnO_4$ per liter.

The oxidation number of Mn changes from $+7$ in MnO_4^- to $+2$ in Mn^{++}. This represents an oxidation number change of 5.

$$\text{redox equivalent weight} = \frac{\text{formula weight}}{\text{oxidation number change}}$$

$$= \frac{158}{5} = 3.16$$

0.100 equivalent = 3.16 g   concentration = 3.16 g/liter

15.35 How many grams of $KClO_3$ will be required for the preparation of 400 ml of 0.20 N $KClO_3$ for use in the reaction,

$$\text{ClO}_3^- + 3 \text{ H}_2\text{SO}_3 = \text{Cl}^- + 3 \text{ SO}_4^{--} + 6 \text{ H}^+ ?$$

15.36 How many milliliters of 0.50 N H_2SO_3 will be required to reduce 120 ml of 0.40 N $K_2Cr_2O_7$?

$$\text{Cr}_2\text{O}_7^{--} + 3 \text{ H}_2\text{SO}_3 + 2 \text{ H}^+ = 2 \text{ Cr}^{+++} + 3\text{SO}_4^{--} + 4 \text{ H}_2\text{O}$$

Solution: ml × normality = ml × normality.

15.37 How many grams of $FeSO_4$ will be oxidized by 24 ml of 0.25 N $KMnO_4$ in the following reaction?

$$\text{MnO}_4^- + 5 \text{ Fe}^{++} + 8 \text{ H}^+ = \text{Mn}^{++} + 5 \text{ Fe}^{+++} + 4 \text{ H}_2\text{O}$$

15.38 $Cr_2O_7^{--}$ will oxidize NO_2^- to NO_3^- in acid solution, the $Cr_2O_7^{--}$ being reduced to Cr^{+++}.

In one student's experiment 20 ml of 0.100 F $K_2Cr_2O_7$ solution reacted with 1.020 g of a mixture of KNO_2 and KNO_3. For this experiment calculate:

(a) The normality of the $K_2Cr_2O_7$.

(b) The number of equivalents of $K_2Cr_2O_7$ used.

(c) The number of equivalents of KNO_2 present in the mixture.

(d) The gram-equivalent weight of KNO_2.

(e) Grams of KNO_2 in the mixture.

(f) Percent of KNO_2 in the mixture.

**Faraday's law and electrochemical
equivalence**

*One faraday of electricity, when passed through a solution of an electrolyte,
will cause one gram-equivalent weight of substance to react, be deposited, or be
liberated at each electrode.* This important generalization is a part of a broader
generalization known as *Faraday's law.*

It should be noted that, since 1 gram-equivalent weight of a substance
such as Ag^+ contains 6.023×10^{23} ions, 6.023×10^{23} electrons will be
required to electrodeposit 1 gram-equivalent weight of silver according to
the reaction, $Ag^+ + e^- = Ag$. Since 1 faraday of electricity will deposit
1 gram-equivalent weight of silver, 1 faraday must represent 6.023×10^{23}
electrons. This is the basis for stating that 1 faraday is a *mole of electrons.*

One faraday of electricity is 96,500 coulombs. (The more exact value,
to 5 significant figures, is 96,489.) One coulomb is the charge that is carried
when one ampere of current flows for one second. Therefore, 1 faraday =
96,500 ampere-seconds = 26.8 ampere-hours. It will be assumed in the problems that follow that the efficiency of the process (the current efficiency) is
100% unless stated otherwise.

PROBLEMS

15.39 Electricity was allowed to flow until 20 g of copper were
deposited from a solution of $CuSO_4$. How many coulombs of electricity
passed through the solution?

Solution: Since the change in oxidation number when Cu^{++} is
reduced to Cu is 2 the equivalent weight of copper is $\frac{1}{2}$ its atomic
weight, namely 31.8.

$$\frac{20.0 \text{ g of Cu}}{31.8 \text{ g of Cu/equivalent}} \times \frac{96,500 \text{ coulombs}}{1 \text{ equivalent}} = 60,700 \text{ coulombs}$$

15.40 A current of 2.00 amp was allowed to flow through a solution
of $AgNO_3$ for 6.00 hr. How many grams of silver were deposited?

Solution: 6.00 hr = 21,600 sec. Two amp of current flowing for
21,600 sec is 43,200 coulombs of electricity. The equivalent weight
of Ag in $AgNO_3$ is 107.9.

$$\frac{43,200 \text{ coulombs}}{96,500 \text{ coulombs/equivalent}} \times \frac{107.9 \text{ g of Ag}}{1 \text{ equivalent}} = 48.4 \text{ g of Ag}$$

15.41 A certain amount of electricity deposited 50.0 g of silver from
a solution of $AgNO_3$. How many grams of copper will this same amount of
current deposit from a solution of $CuSO_4$?

Solution: Since 1 faraday of electricity will deposit 1 equivalent of any element, it follows that the weight of one element deposited by a given amount of electricity will be to the weight of another element deposited by the same amount of electricity as the equivalent weight of the first element is to the equivalent weight of the second element.

$$\frac{\text{weight of copper deposited}}{\text{weight of silver deposited}} = \frac{\text{equivalent weight of copper in CuSO}_4}{\text{equivalent weight of silver in AgNO}_3}$$

The equivalent weight of Ag in $AgNO_3$ is 107.9. The equivalent weight of Cu in $CuSO_4$ is 31.8. Therefore,

$$\frac{\text{grams of copper}}{50.0 \text{ g of Ag}} = \frac{31.8}{107.9}$$

$$\text{grams of copper} = \frac{31.8}{107.9} \times 50.0 \text{ g} = 14.7 \text{ g}$$

15.42 How many grams of cobalt will be deposited from a solution of $CoCl_2$ by 40,000 coulombs of electricity?

15.43 How many coulombs of electricity will be required to deposit 100 g of chromium from a solution of $CrCl_3$?

15.44 How many grams of zinc would be deposited from a solution of $ZnCl_2$ by a current of 3.00 amp flowing for 20.0 hr?

15.45 A quantity of electricity which deposited 70 g of nickel from a solution of $NiCl_2$ will deposit how many grams of hydrogen from a solution of HCl?

15.46 A solution of $CuSO_4$ contains Cu^{++}. A solution of $Na_2Cu(CN)_3$ contains Cu^+ ions in equilibrium with $Cu(CN)_3^{--}$ ions and CN^- ions. A quantity of electricity which deposits 12 g of copper from a solution of copper sulfate will deposit how many grams of copper from a solution of $Na_2Cu(CN)_3$?

15.47 How many grams of nickel will be deposited from a solution of $NiCl_2$ by 4.00 amp of current flowing for 24.0 hr if the current efficiency of the process is 96.0%?

15.48 Calculate the charge, in coulombs, on an electron.

15.49 A current of 2.0 amp was passed through a solution of H_2SO_4 for 20 min. How many milliliters of O_2 gas, measured at STP, were liberated?
Solution hint: 1 mole of O_2 gas is how many equivalents?

15.50 A solution of $CuSO_4$ was electrolyzed, using platinum electrodes. A volume of 6.0 liters of O_2 gas, measured at STP, was liberated at the positive electrode. How many grams of copper were deposited at the negative electrode?

Solution: Number of equivalents of O_2 liberated at the positive electrode equals number of equivalents of Cu deposited at the negative electrode.

I mole of O_2 = 4 equivalents

$$6 \text{ liters of } O_2 = 6/22.4 \text{ mole} = \frac{6 \times 4}{22.4} \text{ equiv}$$

$$\frac{6 \times 4}{22.4} \text{ equiv of Cu} \times 31.8 \text{ g of Cu/equiv} = 34 \text{ g of Cu deposited}$$

➤ **15.51** To electrodeposit all of the Cu and Cd from a solution of $CuSO_4$ and $CdSO_4$ in water required 1.20 faradays of electricity. The mixture of Cu and Cd that was deposited weighed 50.36 g. How many grams of $CuSO_4$ were there in the solution?

➤ **15.52** A certain solution of K_2SO_4 was electrolyzed using platinum electrodes. The combined volume of the dry gases that were evolved was 67.2 cc at STP. Assuming 100 percent current efficiency and no loss of gases during measurement, how many coulombs of electricity were consumed?

➤ **15.53** A current of 2.0 amp was passed through a cell containing 800 ml of 1.0 F H_2SO_4. The following electrode reactions took place:

$$\text{Anode: } 2 H_2O = O_2 \text{ (g)} + 4 H^+ + 4 e^-$$
$$\text{Cathode: } 2 H^+ + 2 e = H_2 \text{ (g)}$$

(a) What time (in sec) was required to liberate 0.050 mole of O_2?

(b) What volume of H_2 gas, measured over water at 25°C and a barometric pressure of 740 mm, was produced at the same time?

(c) How many electrons were involved in the anode reaction in this experiment?

(d) What volume of 2 F NaOH would be required to neutralize all the acid remaining in the cell at the end of the electrolysis?

➤ **15.54** When 1 g of yttrium metal (at. wt 88.92) is treated with excess H_2SO_4, 378 cc of H_2 gas, measured at STP, are liberated. When the resulting solution of yttrium sulfate is electrolyzed with platinum electrodes, using a steady current of 2 amp, 1 g of pure yttrium is deposited on the negative electrode and O_2 gas is liberated at the positive electrode. Calculate:

(a) the formula for yttrium sulfate,

(b) the number of minutes the electrolysis had to proceed to deposit the 1 g of Y, and

(c) the volume, measured at STP, of the oxygen gas liberated.

➤ **15.55** The atomic weight of metal M is 52.01. When the melted chloride of M is electrolyzed, for every gram of metal deposited on the cathode 725 cc of dry chlorine gas, measured at 25°C and 740 mm, are liberated at the anode. Calculate the formula of the chloride of M.

Solution hint: Equivalents of Cl_2 = equivalents of M. Atomic weight of M/equivalent weight of M = valence of M. This is equivalent to stating that atomic weight of M divided by the equivalent weight of M equals the change in oxidation number of M when it is electrodeposited.

➡ **15.56** A metal M is known to form the fluoride MF_2. When 3300 coulombs of electricity are passed through the molten fluoride 1.950 g of M are plated out. What is the atomic weight of M?

➡ **15.57** When a solution of KI is electrolyzed using porous silver electrodes, H_2 gas is liberated at the negative electrode (cathode) and insoluble AgI is deposited in the pores of the positive electrode (anode). All of the AgI that is formed remains in the pores of the anode. At the conclusion of the experiment the anode had increased in weight 5.076 g and 530 cc of dry H_2 gas, measured at 27°C and 720 mm, had been liberated. Calculate the atomic weight of iodine.

➡ **15.58** An aqueous solution of the soluble salt MSO_4 is electrolyzed between inert (platinum) electrodes until 0.0327 g of metal, M, are deposited on the negative electrode (cathode). To neutralize the solution that was formed in the electrolytic cell required 50 ml of 0.020 F KOH. Calculate the atomic weight of metal M.

➡ **15.59** A potassium salt of a ternary acid of molybdenum (Mo, at wt 95.95) has the formula K_2MoO_x. When an acidified solution of K_2MoO_x is electrolyzed between platinum electrodes, only oxygen gas is liberated at the positive electrode and only molybdenum metal is deposited at the negative electrode. When electrolysis is continued until 0.3454 g of molybdenum are deposited 121.0 cc of O_2 gas, measured at STP, are liberated. Calculate the formula of the salt.

Solution: If we know the oxidation number of Mo in MoO_x^{--} we can then calculate the value of x. Since MoO_x^{--} is reduced to Mo metal, the *change* in the oxidation number of Mo in the course of the reaction is equal to the oxidation number of Mo in MoO_x^{--}. The equivalent weight of Mo in this reaction is its atomic weight divided by the change in oxidation number when it is formed from MoO_x^{--}. Therefore, if we know the equivalent weight of Mo in the above reaction we can, following the argument outlined above, calculate the value of x.

Equivalents of Mo deposited = equivalents of O_2 liberated.

$$\text{Equivalents of } O_2 = 4 \times \text{moles of } O_2 = \frac{4 \times 121 \text{ cc}}{22,400 \text{ cc}} = \text{equiv of Mo.}$$

$$\text{Equiv wt of Mo} = 0.3454 \text{ g of Mo} \div \frac{4 \times 121}{22,400} \text{ equiv of Mo}$$
$$= 16.0 \text{ g/equiv.}$$

Change in oxidation number of Mo = At wt/equiv wt = 95.95/16.0 = 6. Therefore, oxidation number of Mo in MoO_x^{--} is 6, $x = 4$, and the formula of the compound is K_2MoO_4.

➡ **15.60** When 0.20 faraday of electricity is passed through a solution of $Pb(NO_3)_2$, a compound containing 20.7 g of Pb is deposited at the positive electrode. What is the oxidation number of lead in the compound that is deposited?

Oxidation potentials

Metals have a tendency to give off one or more electrons to form positive ions. Thus, when zinc is brought in contact with water, or an aqueous solution of some substance, the reaction

$$Zn = Zn^{++} + 2 e^-$$

tends to take place.

Likewise, the reactions $Mg = Mg^{++} + 2 e^-$, $Mn = Mn^{++} + 2 e^-$, $Cu = Cu^{++} + 2 e^-$, $Ag = Ag^+ + e^-$, and so on, tend to occur. The energy or *potential* with which the electrons are expelled, that is the energy with which the metal reacts, varies from one metal to another.

If we list the metals in the order of the *potential* with which the electrons are given off, and include in this list other substances that give off electrons when they react, we obtain the sort of arrangement found in Table 5, page 279. The substance with the strongest tendency to lose electrons is at the top, the one with the weakest tendency is at the bottom.

In arriving at this table the potential of the reaction, H_2 (gas) = $2 H^+ + 2 e^-$, is used as a standard of reference.

Potential is expressed in units of *volts*. The voltage of the reaction, $H_2 = 2 H^+ + 2 e^-$, is arbitrarily assigned a value of zero. That is,

$$H_2 = 2 H^+ + 2 e^- \qquad \text{potential} = 0.00 \text{ V}$$

Any substance which has a stronger tendency to give off electrons than does H_2 has a positive voltage and lies above H_2 in the table; any substance whose tendency to lose electrons is less than that of H_2 has a negative voltage and falls below hydrogen. The voltage listed in the table gives the relative numerical value of this tendency. Thus, for the reaction, $Zn = Zn^{++} + 2 e^-$, the potential is 0.763 V greater than that for H_2 while the potential of the reaction, $Cu = Cu^{++} + 2 e^-$, is 0.337 V less than that for H_2.

When the reaction, $Zn = Zn^{++} + 2 e^-$, occurs, *zinc is oxidized*. The evidence is that its oxidation number is increased from 0 to +2, and two electrons are lost. *Oxidation involves* (1) *an increase in oxidation number of the element oxidized and* (2) *a loss of electrons by the element oxidized.*

In the above reaction zinc functions as a reducing agent. When any substance loses electrons and, for that reason, is oxidized, it functions as a *reducing agent.* We note, in Tables 5 and 6, that all of the substances on the left in the table react by losing electrons. Therefore, all of the substances on the left are reducing agents. The higher the potential of a substance the greater its strength as a reducing agent.

The reaction, $Zn^{++} + 2\,e^- = Zn$, is the exact reverse of the reaction given in the previous paragraph. In this reaction zinc is reduced from Zn^{++} to Zn metal; Zn^{++} functions as an oxidizing agent.

If we examine the reactions in Tables 5 and 6 we will find that, in every instance when the reaction proceeds from right to left the substance on the right functions as an *oxidizing agent*; in the course of the reaction the substance on the right is reduced.

Just as the substance on the left with the *highest potential* is the *strongest reducing agent* so the substance on the right with the *lowest potential* is the *strongest oxidizing agent.* The magnitude of the potential is thus a measure of the relative oxidizing strength of an oxidizing agent as well as the relative reducing strength of a reducing agent.

The reaction, $Zn = Zn^{++} + 2\,e^-$, will not go on by itself. It must be paired up with another reaction, one that will accept the two electrons. The same thing is true of every reaction in this table. Each is only a *half-reaction* and must be coupled with another half-reaction.

When zinc metal is added to a solution of copper sulfate the reaction

$$Zn + Cu^{++} = Zn^{++} + Cu$$

occurs. This overall reaction is the sum of the two half-reactions.

$$Zn = Zn^{++} + 2\,e^- \qquad 0.763\ V$$
$$Cu = Cu^{++} + 2\,e^- \qquad -0.337\ V$$

The substance with the higher potential (Zn) will function as the reducing agent while the substance with the lower potential (Cu^{++}) will function as the oxidizing agent. In effect, the second reaction is reversed and the overall reaction is the sum of the two half-reactions. Likewise, the potential of the overall reaction is the sum of the potentials of the two half-reactions. Since the second reaction is reversed, the sign of its potential is changed from -0.337 to $+0.337$. The potential of the overall reaction is therefore $1.100\ V$.

$Zn = Zn^{++} + 2\,e^-$	$0.763\ V$
$Cu^{++} + 2\,e^- = Cu$	$0.337\ V$
$Zn + Cu^{++} = Zn^+ + Cu$	$1.100\ V$

That the potential of this reaction is, in fact, 1.100 V, can be demonstrated by the following experiment.

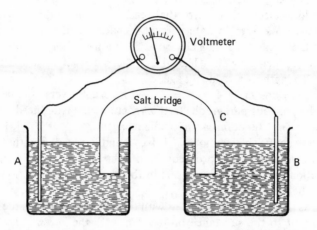

Beaker A in the figure contains a strip of Zn metal in a 1-molal solution of $ZnSO_4$. Beaker B contains a strip of Cu metal in 1-molal $CuSO_4$. The U-shaped tube C contains a gel prepared with a solution of K_2SO_4. It is called a *salt bridge* and serves as a conducting medium between the two solutions. The strips of Cu and Zn are connected by copper wires through a voltmeter.

When the circuit is closed, the voltmeter registers 1.100 V. This proves that *the potential of the cell is the sum of the potentials of the two half-reactions.*

A cell, made up as described above, is commonly designated by the notation $Zn \,|\, 1 \, M \, Zn^{++} \,|\, 1 \, M \, Cu^{++} \,|\, Cu$.

If a hydrogen gas electrode, prepared by keeping a piece of platinum immersed in $1 \, M \, H^+$ and continually enveloped in hydrogen gas, is substituted for the Cu-$CuSO_4$ half cell, the voltage is 0.763. This illustrates one method by which voltages in Tables 5 and 6 were determined.

It should be emphasized that the voltages given in Tables 5 and 6 are for systems in which the concentrations of all substances in solution are 1-molal. (Since molar and molal concentration of aqueous solution are very nearly the same, particularly for dilute solutions, it is common practice to consider that these solutions are $1 \, M$.) The temperature is 25°C, and gases are at a pressure of 1 atm. Under these conditions the potentials are referred to as *standard electrode potentials* and are designated by the symbol $E°$.

To determine the voltage for a pair of half-reactions such as

$$Fe^{++} \rightleftharpoons Fe^{+++} + e^- \qquad\qquad -0.771 \text{ V}$$

$$2 \, Cr^{+++} + 7 \, H_2O \rightleftharpoons Cr_2O_7^{--} + 14 \, H^+ + 6 \, e^- \qquad -1.33 \text{ V}$$

beaker A in the figure on page 224 contains a platinum electrode immersed in a solution 1 M in Fe^{++} and 1 M in Fe^{+++} while beaker B contains a platinum electrode immersed in a solution 1 M in $Cr_2O_7^{--}$, 1 M in Cr^{+++}, and 1 M in H^+.

If a battery is made up with a strip of Cu in a solution of $CuSO_4$ in one cell and a strip of Ag in Ag_2SO_4 in the other cell, its voltage is 0.462 V calculated as follows:

$$Cu = Cu^{++} + 2\,e^- \qquad\qquad -0.337\ V$$
$$2\,Ag^+ + 2\,e^- = 2\,Ag \qquad\qquad +0.799\ V$$
$$\overline{Cu + 2\,Ag^+ = Cu^{++} + 2\,Ag \qquad +0.462\ V}$$

This is referred to as a $Cu\,|\,Cu^{++}\,|\,Ag^+\,|\,Ag$ cell.

Note, in this example, that doubling the number of moles of Ag^+ in the reaction, $2\,Ag^+ + 2\,e^- = 2\,Ag$, does not alter the voltage. The reason is that the voltage is a measure of the *work per electron*.

Note, also, that the substance with the lowest (most negative) potential, Ag^+ in this case, functions as the oxidizing agent. The Ag^+ half-reaction has accordingly been reversed, resulting in its voltage being changed from -0.799 to $+0.799$. As a result, the sum of the voltages is positive ($+0.462$). This illustrates how one can determine whether or not a certain oxidizing agent will react with a certain reducing agent. The rule is this: If the sum of the two potentials, written with the proper signs (sign of the potential of the oxidizing agent is reversed), is *positive*, the two substances *will* react with each other, if it is negative, they will not react.

Will Cu reduce Sn^{++++} to Sn^{++}, being itself oxidized to Cu^{++}?

$$Cu = Cu^{++} + 2\,e^- \qquad\qquad -0.337\ V$$
$$Sn^{++++} + 2\,e^- = Sn^{++} \qquad\qquad +0.15\ V$$
$$\overline{Cu + Sn^{++++} = Cu^{++} + Sn^{++} \qquad -0.19\ V}$$

The sum is negative. Reaction will not occur.

Application of the above rule to Tables 5 and 6 reveals these simple and useful relationships:

Any reducing agent of higher potential will reduce any oxidizing agent of lower potential. Any oxidizing agent of lower potential will oxidize any reducing agent of higher potential.

Any reducing agent on the left *will* reduce (be oxidized by) any oxidizing agent *below* itself and on the right. It *will not* be oxidized by an oxidizing agent above itself and on the right.

Any oxidizing agent on the right will oxidize (be reduced by) any reducing agent *above* itself and on the left. It *will not* oxidize a reducing agent *below* itself on the left.

In the reaction, $Zn + Cu^{++} = Zn^{++} + Cu$, Cu^{++} functions as the oxidizing agent and the half-reaction, $Cu = Cu^{++} + 2e^-$, proceeds to the left. In the reaction, $Cu + 2Ag^+ \rightleftharpoons Cu^{++} + 2Ag$, Cu is the reducing agent and the half-reaction, $Cu = Cu^{++} + 2e^-$, proceeds to the right. This means that the reaction, $Cu = Cu^{++} + 2e^-$, is reversible, and, being reversible, it can reach a state of equilibrium.

What is true of the reaction, $Cu = Cu^{++} + 2e^-$, is true of all redox (oxidation-reduction) half-reactions. They are reversible and can reach a state of equilibrium. For reactions with a high positive potential, indicating that the substance is a strong reducing agent, the equilibrium is far to the right. For reactions with a low potential (a large negative potential), meaning that the substance is a strong oxidizing agent, the equilibrium is far to the left.

Since the reactions are all reversible and can reach a state of equilibrium, they are written with conventional double arrows, $Cu \rightleftharpoons Cu^{++} + 2e^-$.

PROBLEMS

15.61 Calculate the theoretical voltage of each of the following cells, assembled in the manner shown in the figure on page 224.

(a) $Zn \mid 1\ M\ Zn^{++} \mid 1\ M\ Ni^{++} \mid Ni$

(b) $Al \mid 1\ M\ Al^{+++} \mid 1\ M\ Cu^{++} \mid Cu$

(c) $Cu \mid 1\ M\ Cu^{++} \mid 1\ M\ Hg^{++} \mid Hg$

(d) $Pt \mid 1\ M\ Fe^{+++}, 1\ M\ Fe^{++} \mid 1\ M\ MnO_4^-, 1\ M\ Mn^{++} \mid Pt$

15.62 Calculate the theoretical voltage of each of the following cells with alkaline solutions.

(a) $Zn_{(s)} \mid ZnS_{(s)} \mid 1\ M\ S^{--} \mid S_{(s)} \mid Pt_{(s)}$

(b) $Pt_{(s)} \mid H_2\ (1\ atm) \mid 1\ M\ OH^-, 1\ M\ Zn\ (OH)_4^{--} \mid Zn_{(s)}$

(c) $Ni_{(s)} \mid Ni\ (OH)_{2\ (s)} \mid 1\ M\ OH^- \mid O_2\ (1\ atm) \mid Pt_{(s)}$

15.63 State whether or not a reaction will occur when the following are brought together in acidic 1 M solution at 25°C.

(a) MnO_4^- and I^-

(b) Cr^{++} and Cu^{++}

(c) Sn^{++} and H_3PO_4

(d) H_2S (at 1 atm) and Fe^{+++}

(e) Sn and Sn^{++++}

15.64 State whether or not a reaction will occur when the following are brought together in alkaline 1 M solution at 25°C.

(a) MnO_4^- and ClO^-

(b) Cl^- and BrO^-

(c) PH_3 and $Sn(OH_4)^{--}$

(d) I^- and Br^-

(e) H_2O_2 and MnO_4^-

Effect of change of concentration

Since each half-reaction reaches a state of equilibrium, it follows that the equilibrium, and, hence, the voltage, will change when the concentration of a reactant or product is changed. Thus for the reaction

$$Fe^{++} \rightleftarrows Fe^{+++} + e^- \quad (-0.771 \text{ V})$$

an increase in concentration of Fe^{++} will shift the equilibrium to the right, thereby increasing the positive potential (decreasing the negative potential). If the concentration of Fe^{++} is reduced, the equilibrium will shift to the left and the potential will decrease. If the concentration of Fe^{+++} is increased, the equilibrium will shift to the left and the potential will decrease while, if the concentration of Fe^{+++} is reduced, the equilibrium will shift to the right and the potential will increase.

The quantitative effect of change in concentration can be calculated by the use of the *Nernst equation*,

(1) $$E = E° - \frac{2.303 \, RT}{nF} \log \frac{[product]}{[reactant]}$$

In this formula $E°$ is the standard potential of the half-reaction, E is its potential under the particular condition of the experiment, 2.303 is the *constant* for conversion from natural logarithms to logarithms to the base 10, R is the gas *constant* (with a value of 1.98 cal/mole $\times$ deg), F is the *faraday* (with a value of 23,060 cal/volt), n is the number of electrons transferred in the half reaction, and T is the absolute temperature. Therefore, at constant temperature, 2.303 RT/F will be constant for all reactions; its value, at 25°C, is 0.059 volts/mole. Thus Equation (1) takes the simplified form

$$E = E° - \frac{0.059}{n} \log \frac{[product]}{[reactant]}$$

Since the product, Fe^{+++} in the above example, is in the oxidized state and the reactant, Fe^{++}, is in the reduced state, the formula is commonly written

$$E = E° - \frac{0.059}{n} \log \frac{[oxidized \, state]}{[reduced \, state]}$$

Suppose the concentration of Fe^{++} is 0.1 M and the concentration of Fe^{+++} is 1.0 M:

$$E = -0.771 - \frac{0.059}{1} \log \frac{1.0}{0.1} = -0.771 - 0.059 = -0.830 \text{ V}$$

If more than one reactant or more than one product is involved, each species is included in the concentration term and, as in the standard equilibrium formula, each is raised to a power equal to the number of moles that react. Solid or liquid species are not included in the formulation and the concentration of H_2O is constant at 55.6 moles per liter. Thus, for the half-reaction,

$$Cl^- + 3\,H_2O \rightleftharpoons ClO_3^- + 6\,H^+ + 6\,e^- \qquad (-1.45\ V)$$

$$E = -1.45 - \frac{0.059}{6} \log \frac{[ClO_3^-][H^+]^6}{[Cl^-]}$$

and for the reaction,

$$Zn\,(s) \rightleftharpoons Zn^{++} + 2\,e^- \qquad (0.763\ V)$$

$$E = 0.763 - \frac{0.059}{2} \log [Zn^{++}]$$

Note that, if all concentrations are 1 M, the term [oxid]/[red] = 1. Since log 1 = 0, the entire term,

$$\frac{-0.059}{n} \log \frac{[oxid]}{[red]}$$

equals zero, and $E = E°$.

If [oxid] is less than 1 or [red] is greater than 1, making [oxid]/[red] less than 1, the log will be negative. As a result, the term,

$$\frac{-0.059}{n} \log \frac{[oxid]}{[red]}$$

will be positive and E will be more positive than $E°$. This is exactly the conclusion that was reached in our qualitative inspection. If the ratio of [oxid] to [red] is greater than 1, the log will be positive, the term,

$$\frac{-0.059}{n} \log \frac{[oxid]}{[red]}$$

will be negative, and E will be less positive than $E°$.

PROBLEMS

15.65   Calculate the potential of the half-reaction, $Zn = Zn^{++} + 2\,e^-$, when the concentration of the Zn^{++} ion is 0.10 M.

15.66   Calculate the potential of the half-reaction, $Fe^{++} \rightleftharpoons Fe^{+++} + e^-$, when Fe^{++} and Fe^{+++} are, respectively, 0.40 M and 1.60 M.

15.67   Calculate the potential of the following cells.

(a) $Zn_{(s)} \,|\, 0.200\ M\ Zn^{++} \,||\, 0.100\ M\ Cu^{++} \,|\, Cu_{(s)}$

(b) $Zn_{(s)} \,|\, 0.200\ M\ Zn^{++} \,||\, 1.00 \times 10^{-4}\ M\ Cu^{++} \,|\, Cu_{(s)}$

(c) $Zn_{(s)} | 0.200\ M\ Zn^{++} || 1.00 \times 10^{-8}\ M\ Cu^{++} | Cu_{(s)}$

(d) $Zn_{(s)} | 1.00 \times 10^{-8} M\ Zn^{++} || 0.200\ M\ Cu^{++} | Cu_{(s)}$

15.68 Calculate the potential of the following cells:

(a) $Zn_{(s)} | ZnS_{(s)} | 0.0400\ M\ S^{--} || 2.00\ M\ CN^-,\ 1\ M\ Zn(CN)_4{}^{--} | Zn_{(s)}$

(b) $Zn_{(s)} | ZnS_{(s)} | 2.00\ M\ S^{--} || 0.04\ M\ CN^-,\ 1\ M\ Zn(CN)_4{}^{--} | Zn_{(s)}$

15.69 Calculate the potential of the cell $Pt_{(s)} | Cl_{2\,(g)}$ (1 atm) $| 1.00\ M$ $Cl^- || 0.0100\ M\ Mn^{++},\ 2.00\ M\ MnO_4{}^- | Pt_{(s)}$

(a) at pH $= 1.00$

(b) at pH $= 4.00$

➡ **15.70** Calculate the $[Ni^{++}]$ required for the potential of the cell $Ni_{(s)} | Ni^{++} || 0.001\ M\ Sn^{++} | Sn_{(s)}$ to be 0.0 V.

Calculation of equilibrium constants

The Nernst formula enables us to calculate the equilibrium constant for the overall oxidation-reduction reaction. Thus, suppose we set up a cell like the figure on page 224 with $Br_2 - Br^-$ in one beaker and $I_2 - I^-$ in the other, close the circuit, and allow the system to stand until the voltage drops to zero. The reaction, $Br_2 + 2\ I^- \rightleftharpoons I_2 + 2\ Br^-$, will then have reached a state of equilibrium. When this point is reached, the potential generated by the reaction, $2\ Br^- = Br_2 + 2\ e^-$, will exactly equal the potential generated by the reaction, $2\ I^- = I_2 + 2\ e^-$.

But

$$E^{Br} = -1.065 - \frac{0.059}{2} \log \frac{[Br_2]}{[Br^-]^2}$$

and

$$E^{I} = -0.536 - \frac{0.059}{2} \log \frac{[I_2]}{[I^-]^2}$$

Then

$$-1.065 - \frac{0.059}{2} \log \frac{[Br_2]}{[Br^-]^2} = -0.536 - \frac{0.059}{2} \log \frac{[I_2]}{[I^-]^2}$$

$$\frac{0.059}{2} \left(\log \frac{[I_2]}{[I^-]^2} - \log \frac{[Br_2]}{[Br^-]^2} \right) = 1.065 - 0.536 = 0.529$$

$$\frac{0.059}{2} \log \frac{\dfrac{[I_2]}{[I^-]^2}}{\dfrac{[Br_2]}{[Br^-]^2}} = 0.529$$

$$\log \frac{[I_2] \times [Br^-]^2}{[I^-]^2 \times [Br_2]} = \frac{0.529 \times 2}{0.059} = 18$$

But $\dfrac{[I_2] \times [Br^-]^2}{[I^-]^2 \times [Br_2]}$ is the equilibrium constant, K, for the reaction $Br_2 + 2\,I^-$
$\rightleftharpoons 2\,Br^- + I_2$.

$$\log K = 18$$
$$K = 10^{18}$$

This means that the above reaction is practically complete to the right.

If we examine the calculations given above we will find that, for a reaction at equilibrium,

$$\log K = \frac{n\,(\text{difference in the standard potentials of the two half-reactions})}{0.059}$$

and

$$K = 10^{n\Delta E^\circ / 0.059}$$

where n is the number of electrons gained or lost in each balanced half-reaction and ΔE° is the difference between the two standard potentials.

PROBLEM

15.71 Calculate the equilibrium constant for each of the following reactions.

(a) $Cl_2 + 2\,Br^- \rightleftharpoons 2\,Cl^- + Br_2$

(b) $2\,Fe^{+++} + 2\,I^- \rightleftharpoons 2\,Fe^{++} + I_2$

(c) $Cr_2O_7^{--} + 3\,Sn^{++} + 14\,H^+ \rightleftharpoons 2\,Cr^{+++} + 3\,Sn^{4+} + 7\,H_2O$

(d) $Zn + Cu^{++} \rightleftharpoons Zn^{++} + Cu$

**Calculation of the potential of a half-reaction
from the potentials of other half-reactions**

We have learned that when two half-reactions are combined to give a *complete chemical reaction*, the voltage of the couple is the sum of the two voltages, due regard being given to the sign of each voltage.

When the potential of a *half-reaction* is calculated from two other half-reactions we must keep in mind that (1) the expulsion of electrons requires energy, (2) that, since the potential of a reaction is a measure of the work per unit charge, the total energy in electron-volts will be the product of volts × electrons, and that (3) in any series of reactions the net change in energy is the algebraic sum of the energy changes in all steps.

Thus, suppose we wish to calculate the potential of the half-reaction, $BrO^- + 4\,OH^- = BrO_3^- + 2\,H_2O + 4\,e^-$, from the two half-reactions, (1) and (2).

Reaction		n	$E°$	Total Energy
(1) $6 OH^- + Br^-$	$= BrO_3^- + 3 H_2O + 6 e^-$		-0.61 V	$6 \times -0.61 = -3.66$ eV
(2) $2 OH^- + Br^-$	$= BrO^- + H_2O + 2 e^-$		-0.76 V	$2 \times -0.76 = -1.52$ eV
(3) $4 OH^- + BrO^-$	$= BrO_3^- + 2 H_2O + 4 e^-$		-0.535 V	-2.14 eV

If we subtract reaction (2) from reaction (1), taking proper note of signs, we obtain reaction (3). Note that, in reaction (3), the total energy is -2.14 electron volts. Since 4 electrons are given off per molecule of BrO^-, the potential, $E°$, in volts will be $-2.14 \div 4$ or -0.535.

When a balanced equation for a chemical reaction is obtained by combining two half-reactions, the number of moles of each reactant must be chosen so that the total number of electrons lost by the oxidized substance equals the total number of electrons gained by the reduced substance; as a result, the electrons cancel and no electrons appear in the final balanced equation. In contrast, when a *half-reaction is calculated from two half-reactions,* the electrons do not cancel. Instead, a chemical species common to the two parent half-reactions is canceled. (In the example given above Br^- ions are canceled.) Every half-reaction must include electrons; if electrons are not included in the balanced equation it is not a half-reaction; it is a balanced equation for a chemical reaction. If electrons appear in the balanced equation it is a half-reaction, not a chemical reaction; like all half-reactions, it will not occur unless it is combined with another appropriate half-reaction.

PROBLEMS

15.72 Given the half-reactions:

$$S + 3 H_2O = H_2SO_3 + 4 H^+ + 4 e^- \qquad E° = -0.45 \text{ V}$$
$$H_2SO_3 + H_2O = SO_4^{--} + 4 H^+ + 2 e^- \qquad E° = -0.17 \text{ V}$$

Calculate $E°$ for the half-reaction,

$$S + 4 H_2O = SO_4^{--} + 8 H^+ + 6 e^-$$

Solution: Add the two half-reactions, including the total energies, cancel the H_2SO_3, and divide the total energy in electron volts by the number of electrons in the final half-reaction.

15.73 Given the half-reactions:

$$Cl^- + 3 H_2O = ClO_3^- + 6 H^+ + 6 e^- \qquad E° = -1.45 \text{ V}$$
$$2 Cl^- = Cl_2 + 2 e^- \qquad E° = -1.36 \text{ V}$$

Calculate $E°$ for the half-reaction, $\frac{1}{2} Cl_2 + 3 H_2O = ClO_3^- + 6 H^+ + 5 e^-$.

Solution: Divide the second equation by 2 to give

$$Cl^- = \tfrac{1}{2} Cl_2 + 1\, e^- \qquad E° = -1.36\ V$$

Then subtract this equation from the first, including the total energies.

Finally, divide the total final energy in electron volts by the number of electrons in the final half-reaction.

15.74   Given the standard potentials ($E°$) of the following half-reactions:

Half-Reaction	$E°$
(1) $M = M^{++} + 2\, e^-$	+0.80 V
(2) $M = M^{+++} + 3\, e^-$	+0.70 V
(3) $M^{+++} + H_2O = MO^{++} + 2\, H^+ + e^-$	+0.30 V
(4) $M^{++} + 2\, H_2O = MO_2{}^+ + 4\, H^+ + 3\, e^-$	+0.20 V

Calculate the standard potential of the half-reaction in which M^{++} is oxidized to MO^{++}.

Solution: In this instance there are no two equations one of which involves M^{++}, the other MO^{++}, each of which contains the same species. So we must try to create two such equations. Inspection tells us that if we subtract (1) from (2) we will obtain the half-reaction in which M^{++} is converted to M^{+++}. We note that, in Equation (3) M^{+++} is converted to MO^{++}. Therefore, if we add our newly-created equation to (3) we will obtain the desired half-reaction.

15.75   Given the following standard potentials in basic solution:

$$BrO^- + 4\, OH^- = BrO_3{}^- + 2\, H_2O + 4\, e^- \qquad E° = -0.54\ V$$

$$Br^- + 2\, OH^- = BrO^- + H_2O + 2\, e^- \qquad E° = -0.76\ V$$

Will BrO^- disproportionate to give Br^- and $BrO_3{}^-$? If the answer is "yes," write the equation for the disproportionation reaction.

Solution: Disproportionation occurs when a substance undergoes a redox reaction with itself. Some of the substance acts as a reducing agent, some as an oxidizing agent. For disproportionation to occur the potential of the half-reaction in which the substance functions as a reducing agent must be higher than the potential of the half-reaction in which it functions as an oxidizing agent. Inspection of the above two equations shows that this requirement is satisfied; therefore, disproportionation will occur.

The balanced equation for the reaction is obtained in the conventional manner by multiplying the lower equation by 2 (to give 4 electrons), then subtracting it from the upper equation to give

$$3 \, BrO^- = 2 \, Br^- + BrO_3^-$$

15.76 Given the following half-reactions in acid solution:

Half-Reaction	$E°$
$H_2S = S + 2 \, H^+ + 2 \, e^-$	-0.14 V
$H_2SO_2 + H_2O = H_2SO_3 + 2 \, H^+ + 2 \, e^-$	-0.40 V
$2 \, S + 3 \, H_2O = S_2O_3^{--} + 6 \, H^+ + 4 \, e^-$	-0.50 V
$S + 4 \, H_2O = SO_4^{--} + 8 \, H^+ + 6 \, e^-$	-0.36 V
$S_2O_3^{--} + 3 \, H_2O = 2 \, H_2SO_3 + 2 \, H^+ + 4 \, e^-$	-0.4025 V

Will H_2SO_3 disproportionate to give S and SO_4^{--}? Give the data upon which your answer is based. If the answer is "yes," write the equation for the disproportionation reaction.

15.77 You are given the following half-reactions in alkaline solution:

Half-Reaction	$E°$
$ClO_2^- + 2 \, OH^- = ClO_3^- + H_2O + 2 \, e^-$	-0.33 V
$ClO^- + 2 \, OH^- = ClO_2^- + H_2O + 2 \, e^-$	-0.66 V
$Cl^- + 2 \, OH^- = ClO^- + H_2O + 2 \, e^-$	-0.89 V
$Cl^- = \frac{1}{2} \, Cl_2 + 1 \, e^-$	-1.36 V

(a) Calculate the potential, $E°$, of the half-reaction, $\frac{1}{2} \, Cl_2 + 6 \, OH^- = ClO_3^- + 3 \, H_2O + 5 \, e^-$.

(b) Will ClO^- disproportionate to give Cl^- and ClO_3^-? Show the calculations that lead to your conclusion. If the answer is "yes," write the equation for the disproportionation reaction.

15.78 In acid solution will H_3PO_2 disproportionate to give PH_3 and H_3PO_4? (See Table 5.)

15.79 Calculate the potential of the half-reaction,

$$PH_3 + 3 \, H_2O \rightleftharpoons H_2PO_3 + 7 \, H^+ + 7 \, e^-$$

15.80 What oxidation state of chlorine will be formed when 1.0 M ClO_4^- is reduced with excess solid $Mn(OH)_2$ in alkaline solution? (See Table 6.)

15.81 Calculate the potential of the half-reaction,

$$Cl^- + 8\,OH^- \rightleftharpoons ClO_4^- + 4\,H_2O + 8\,e^-$$

15.82 The metal M has five oxidation states: 0, 2, 3, 4, and 5, which are related by the redox potentials given below; the potentials for zinc, iron, and tin are also included. Assume that all possible reactions are very rapid.

	Half-Reaction	$E°$
A	$M = M^{++} + 2\,e^-$	+0.80
	$Zn = Zn^{++} + 2\,e^-$	+0.76
B	$M = M^{+++} + 3\,e^-$	+0.70
C	$M + H_2O = MO^{++} + 2\,H^+ + 4\,e^-$	+0.60
D	$M^{++} = M^{+++} + e^-$	+0.50
	$Fe = Fe^{++} + 2\,e^-$	+0.45
E	$M + 2\,H_2O = MO_2^+ + 4\,H^+ + 5\,e^-$	+0.44
F	$M^{++} + H_2O = MO^{++} + 2\,H^+ + 2\,e^-$	+0.40
G	$M^{+++} + H_2O = MO^{++} + 2\,H^+ + e^-$	+0.30
H	$M^{++} + 2\,H_2O = MO_2^+ + 4\,H^+ + 3\,e^-$	+0.20
	$Sn = Sn^{++} + 2\,e^-$	+0.14
I	$M^{+++} + 2\,H_2O = MO_2^+ + 4\,H^+ + 2\,e^-$	+0.05
J	$MO^{++} + H_2O = MO_2^+ + 2\,H^+ + e^-$	−0.20

(a) What is the potential of the reaction

$$3\,Sn^{++} + 2\,M^{++} + 4\,H_2O = 3\,Sn + 2\,MO_2^+ + 8\,H^+$$

(b) Will excess metallic Zn reduce M^{++} to M?

(c) Will excess metallic Zn reduce M^{+++} to M^{++}?

(d) Will excess metallic Zn reduce M^{+++} to M?

(e) What oxidation state of M is produced when 1 molal MO_2^+ is treated with
    (1) excess metallic zinc?
    (2) excess metallic tin?
    (3) excess metallic iron?

(f) Will oxidation state +3 disproportionate to +2 and +5?

(g) Show how $E°$ for half-reaction **H** can be calculated by appropriate combination of $E°$'s for
    (1) half-reactions **A** and **E**.
    (2) half-reactions **D**, **G**, and **J**.
    (3) half-reactions **J** and **F**.
    (4) half-reactions **A**, **C**, and **J**.

➡ **15.83** Calculate K for the dissociation of $Zn\,(NH_3)_4^{++}$ ion using the data of Tables 5 and 6.

Solution hint: Use reaction 6 of Table 5 and reaction 13 of Table 6.

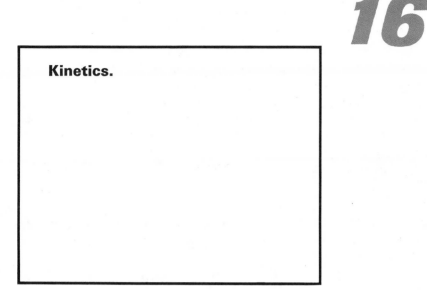

Kinetics.

The study of chemical thermodynamics and equilibrium deals with the direction of spontaneous change and the eventual most stable state of a system, but gives us no information on the time necessary for chemical change to occur. Kinetics deals with the rate at which a process occurs and the way in which the rate is affected by various factors. Two of the most important of these factors are concentration and temperature.

Since the rate of a reaction, in general, depends on the concentration of the constituents of the reaction and since these concentrations are changing as the reaction proceeds, the rate of the reaction is also changing as the reaction proceeds. In order to write an expression for the rate of the reaction we must take into account its dependence on time. One approach is to define the average rate over a period of time. For example suppose we start with a solution where $[A] = 1.00\ M$ and allow the reaction $A \longrightarrow B$ to proceed for 100 sec, at the end of which time we find that $[A] = 0.75\ M$ and $[B] = 0.25\ M$. The average rate for this time can be expressed as 0.25 mole of B per liter of solution is formed in 100 sec, that is $\Delta[B] = 0.25\ M$ and $\Delta t = 100$ sec. The average rate is

$$\frac{\Delta[B]}{\Delta t} = \frac{0.25}{100} = 2.5 \times 10^{-3}\ \frac{\text{mole}}{\text{liter-sec}}$$

We could also express the rate in terms of $\Delta[A]$ which is $0.75 - 1.00 =$

$-0.25\ M$. The average rate is

$$-\frac{\Delta[A]}{\Delta t} = -\frac{(-0.25)}{100} = 2.5 \times 10^{-3}\ \frac{\text{mole}}{\text{liter-sec}}$$

by using a minus sign for the decreasing concentration the value of the average rate is always positive. The stoichiometry of the reaction $A \longrightarrow B$ indicates that for every mole of B which is formed a mole of A has reacted and it therefore follows that $\Delta[B]/\Delta t = -\Delta[A]/\Delta t$. For reactions which are not simply mole-for-mole conversions the coefficients of the chemical equation must be taken into account in writing expressions for the rate and in relating them. For the process

$$H_{2(g)} + I_{2(g)} \longrightarrow 2\ HI_{(g)}$$

every time one mole of H_2 reacts one mole of I_2 also reacts and two moles of HI are produced. Therefore the average rate can be expressed by any one of the following

$$-\frac{\Delta[H_2]}{\Delta t} = -\frac{\Delta[I_2]}{\Delta t} = \frac{1}{2}\frac{\Delta[HI]}{\Delta t}$$

For the general reaction

$$aA + bB \longrightarrow cC + dD$$

where the lowercase letters represent coefficients, the expressions for the average rate are

$$-\frac{1}{a}\frac{\Delta[A]}{\Delta t} = -\frac{1}{b}\frac{\Delta[B]}{\Delta t} = \frac{1}{c}\frac{\Delta[C]}{\Delta t} = \frac{1}{d}\frac{\Delta[D]}{\Delta t}$$

As we have seen, expressions involving the use of Δ just give us the average rate over a time interval. Suppose however that we would like to know what the rate is after exactly 50 seconds for the above hypothetical reaction $A \longrightarrow B$. In general the value of the average rate we found, 2.5×10^{-3} mole/liter-sec, will be only a fair approximation to the rate at any specified time, which is called the instantaneous rate. It can also be seen that the approximation will get better as the time interval gets smaller. Thus if the average rate is measured for the time interval from 40 to 60 seconds, it will be closer to the instantaneous rate at 50 seconds. If the time interval is from 49 to 51 seconds, it will be better still and the average rate will be exactly the instantaneous rate if the time interval, Δt, around 50 seconds becomes infinitely small. This can be expressed as the instantaneous rate and is equal to

$$\lim_{\Delta t \to 0} \frac{\Delta[B]}{\Delta t} \quad \text{or} \quad -\lim_{\Delta t \to 0} \frac{\Delta[A]}{\Delta t}$$

or using the usual notation, the instantaneous rate is $d[B]/dt$ or $-d[A]/dt$. The relationship of instantaneous rates to stoichiometry is the same as that of average rates to stoichiometry so that for the general reaction written above

$$-\frac{1}{a}\frac{d[A]}{dt} = -\frac{1}{b}\frac{d[B]}{dt} = \frac{1}{c}\frac{d[C]}{dt} = \frac{1}{d}\frac{d[D]}{dt}$$

The experimental approach to the study of the rate of a reaction is to measure the concentration of one or more species at different times as the reaction proceeds. Sometimes instead of measuring concentrations directly some other quantity directly related to concentration is measured. There are a number of such quantities; pressure in the case of gases, intensity of light absorption in the case of colored materials, or radioactivity levels in the case of radioactive substances are examples. In order to calculate values for the average rate all that is necessary is to calculate from the data the changes in concentration or the measured related quantity during measured time intervals. If the time intervals are fairly small, the values of the average rates will be good approximations of the instantaneous rates at times within the intervals. However, in order to calculate instantaneous rates from experimental data it is necessary to know exactly how the rate changes with changes in concentration.

The relationship betweeen the rate of a reaction and the concentrations of the species in the system is called the differential rate law. In general it is *not* possible just to look at a chemical equation and know what the differential rate law will be. Either it must be experimentally determined or more must be known about the details of how the reactants become the products, that is, the mechanism of the reaction. For the reaction A $\longrightarrow$ B it might be found that the rate is proportional to [A]. The differential rate law is $-d[A]/dt$ or $d[B]/dt = k[A]$ where k is a proportionality constant called the specific rate constant. But it might also be found that the differential rate law is

$$\frac{-d[A]}{dt} = k[A]^2 \quad \text{or} \quad \frac{-d[A]}{dt} = k[A]^{1/2}$$

or a number of other possibilities. For the general reaction aA + bB $\longrightarrow$ cC + dD many different rate laws are possible. For example

$$-\frac{1}{a}\frac{d[A]}{dt} = k[A][B] \quad \text{or} \quad k[A]^2[B] \text{ or } k[A][B]^2$$

or in general $k[A]^m[B]^n$ where m and n are most commonly zero, integers, or half-integers. The *order* of a reaction is described in terms of the exponents of the concentrations in the rate laws. The overall order is $m + n$, the order with respect to A is m, and with respect to B is n.

The experimental determination of the rate law typically involves measurement of the change in rate with change in concentration. For example if a reaction is first order in [A], doubling the concentration will double the rate; if it is second order in [A], doubling the [A] should quadruple the rate, and so on. It is also possible to determine the rate law if the mechanism of a reaction is known. In particular if a process is known to be a single-step or elementary process the rate law can be written from the stoichiometry by using the coefficients of the chemical equation as exponents in the rate. If A $\rightarrow$ B in a single step, the reaction is first order in [A]. If $2A \rightarrow C$ in a single step, the reaction is second order in [A]. If A + B $\rightarrow$ D in a single step, the reaction is first order in [A], first order in [B], and second order overall. Single-step processes are never higher than third order overall. It should be emphasized that this approach works only if the process is known to be an elementary one and that many processes occur by a series of steps.

Since the differential rate laws relate the instantaneous rate to concentration the instantaneous rate at any concentration can be calculated provided the value of k, the specific rate constant, is known. The value of k can be calculated from rate data in several ways. A differential rate law can be expressed in terms of average rates by defining an average concentration for the time interval of the average rate. For the example

$$A \rightarrow B \text{ given above the } [A]_{ave} \text{ is } \frac{1.00 + 0.75}{2} = 0.88 \ M$$

calculated in the usual way that averages are calculated. If this is a first order reaction then $-\Delta[A]/\Delta t = k[A]_{ave}$; if it is a second order reaction then $-\Delta A/\Delta t = k[A]_{ave}^2$. Expressions such as these can be used to calculate an approximate value of k.

A better value of k can be calculated if we know exactly how concentrations depend on time. This dependence can be obtained by performing the mathematical operation called integration on the differential rate law. In many cases this is very complicated but in two cases the integrated rate laws are of simple form. The first order rate law $-d[A]/dt = k[A]$ has the integrated form

(1) $$2.3 \log \frac{[A_0]}{[A]} = kt$$

where $[A_0]$ is the initial concentration of A when the time, t, is 0, [A] is the concentration at any time measured after the initial time, and k is the specific rate constant. The second order rate law $-d[A]/dt = k[A]^2$ has the integrated form

(2) $$\frac{1}{[A]} - \frac{1}{[A_0]} = kt$$

If the reaction under investigation has one of these two rate laws, values of k can be obtained by substitution of the data into the appropriate integrated rate law.

It is a well known phenomenon that qualitatively the rates of most processes are accelerated when the temperature is raised. Quantitatively the effect of temperature on rate depends on the value of a quantity called the activation energy, E_a. The activation energy is the difference in energy between the reactants and the highest energy state between the reactants and products, which is called the transition state. The relationship between a specific rate constant k_1 at a temperature T_1 and another specific rate constant k_2 at another temperature T_2 is

$$(3) \qquad 2.3 \log \frac{k_2}{k_1} = -\frac{E_a}{R}\left(\frac{1}{T_2} - \frac{1}{T_1}\right)$$

$$(4) \qquad \qquad\qquad = -\frac{E_a}{R}\left(\frac{T_1 - T_2}{T_1 T_2}\right)$$

PROBLEMS

16.1 A large sample of $CaCO_{3(s)}$ is placed in an evacuated container and heated to 540°C, to bring about the reaction

$$CaCO_{3(s)} \rightarrow CaO_{(s)} + CO_{2(g)}$$

and the pressure of CO_2 is monitored. The pressure is found to be 100 mm after 5 minutes, and 200 mm after 10 minutes. Calculate the average rate of appearance of CO_2 during each 5-minute interval and during the 10-minute interval.

Solution: The average rate of appearance of CO_2 is $\Delta P_{CO_2}/\Delta t$. Initially $P_{CO_2} = 0$, after 5 minutes $P_{CO_2} = 100$ mm. So

$$\frac{\Delta P_{CO_2}}{\Delta t} = \frac{100 - 0}{5 - 0} = 20 \frac{mm}{min}$$

During the second 5-minute period P_{CO_2} increases from 100 mm to 200 mm and the time increases from 5 minutes to 10 minutes

$$\frac{\Delta P_{CO_2}}{\Delta t} = \frac{200 - 100}{10 - 5} = 20 \frac{mm}{min}$$

For the 10-minute period

$$\frac{\Delta P_{CO_2}}{\Delta t} = \frac{200 - 0}{10 - 0} = 20 \frac{mm}{min}$$

16.2 It is found that $N_2O_{3(g)}$ decomposes according to $N_2O_{3(g)} \rightarrow NO_{2(g)} + NO_{(g)}$. A bulb is filled with N_2O_3 to a pressure of 100 mm and the decomposition is monitored by measuring the total pressure. After 75 seconds the pressure is measured to be 150 mm and after 150 seconds the pressure is measured to be 175 mm. Calculate the average rate of disappearance of $N_2O_{3(g)}$ during these time intervals.

Solution: The rate of disappearance of $N_2O_{3(g)}$ is given by $-\dfrac{\Delta P_{N_2O_3}}{\Delta t}$.

Thus we must calculate $P_{N_2O_3}$ after 75 seconds and after 150 seconds from the data given on the total pressure. It should be noted that the pressure increases as the reaction proceeds because each mole of $N_2O_{3(g)}$ which decomposes produces 2 moles of gaseous products. If we let $X = $ mm of $N_2O_{3(g)}$ which decomposes after 75 seconds, then at this time $P_{N_2O_3} = 100 - X$ $P_{NO} = P_{NO_2} = X$. Since $P = P_{N_2O_3} + P_{NO} + P_{NO_2}$ then

$$150 = (100 - X) + X + X$$

$$X = 50 \text{ mm and } P_{N_2O_3} = 100 - 50 = 50 \text{ mm}$$

$$\frac{-\Delta P_{N_2O_3}}{\Delta t} = -\frac{50 - 100}{75} = 0.67 \frac{\text{mm}}{\text{sec}}$$

After 150 seconds, let $Y = $ the mm of N_2O_3 which has decomposed. Then $175 = (100 - Y) + Y + Y$, $Y = 75$ mm, and $P_{N_2O_3} = 100 - 75 = 25$ mm.

$$\frac{-\Delta P_{N_2O_3}}{\Delta t} = -\frac{25 - 50}{150 - 75} = 0.33 \frac{\text{mm}}{\text{sec}}$$

for the interval from 75 to 150 seconds. For the entire time interval

$$\frac{-\Delta P_{N_2O_3}}{\Delta t} = -\frac{25 - 100}{150 - 0} = 0.50 \frac{\text{mm}}{\text{sec}}$$

16.3 Calculate the average rate of appearance of Br^- in the reaction $CH_3Br + I^- \rightarrow CH_3I + Br^-$ during each 5-minute time interval from the following data:

[I$^-$] moles/liter	1.00	0.950	0.902	0.857
t minutes	0	5	10	15

16.4 Calculate the average rate of disappearance of $NO_{2(g)}$ during each time interval in the reaction $2NO_{2(g)} \rightarrow N_2O_{4(g)}$, from the following measurements of total pressure:

P_T mm	250	238	224	210
t sec	0	200	500	900

16.5 Calculate the average rate of appearance of $N_2O_{4(g)}$ during the indicated time intervals from the data in Problem 16.4.

Solution: The average rate of appearance of $N_2O_{4(g)}$ is one half the average rate of disappearance of $NO_{2(g)}$ since one mole of $N_2O_{4(g)}$ forms every time two moles of $NO_{2(g)}$ react.

16.6 For a certain time interval it is found that for the process $4 NH_{3(g)} + 5 O_{2(g)} \rightarrow 4 NO_{(g)} + 6 H_2O_{(g)}$ the average rate of appearance of water is 21.3 mm/min. Calculate the average rate of appearance of $NO_{(g)}$ and the average rates of disappearance of $NH_{3(g)}$ and $O_{2(g)}$.

16.7 The reaction $2 HI_{(g)} \rightarrow H_{2(g)} + I_{2(g)}$ is followed by monitoring the intensity of the purple color of $I_{2(g)}$ as it appears. The intensity of the purple color, as measured by a quantity called the optical density, changes during a 10-minute interval from 5.24 to 5.68. In another experiment it is found that the optical density of the I_2 is related to the P_{I_2} by $P = 15.2$ O.D. Calculate the average rate of disappearance of HI during the 10-minute interval.

16.8 The reaction $2 NO_{(g)} + O_{2(g)} \rightarrow 2 NO_{2(g)}$ is found to follow the differential rate law $-dP_{O_2}/dt = k P_{O_2} P_{NO}^2$. What is the overall order of the reaction and the order with respect to O_2 and NO?

16.9 The reaction $H_{2(g)} + Br_{2(g)} \rightarrow 2 HBr_{(g)}$ is found to follow the differential rate law $d[HBr]/dt = k[H_2][Br_2]^{1/2}$. What is the overall order of the reaction and the order with respect to $H_{2(g)}$ and $Br_{2(g)}$?

16.10 The reaction $C_4H_9I + H_2O \rightarrow C_4H_{10}O + H^+ + I^-$ is first order in C_4H_9I. Calculate the value of k, the specific rate constant, from the following data obtained by starting with $[C_4H_9I] = 0.200 \ M$.

[H⁺] moles/liter	~0	0.0587	0.100	0.130	0.150
t min	0	30	60	90	120

Solution: Since the reaction is first order in C_4H_9I we can write

$$\frac{-\Delta[C_4H_9I]}{\Delta t} = k[C_4H_9I]_{ave} = \frac{\Delta[H^+]}{\Delta t}$$

For each 30-minute time interval we can calculate the $[C_4H_9I]_{ave}$ since $[C_4H_9I] + [H^+] = 0.200 \ M$. We can also calculate the average rate as in Problem 16.1. At $t = 0$, $[C_4H_9I] = 0.200 \ M$. At $t = 30$ min, $[C_4H_9I] = 0.2 - [H^+] = 0.141 \ M$; $[C_4H_9I]_{ave} = (0.200 + 0.141)/2 = 0.17 \ M$. At $t = 60$ min, $[C_4H_9I] = 0.200 - 0.100 = 0.100 \ M$ and $[C_4H_9I]_{ave}$ from 30 to 60 min $= (0.100 + 0.141)/2 = 0.121 \ M$. By the same method $[C_4H_9I]_{ave} = 0.850 \ M$ from 60 to 90 min and $= 0.60 \ M$ from 90 to 120 min.

The average rates for the successive time intervals can be calculated directly from $[H^+]$:

$$\frac{\Delta[H^+]}{\Delta t} = \frac{0.0587 - 0}{30} = 1.96 \times 10^{-3} \ \frac{moles}{liter\text{-}min}$$

for the first 30 minutes;

$$= \frac{0.100 - 0.0587}{30} = 1.38 \times 10^{-3} \ \frac{moles}{liter\text{-}min}$$

for the second 30 minutes; and so on.

To calculate k we substitute each pair of values of $[C_4H_9I]_{ave}$ and $\dfrac{\Delta[H^+]}{\Delta t}$. Thus $1.96 \times 10^{-3} = k(0.171)$, $k = 0.0115$ min^{-1}. (Note that the units moles/liter appear on both sides of the equation and therefore cancel each other.) For the second 30-minute interval $1.38 \times 10^{-3} = k(0.121)$, $k = 0.0114$ min^{-1}. This process can be repeated to obtain two more values of k and the best value of k can be obtained by averaging all the values so obtained. It should be noted that in order for the values of k obtained from each interval to be close to the correct one it is necessary to know the order of the reaction. Often it is possible to deduce the order of a reaction by testing the data in the way done here for different orders and seeing which order gives the best agreement between the calculated values of k for each time interval.

16.11 Using the data of Problem 16.3, and assuming the reaction is run in such a way as to be first order in $[I^-]$ and independent of $[CH_3Br]$ calculate the value of k.

16.12 Use the data from Problem 16.4 to calculate k, given that the reaction is second order in NO_2.

16.13 The reaction $2A \rightarrow B$ is either first or second order in A. Deduce the order of the reaction and then calculate k from the following data:

[A] moles/liter	1.00	0.862	0.758	0.678	0.609
t sec	0	500	1000	1500	2000

16.14 Calculate the specific rate constant from the data of Problem 16.10 using the integrated form of the first order rate equation.

Solution: In the form of the integrated rate equation

$$2.3 \log \frac{[A]}{[A_0]} = -kt,$$

$[A]$ and $[A_0]$ refer to C_4H_9I. Since the $[C_4H_9I]$ is given to be 0.200 M initially, the $[C_4H_9I]$ at any time is $0.200 - [H^+]$. The calculation of k simply involves substitution of the appropriate pairs of values into the equation. For example at 30 minutes $[C_4H_9I] = 0.200 - 0.0587 = 0.1413$. Substituting

$$2.3 \log \frac{(0.1413)}{0.200} = -k(30)$$

$$k = 1.16 \times 10^{-2} \text{ min}^{-1}$$

At 60 minutes $[C_4H_9I] = 0.200 - 0.100 = 0.100$. Substituting

$$2.3 \log \frac{(0.100)}{(0.200)} = -k60$$

$$k = 1.16 \times 10^{-2} \text{ min}^{-1}$$

This process can be repeated for each measurement and the values of k so obtained averaged to give the best value.

16.15 Use the integrated first order rate equation to calculate k for the reaction A $\longrightarrow$ B from the following data:

[A] moles/liter	1.0	0.80	0.63	0.40
t hr	0	100	200	400

16.16 The reaction $N_2O_{5(g)} \longrightarrow 2\,NO_{2(g)} + \frac{1}{2}\,O_{2(g)}$ is first order in $N_2O_{5(g)}$. The reaction is monitored by measuring the total pressure and the following data are obtained:

P_T mm	350	490	592	718	788
t min	0	10	20	40	60

Use the integrated first order rate equation to calculate the value of k.

Solution hint: Use a similar method to Problem 16.2 to calculate $P_{N_2O_5}$ from P_T.

16.17 Use the data of Problem 16.10 to find how much time is required for the $[C_4H_9I]$ to be reduced from 0.200 M to 0.125 M.

Solution: Since this reaction is known to be first order in $[C_4H_9I]$ the integrated first order rate equation may be used.

$$t = \left(2.3 \log \frac{A}{A_0}\right)\left(-\frac{1}{k}\right)$$

$$= \left(2.3 \log \frac{0.125}{0.200}\right)\left(-\frac{1}{1.16 \times 10^{-2}}\right)$$

$$= 40.5 \text{ min}$$

16.18 The reaction $SO_2Cl_{2(g)} \longrightarrow SO_{2(g)} + Cl_{2(g)}$ is first order, $k = 2.20 \times 10^{-5}$ sec^{-1}. If 500 mm of SO_2Cl_2 are introduced into a bulb, how long will it take for the $P_{SO_2Cl_2}$ to drop to 100 mm?

16.19 The specific rate constant for the first order reaction $N_2O_{3(g)} \longrightarrow NO_{2(g)} + NO_{(g)}$ at a certain temperature is 8.93×10^{-3} min^{-1}. Calculate how long it would take the pressure to reach 135 mm if 100 mm of $N_2O_{3(g)}$ is initially introduced into an evacuated bulb.

16.20 Referring to the reaction of Problem 16.18 calculate the $P_{SO_2Cl_2}$ after 5 hr; calculate the total pressure at this time.

16.21   The following rate data are obtained for the reaction $2A_{(g)} \rightarrow$ $B_{(l)}$, which is second order.

P_A mm	300	200	151	99
t sec	0	250	500	1000

Calculate the specific rate constant.

Solution: Since the initial concentration and the concentration at different times are given, k can be calculated by substitution into the integrated second order rate equation.

$$\frac{1}{A} - \frac{1}{A_0} = kt \text{ for } t = 250 \text{ sec}$$

$$\frac{1}{200} - \frac{1}{300} = k(250) \qquad k = 6.67 \times 10^{-6} \frac{\text{liters}}{\text{mole-sec}}$$

Repeating for another time: $\frac{1}{151} - \frac{1}{300} = k(500)$

$k = 6.58 \times 10^{-6}$ liters/mole-sec. The best value of k is then obtained by averaging the value of k obtained from each measurement.

16.22   At a certain temperature the following data are obtained for the second order reaction $2C_4H_6 \rightarrow C_8H_{12}$

$[C_4H_6]$ moles/liter	0.120	0.111	0.104	0.092
t min	0	200	400	800

16.23   Use the data of Problem 16.4 and the integrated second order rate equation to calculate k for the reaction.

16.24   For the reaction of Problem 16.21, how long does it take for P_A to drop to 50 mm?

16.25   For the reaction of Problem 16.21, what will be the value of P_A after 5000 sec?

16.26   Used the integrated form of the first order rate equation to show that the time necessary for the initial concentration of a substance which reacts in a first order process to be reduced by half is independent of the value of the initial concentration. Find the relationship between this time, which is called the half-life, and the specific rate constant.

Solution: If we start initially with $[A] = A_0$, when half of A_0 reacts $[A] = A_0/2$. Substituting gives

$$2.3 \log \frac{A_0/2}{A_0} = -kt \text{ or } 2.3 \log \frac{1}{2} = -kt; \ t = \frac{0.693}{k}$$

16.27 Show that for a first order reaction the time required for 99.9% of the reaction to take place is ten times that required for 50% to take place.

16.28 A compound decomposes by a first order reaction and it is found that after 6 hours, 20% of the original material remains. Calculate the specific rate constant, k, the half-life, and the time it took for the first 20% of the material to react.

➡ **16.29** A certain substance A decomposes. It is found that 50% of A remains after 100 seconds. How much A remains after 200 seconds if the reaction with respect to A is (a) zero order, (b) first order, and (c) second order?

Solution hint: For a second order reaction the half-life depends on A_0. Thus use the integrated equation to calculate k in terms of A_0 and then calculate [A] at 200 sec in terms of A_0.

16.30 A reaction has an activation energy, $E_a = 20$ kcal and a specific rate constant, $k = 5.1 \times 10^{-2}$ sec^{-1} at 300°K. Calculate the specific rate constant at 400°K.

Solution: Substitute in Equation 16.4.

16.31 A certain reaction doubles its velocity when the temperature is raised from 298°K to 310°K. Calculate the activation energy.

16.32 Consider the reaction A + B $\rightleftharpoons$ C + D which is first order in A and first order in B for the forward reaction and first order in C and first order in D for the back reaction. Let k_1 and k_{-1} be the specific rate constants for the forward and back reactions respectively. Derive a relationship between these specific rate constants and the equilibrium constant K for the reaction.

17

Quantum theory and the structure of atoms.

The study of the interaction between electromagnetic radiation and matter has played a prominent role in the development of our current ideas on the structure of matter. All electromagnetic radiation can be described as waves moving with a velocity, $c = 3 \times 10^{10}$ cm/sec. A wave is a regular succession of maxima and minima or peaks and valleys. The distance between two consecutive maxima is called the wavelength, λ. The frequency, ν, is the number of maxima which pass a fixed point in unit time, usually one second. From these definitions it follows that $c = \lambda\nu$, that is wavelength and frequency are inversely proportional.

The energy associated with electromagnetic radiation is carried as discrete particles sometimes called quanta. The energy of one of these particles is proportional to the frequency of the wave, $E = h\nu$, where E is the energy in ergs, ν is frequency in sec^{-1}, and h is a proportionality constant called Planck's constant which has the value 6.625×10^{-27} erg-sec.

One of the classic experiments that illustrates this idea is the photoelectric effect. It is observed that when light impinges on a metal, electrons may be emitted depending only on the frequency of the light. Only when the frequency of light is equal to or greater than a value characteristic for each metal will electrons be emitted. This value is called the threshold frequency, ν_0, and the energy of a single quantum is given by $E = h\nu_0$. This must also be the energy E, called the binding energy, with which a single electron is held by the metal. When the frequency of light impinging on the metal is greater than the threshold frequency, only a part of the available energy is needed to overcome the attraction between the metal and the electron. The excess energy results *not* in the emission of more electrons but in excess kinetic energy of the electron being emitted. The number of electrons emitted depends rather on the intensity, that is the number of quanta, of the incident light.

These relationships are given by

(1) $$h\nu = E + \tfrac{1}{2}mv^2$$

where ν is the frequency of the incident light and $\tfrac{1}{2}mv^2$ is the excess kinetic energy.

When a pure element is heated, for example hydrogen in stars, radiation is emitted. It is found that this emitted radiation consists of only a relatively small number of wavelengths, so that when it is passed through a prism it appears as a number of lines, called the atomic spectrum, characteristic for each element. It was found that the measured wavelengths of these lines for hydrogen obey a surprisingly simple numerical relationship.

(2) $$\bar{\nu} = \frac{1}{\lambda} = R_H\left(\frac{1}{n_L^2} - \frac{1}{n_H^2}\right)$$

where λ is the wavelength of the line, R_H is the Rydberg constant with a value of 1.10×10^5 cm^{-1}, and n_L and n_H are arbitrarily assigned small integers restricted only by $n_H > n_L$. The symbol $\bar{\nu}$ is called the wave number and is defined as $1/\lambda$. Every time two small integers are substituted into this equation a calculated wavelength is obtained which will correspond to that of a line observed in the atomic spectrum of hydrogen.

Observations such as these eventually led to the quantum description of the atom in which only certain energies are possible for the electrons. The emission of a quantum of radiation, $h\nu$, thus corresponds to an electron going from a high energy to a low energy level. The difference in energy between the two levels, ΔE, is equal to $h\nu$. Similarly when a quantum of radiation is absorbed it causes an electron to go from a low energy level to a higher one. Again $\Delta E = h\nu$.

In order to understand the behavior of electromagnetic radiation it is necessary to think in terms of a wave-particle duality. That is, radiation has some wave characteristics and some particle characteristics. It was suggested that the properties of matter also require a wave-particle duality, that is, associated with a moving particle will be a wavelength defined by $\lambda = \dfrac{h}{mv}$, where λ is the wavelength, h is Planck's constant, m is the mass of the particle, and v is its velocity. The quantity mv is sometimes written as p, the momentum. For macroscopic particles such as a baseball, m is so large that λ is inconsequentially small. However for submicroscopic particles such as an electron, m is so small that λ has an appreciable value.

PROBLEMS

17.1 What is the frequency of green light whose wavelength is 5200 Å?

Solution: Substitute into $\nu = c/\lambda$ using the conversion Å $= 10^{-8}$ cm.

17.2 What is the wavelength of a radio wave whose frequency is 6×10^{11} sec^{-1}?

17.3 What is the velocity of a sound wave whose frequency is 1000 sec^{-1} and whose wavelength is 33 cm?

17.4 What is the energy of a quantum of gamma radiation whose frequency is 4.0×10^{19} sec^{-1}?

17.5 What is the energy of a quantum of ultraviolet radiation whose wavelength is 3000 Å?

17.6 How many quanta of radiation whose frequency is 1×10^{15} sec^{-1} will be required for an energy of 1 erg?

➡ **17.7** The value of C_p for $H_2O_{(l)}$ is 1.0 cal/deg-g. Water is exposed to infrared radiation, $\lambda = 2.8 \times 10^{-4}$ cm. Assume that all the radiation is absorbed and converted to heat. How many quanta will be required to raise the temperature of one gram of the water 2°C? (1 cal $= 4.18 \times 10^7$ ergs.)

17.8 What is the binding energy of an electron in a metal whose threshold frequency for the photoelectric effect is 5×10^{14} sec^{-1}? What is it for one mole of electrons in kcal? (1 erg/electron $= 1.4 \times 10^{13}$ kcal/mole.)

17.9 What will be the kinetic energy of an electron emitted from the metal in Problem 17.8 when it is exposed to radiation of frequency 2×10^{15} sec^{-1}?

17.10 Calculate the threshold frequency necessary for the photoelectric effect to occur in a metal which is found to emit electrons with kinetic energy of 2.2×10^{-22} erg/electron when exposed to radiation of frequency 1×10^5 sec^{-1}.

➡ **17.11** A certain metal is found to have a threshold frequency of 4.5×10^{14} sec^{-1} for the photoelectric effect. The mass of an electron is 9.1×10^{-28} g. Calculate the velocity of an electron emitted by this metal when it is exposed to radiation of wavelength 5500 Å.

17.12 A series of four lines in the atomic spectrum of hydrogen will correspond to an electron in the fifth energy level going down to each of the lower energy levels. Calculate the wavelength of each line.

Solution: In Equation (2) n_H represents the higher energy level and n_L the lower energy level involved in the emission of radiation. Thus $n_H = 5$ and $n_L = 1, 2, 3$, and 4. Substitution of these values into Equation (2) will give the appropriate wavelengths.

17.13 When hydrogen is at relatively low temperatures its electron is in the first energy level. This is called the ground state. What is the longest wavelength of radiation which can be absorbed by the hydrogen?

Solution: Equation (2) can also be used to correlate absorption which involves an electron going from a low to a high energy level.

17.14 What is the difference in energy between the second and third level of hydrogen?

Solution: Calculate the wavelength of the radiation corresponding to this transition and then the frequency and corresponding energy of this radiation.

➡ **17.15** What is the energy required to remove an electron from hydrogen in its ground state?

17.16 An electron at 300°K is found to have a wavelength of 60 Å. Its mass is 9.1×10^{-28} g. What is its velocity?

17.17 A xenon atom travels with a velocity of 2.4×10^4 cm/sec at 300°K. What is its wavelength?

17.18 A baseball weighing 100 g is thrown with a velocity of 90 mph. What is its wavelength? (1 mile $= 1.61 \times 10^5$ cm.)

Electronic configurations of atoms

Just as is the case for the hydrogen atom, so it is true for multielectron atoms that only certain energy levels are allowed for the electrons. The ground state electronic configuration of an atom, which is usually the one of interest, will involve filling the lowest energy levels possible with the available electrons. In order to describe an electronic configuration we first describe the allowed energy levels and then specify which ones contain electrons.

The description of energy levels is based on the mathematical solutions of the Schrödinger wave equation for hydrogen, which relate the allowed energies to the mass and wave motion of a particle and whose solution exactly describes the hydrogen atom. The approximation is made that the energy levels of a multielectron atom are like those of hydrogen. An allowed energy level is described as a region of space enclosed by an imaginary surface whose shape and location are obtained from the solution to the wave equation. This surface is called an orbital and it can be specified by a set of three integers called quantum numbers which specify its energy, shape, and direction.

The first of these integers is called the principal quantum number and is represented by n. The value of n specifies almost completely the energy associated with the orbital. The principal quantum number may have any integral value $\geqslant 1$. The lower its value the lower the energy. The second of these integers is called the angular momentum quantum number, l. The value of l is specified after the value n and is restricted by the value of n. The l may have any integral value from 0 to $n - 1$. The value of l further specifies the energy of the orbital and also gives information about its shape. The third of these integers is called the magnetic quantum number m_l. Its values are restricted by the values of l. It can equal $-l, -l + 1, \ldots, 0, \ldots, l - 1, l$. It serves to specify the directional properties of the orbital but is not related to its energy.

By specifying, within the given restrictions, values for the three quantum numbers an orbital, or an electron in the orbital is described. However to complete the description of an electron it is also necessary to have information about its relative spin. This is done with the spin quantum number, m_s, which can have a value of $+\frac{1}{2}$ or $-\frac{1}{2}$. This number just indicates relative spin. Thus if two electrons both have a spin of $+\frac{1}{2}$ or of $-\frac{1}{2}$ they are spinning the same way and are said to be unpaired or parallel; if one is $+\frac{1}{2}$ and one is $-\frac{1}{2}$ they are spinning in opposite ways and are said to be paired.

By convention certain letters are used to specify orbitals or electrons in place of the numerical values of the l and m_l quantum numbers. Thus when $l = 0$, this corresponds to a spherical orbital called an s orbital. When $l = 1$ this is a two-lobed orbital called a p orbital. When $l = 1$, m_l can have the values $-1, 0, +1$ which correspond to three p orbitals, differing in their direction but not their energy, represented as p_x, p_y, and p_z. When $l = 2$ this denotes a d orbital. When $l = 2$, m_l can have the values $-2, -1, 0, +1, +2$ which correspond to five d orbitals differing in their direction but not their energy, represented as d_{xy}, d_{xz}, d_{yz}, d_{z^2}, and $d_{x^2-y^2}$. When $l = 3$ this denotes an f orbital of which there are seven, differing in direction but not energy. The value of the principal quantum number precedes the letter designations. Thus the hydrogen-like set of allowed orbitals for a multielectron atom begins: $1s, 2s, 2p_x, 2p_y, 2p_z, 3s, 3p_x, 3p_y, 3p_z, 3d_{xy}, 3d_{xz}, 3d_{yz}, 3d_{z^2}, 3d_{x^2-y^2}, \ldots$, etc.

The ground state electronic configuration of an atom is worked out from these orbitals by the Aufbau method in which the electrons are placed into the available orbitals in order of increasing energy subject to certain restrictions. The Pauli exclusion principle states that in a given species no two electrons can have the same four quantum numbers. Thus an orbital can hold a maximum of two electrons and they must spin in opposite directions. In a situation where more than one orbital of equal energy is available to be filled, Hund's rules specify the manner in which the electrons will occupy these orbitals. It is of lower energy to have one electron in each of two different orbitals of the *same* energy, than to have two electrons in the same orbital. When there is only one electron in each of two orbitals it is better to have them spinning the same way.

Electronic configurations are written by indicating the number of electrons in each orbital with a superscript. Thus the ground state electronic configurations of the first few elements are: H, $1s^1$; He, $1s^2$; Li, $1s^2\,2s^1$; Be, $1s^2 2s^2$; and B, $1s^2 2s^2 2p^1$. Relative spin is indicated with arrows pointing up or down. Thus C is $1s^2 2s^2 2p_x \uparrow 2p_y \uparrow$. Both electrons spin the same way. N is $1s^2 2s^2 2p_x \uparrow 2p_y \uparrow 2p_z \uparrow$ and O is $1s^2 2s^2 2p_x^2 2p_y \uparrow 2p_z \uparrow$. When a group of orbitals of the same energy are filled or almost filled it is common to write them together. Thus F is $1s^2 2s^2 2p^5$ and Ne is $1s^2 2s^2 2p^6$.

It should be noted that with neon all the orbitals of principal quantum number 2 are filled and the next element Na begins with $3s$. It is common to abbreviate the electronic configuration of Na as $[Ne]3s^1$ or of Mg as $[Ne]3s^2$ or of Al as $[Ne]3s^2 3p^1$. When argon is reached the $3s$ and $3p$ orbitals are filled: $[Ne]3s^2 3p^6$, and the electronic configurations of elements after argon are written beginning with $[Ar]$. Similarly the $4s$ and $4p$ orbitals are filled with krypton: $[Ar]3d^{10}4s^2 4p^6$, so that electronic configurations of elements after Kr are written beginning with $[Kr]$.

The order in which the orbitals fill follows the regular sequence above until element 19, K. At this point there are so many electrons present that the hydrogen-like approximation starts breaking down, and a number of exceptions must be committed to memory. In general the $4s$ fills before the $3d$; the $5s$ before the $4d$, and the $6s$ before the $4f$ before the $5d$. The following table helps to remember this, the diagonal lines indicating the order of filling:

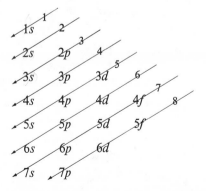

There are numbers of exceptions however. One type of exception seems to reflect stability associated with a half-filled or filled set of orbitals of the same energy. For example Cr has the configuration $[Ar]4s^1 3d^5$ and Cu has $[Ar]4s^1 3d^{10}$.

Another difficulty with the hydrogen-like approximation is seen in the relative ease of removal of electrons from an atom. Contrary to expectation the last electron to be added by the Aufbau procedure is not always the most easily removed. For example it is found that in the series of elements from Sc to Zn the $4s$ electron is more easily removed than the $3d$ electron.

Being able to write ground state electronic configurations is extremely important because the organization of the periodic table is based on columns of elements having the same electronic arrangement of their outermost electrons, called the valence electrons. For example, column I of the periodic table is composed of elements whose valence electron configuration can be represented as ns^1. Thus Na is $[Ne]3s^1$, K is $[Ar]4s^1$, Rb is $[Kr]5s^1$, and Cs is $[Xe]6s^1$. Also it is the valence electron configuration which determines most of the chemical behavior of an element.

PROBLEMS

17.19 Which of the following orbital designations are not possible? $8s$, $2d$, $7p$, $4f$, $1p$, $3f$.

17.20 Write the ground state electronic configuration of the following: (a) Sr (b) Co (c) Se (d) Ba.

17.21 Write the ground state electronic configuration of the following rare gases: (a) Kr (b) Xe (c) Rn.

17.22 Name the element which fits each of the following descriptions:

(a) The lightest atom containing a half-filled p subshell.

(b) The heaviest rare gas that has no electrons in d orbitals.

(c) An element that has two electrons with quantum numbers $n = 4$, $l = 0$ and eight electrons with quantum numbers $n = 3$, $l = 2$.

(d) Two elements with two unpaired $3p$ electrons.

(e) An element with a filled d shell but a half-filled $4s$ shell.

17.23 Without consulting a periodic table predict the atomic numbers of the next three elements in the column under F in the periodic table.

17.24 Without consulting a periodic table predict which element between 32 and 35 will most closely resemble element 52.

17.25 Name an element to fit each of the following:

(a) An element whose dipositive cation has the same electronic configuration as He.

(b) An element whose tetrapositive cation has the same electronic configuration as the monoanion of Cl.

(c) An element with a half-filled $4s$ shell and a half-filled d shell.

(d) An element whose tripositive cation has a half-filled $3d$ shell.

17.26 How many unpaired electrons would be found in the following atoms in their ground states: (a) Mn (b) V (c) Fe (d) Zn?

17.27 The first excited state refers to the electronic configuration closest to but higher than that of the ground state. Write electronic configurations of the first excited state of the following: (a) C (b) Ne (c) K.

17.28 The elements 58–70, called the rare earths, all have very similar properties. Suggest a reason for this based on their electronic configurations.

➡ **17.29** Imagine a universe in which the m_l quantum is only allowed positive values ($m = 0, 1, 2, \ldots, l$). Assuming all other quantum numbers are unaffected, that the Pauli exclusion principle still holds, and that the name corresponding to an element of a given atomic number is the same, answer the following questions:

(a) The new electronic configuration for fluorine will be?

(b) The electronic configuration for the element below fluorine in the new periodic table will be?

(c) The numbers and names of the first three rare gases in the new periodic table will be?

(d) Two elements with three unpaired electrons of principal quantum 3 will be?

➡ **17.30** From its electronic configuration predict which of the first ten elements would be most similar to the as yet unknown element 117.

➡ **17.31** Imagine a universe in which the value of the m_s quantum number can be $+\frac{1}{2}$, 0, and $-\frac{1}{2}$ instead of just $\pm\frac{1}{2}$. Assuming that all of the other quantum numbers can take only the values possible in our world and that the Pauli exclusion principle applies, answer the following questions:

(a) The new electronic configuration for nitrogen will be?

(b) The electronic configuration for the element below nitrogen in the new periodic table will be?

(c) The numbers of the first three rare gases in the new periodic table will be?

The structure of molecules.

The atoms in a molecule are held to each other by chemical bonds which can be thought of as arising when two or more electrons are associated with two or more atoms. Bonds are most commonly classified as ionic or covalent depending on the extent to which the electrons in the bond are associated with each of the atoms bonded together. For example in the molecule H_2 each H atom contributes its $1s$ electron to the bond and the electron is equally associated with or shared between the two H atoms. This is a covalent bond. On the other hand, in a molecule such as CsF, the Cs contributes a $6s$ electron and the fluorine a $2p$ electron to the bond, but the electron pair of the bond is primarily associated with the fluorine. This is an ionic bond. Most chemical bonds fall between these two extremes and are classified as covalent or ionic based on the nature of the atoms involved in the bonds and on observations of chemical properties.

The vast majority of chemical bonds encountered are covalent bonds, both atoms of the bond having a substantial association with the electrons of the bond. It is further convenient to divide covalent bonds into two types: polar and nonpolar. In the latter both atoms in the bond have an equal or almost equal association with the electrons in the bond. For example, homo-nuclear diatomics like Cl_2, F_2, and N_2 have nonpolar bonds; and bonds between C and H or C and C are nonpolar. In many bonds which are covalent one of the two atoms in the bond clearly has a greater association with the shared electrons than the other. This is called a polar bond and common examples are H—O, H—N, and H—halogen bonds.

It is often possible to predict the type of bond by use of a devised property called electronegativity, which is roughly a measure of the relative attraction of a bonded atom for the shared electrons. The greater the difference in electronegativity between the two atoms involved in a bond, the more ionic will be the bond, the smaller the difference in electronegativity the more nonpolar the bond. The electronegativity of an atom is correlated with its position in the periodic table; electronegativity increases in going from left to right and from bottom to top of the table. Thus fluorine is the most electronegative element followed by oxygen, nitrogen, and chlorine, while cesium is the least electronegative common element followed by rubidium, potassium, and barium. A rough quantitative rule is that when the electronegativity difference between the two atoms of the bond is greater than 1.7 the bond is ionic, less than 1.7 it is covalent and polar, and less than 0.5 it is nonpolar.

The stability that results from the formation of bonds can be rationalized primarily in terms of the valence electrons of the atom, which are the electrons of highest principal quantum number or sometimes, when a shell is incomplete, of second highest principal quantum number. The bonding in many compounds can be understood in terms of the stability that seems to be associated with a valence electron configuration of ns^2np^6. That is, if an atom can achieve this electronic configuration by forming bonds in which it loses, gains, or shares electrons it will do so. This is known as the *octet rule* and works well for the lower elements and Groups IA, IIA, VA, VIA, and VIIA of the periodic table, with the exception of hydrogen and helium which can accommodate only two valence electrons. Despite the fact that there is also a great deal of bond formation which violates the octet rule, it is often a very useful starting point for working out the bonding in molecules.

The most common notation chemists use to represent the bonding in covalent molecules is a slight modification of the electron dot picture originally devised by G. N. Lewis. It is based on the following rules:

1. Only the valence electrons will be shown.

2. A *pair* of shared electrons is represented by a dash between the two atoms sharing them. Thus when two atoms are sharing one pair of electrons they will be connected by a single dash called a single bond; when they are sharing two pairs of electrons they will be connected by two dashes called a double bond; and when they are sharing three pairs of electrons they will be connected by three dashes called a triple bond.

3. An unshared or nonbonded pair of electrons will be represented by a pair of dots written next to the atom with which they are associated.

4. All the valence electrons of each atom must be accounted for by either dots or dashes.

Some of the electronic properties of the elements which help in writing Lewis structures are:

1. Hydrogen can accommodate no more than two electrons.
2. In stable compounds C, N, O, and F are *never* surrounded by more than eight electrons (the octet rule) and only very rarely by less.
3. Compounds of many other elements often involve octets around all the atoms but there are many exceptions among elements higher than silicon.
4. Elements of Group IIIA often have incomplete octets.
5. Compounds with an odd number of valence electrons must have at least one atom which violates the octet rule.

In many cases a series of steps can be followed in order to arrive at a correct Lewis structure for a compound. This is illustrated for HNO_3, nitric acid:

(a) Determine the total number of valence electrons available by adding the valence electrons of each atom in the molecule. This is most easily determined by looking to see which column of the periodic table each element is in. Thus, Group IA has 1 valence electron, Group IIA has 2, all the way to Group VIIA which has 7. For the B columns, it is more complex and sometimes requires working out the electronic configuration. HNO_3 has $1 + 5 + 6 \times 3 = 24$ valence electrons.

(b) Connect all the atoms in the molecule using no more than one dash (one electron pair) between two atoms. In order to do this correctly, certain information may be necessary:

(i) Hydrogen can only have one electron pair around it and therefore, only one bond.

(ii) The only compounds in which two oxygens are joined to each other are peroxides. Normally, we do not find oxygen-oxygen bonds.

(iii) Sometimes an experimental observation is necessary, which information must be given. For example, in N_2O, the attachment is found to be N—N—O not N—O—N.

(iv) Most strong oxyacids have the H on oxygen. Thus for HNO_3, we write

because we know (or are told) that since it is a strong acid, the H is attached to O. Note: The octet rule prohibits more than four atoms joined to a central one.

This is the most difficult part of writing these structures, but as one gets more factual information, it becomes easier.

(c) Count the number of electrons used for all the dashes (2 per dash) and subtract this number from the total available to see how many electrons are still left over.

For HNO_3 we have used $4 \times 2 = 8$ and have $24 - 8 = 16$ left over.

(d) Locate those atoms which do not yet have completed octets. Calculate how many more electrons are needed to complete each octet and add these numbers.

For

two of the O's need 6, one of the O's needs 4, and the N needs 2 more electrons to complete their octets. H is complete.

The total electrons needed are $2 \times 6 + 1 \times 4 + 1 \times 2 = 18$.

(e) Compare the numbers from (c) and (d). If they are the same, fill out the octets with pairs of dots so each atom is surrounded by 8 electrons. If they are not the same, (c) may be less than (d), almost always by an even number of electrons. Divide this number by 2. The result gives the number of additional dashes or bonds required. Put in these dashes between two atoms with incomplete octets and then add dots to complete all the octets. Generally, one or two dashes are added where a dash is already found. That is cyclic structures are not used unless some specific information is available indicating they should be. Thus,

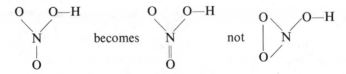

Two or three dashes (2 or 3 pairs) between two atoms are common and are called a double bond and triple bond, respectively.

(f) Use the remaining electrons to complete the remaining octets with dots.

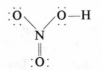

(g) Check to see that the correct number of electrons have been used.

(h) If the number in (c) is less than the number in (d) and the atoms are such that no more dashes may be added, then one or more atoms will have an incomplete octet. This occurs, for example, in BF_3 where the boron, a group IIIA element has an incomplete octet.

(i) If the number in (c) is greater than the number in (d) then one or more of the atoms will be surrounded by more than eight electrons. Such an atom is said to have an expanded octet. In many molecules only one atom will be permitted to expand its octet. For example in ClF_3, it must be the Cl with the expanded octet since F may not do so. The structure is

In other molecules it will be possible to expand the octet of more than one of the atoms. Often it will be the least electronegative atom which is expanded. For example in PCl_5 it is the P with the expanded octet. The structure is

(j) The principle of electroneutrality is often helpful in arriving at the best Lewis structure, especially for compounds of C, N, and O. Normally C forms four bonds, N forms three bonds and has one nonbonded electron pair, and O forms two bonds and has two nonbonded electrons. Often in writing a Lewis structure it will be found that this normal complement of bonds is not obtained. This causes relative destabilization and results in the creation of "formal charges." Thus when an atom with a complete octet forms one more bond than normal, e.g., oxygen with three bonds or nitrogen with four bonds, it is thought of as bearing a formal charge of $+1$. When an atom with a complete octet forms one less bond than normal, e.g., carbon with three bonds or oxygen with one bond, it is thought of as bearing a formal charge of -1. In neutral compounds the sum of any formal charges which exist must be zero. The observation that the existence of formal charges is destabilizing is often helpful in arriving at the best Lewis structure for a compound. For example two choices are apparent for HCN: H—C≡N: and $\overset{(-)}{:C}$≡$\overset{(+)}{N}$—H. The first structure is correct and consistent with avoiding the formal charges of the second. Often there is a choice between a structure with formal charges or with an expanded octet. Again when permitted it is better to expand the octet. For example two choices are apparent for H_2SO_4:

$$H-\overset{\displaystyle \overset{..}{\underset{..}{O}}{}^{(-)}}{\underset{\displaystyle \overset{..}{\underset{..}{O}}{}^{(-)}}{\overset{|}{\underset{|}{S}}{}^{(++)}}}-\overset{..}{\underset{..}{O}}-H \quad \text{and} \quad H-\overset{..}{\underset{..}{O}}-\overset{\displaystyle :O:}{\underset{\displaystyle :O:}{\overset{\|}{\underset{\|}{S}}}}-\overset{..}{\underset{..}{O}}-H$$

Since sulfur is permitted to expand its octet, the second choice is preferred.

PROBLEMS

18.1 List the following compounds in order of increasing ionic bond character: (a) NH_3 (b) H_2O (c) HF (d) CH_4 (e) H_2

Solution hint: The greater the electronegativity difference between two atoms, the more ionic the bond. Use a periodic table.

18.2 List the following compounds in order of increasing ionic bond character: (a) HI (b) I_2 (c) CsI (d) CsF (e) NaI (f) HF (g) NCl_3

18.3 Give the number of an element which fits each of the following descriptions without consulting a periodic table:

(a) The most electronegative element in the first row

(b) The least electronegative element in the third row

(c) The most electronegative element of the group with atomic number: 3, 11, 12, 13, 35

(d) Two third row elements with four valence electrons

(e) The element closest in atomic number to Te (52) which has the same valence shell electronic configuration with a different principal quantum number.

(f) The element of the group with atomic numbers 7, 8, 15, 17 which forms the least polar bond with H.

(g) The two lowest numbered elements which often expand their octets.

(h) The element of the group with atomic number 16, 17, 33, 34, 35 which forms the most ionic compound with lithium.

18.4 What is the correlation between valence electron configuration and the normal number of bonds formed in the series of elements 3–9?

18.5 Which of the following molecules cannot completely satisfy the octet rule?

(a) N_2O (b) NO (c) N_2O_3 (d) NO_2 (e) N_2O_4
(f) Cl_2O (g) N_2O_5

18.6   Which of the following molecules cannot have an expanded octet?
(a) H_2SO_4 (b) H_3PO_4 (c) OF_2 (d) $HClO_4$ (e) HNO_3

18.7   Draw Lewis structures for the following molecules.
(a) CH_2O (b) NF_3 (c) OF_2 (d) N_2

18.8   Draw Lewis structures for (a) HNO_2 (b) N_2O_3 (no N-N bond)
(c) C_2H_2 (d) C_2H_4

18.9   (a) Draw the Lewis structure for CH_4O, methyl alcohol. (b)
Ethyl alcohol, C_2H_6O, is very similar in behavior to methyl alcohol. Draw
the correct Lewis structure of ethyl alcohol based on the structure of methyl
alcohol. (c) Dimethyl ether is a compound which also has the composition
of C_2H_6O. Draw its Lewis structure.

18.10   Draw Lewis structures for the following molecules with incomplete octets: (a) BF_3 (b) $AlCl_3$ (c) NO

18.11   Draw Lewis structures for the following molecules which have
expanded octets: (a) H_3PO_4 (b) $HClO_3$ (c) ICl_5

18.12   Draw Lewis structures for the following ions: (a) CH_3^+ (b)
NH_4^+ (c) ClO^- (d) I_3^-

18.13   Use the principle of electroneutrality to draw Lewis structures
for (a) CNBr (b) HClO (c) SO_3

➡ **18.14**   Draw the Lewis structure of P_4 based on the observations that
no expanded octets and no double bonds are present and that all the P atoms
are equivalent.

18.15   Draw two possible Lewis structures for NH_3O and predict
which represents the more stable compound.

18.16   Draw Lewis structures for (a) $SiCl_4$ (b) PCl_5 (c) SCl_4 and (d)
$BrCl_5$. Which have the same electron configurations as the corresponding
chlorides of the first row elements?

Resonance

During the elaboration of the Lewis structures of many molecules a difficulty
seems to arise in selecting between which two atoms in the structure a multiple bond should be placed. For example in writing the Lewis structure of
HNO_3 a point was reached where it was necessary to put one multiple bond
into

It is possible to write

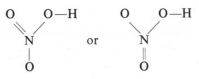

Similarly in writing the Lewis structure of N_2O, two multiple bonds have to be placed in N—N—O. It is possible to write N=N=O, N≡N—O, or N—N≡O.

Whenever such a situation arises, the correct structure of the molecule cannot be represented by a single Lewis structure; and any single Lewis structure which can be written does not correspond to a real species. The correct structure of the molecule is a blend of two or more of these imaginary Lewis structures. They are called contributing structures and the molecule is called a resonance hybrid and is said to be resonance stabilized. It is important to understand that none of the contributing structures ever exist; they are imaginary but are a convenient way of describing a real structure which is a *blend* of them. For example experimental evidence indicates that $NO_3{}^-$, the nitrate ion, has only one type of oxygen and one type of nitrogen-oxygen bond. This is consistent with describing nitrate as an equal blend of the three imaginary contributing structures:

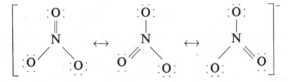

Contributing structures are separated by the double-headed arrow, which should not be confused with the arrows used in chemical equations. It is important to note that contributing structures never involve changes in the original attachments of the atoms but only involve the way the extra electron pairs required for multiple bonding are distributed.

PROBLEMS

18.17 Draw the important contributing structures for the following compounds:

(a) N_2O_4 (b) O_3 (c) CH_3NO_2 (C-N bond)

18.18 Draw the important contributing structures for the following ions:

(a) $CO_3{}^{--}$ (b) $SO_4{}^{--}$ (c) $NO_2{}^-$

18.19   Draw the important contributing structures for the following odd electron molecules:

(a) NO   (b) NO_2

Solution: Extra contributing structures arise here because of the possibility of having the odd electron and incomplete octet on N or O.

18.20   Draw two contributing structures for a molecule with the following properties: formula is C_6H_6, has one type of C, one type of H, one type of carbon-carbon bond intermediate between a single and double bond, and has a cyclic structure.

18.21   Account for the observation that allyl cation

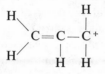

is more stable than the corresponding cation without the double bond.

Hybridization and molecular geometry

Another approach to the description of molecular structure is to think of bonds as arising by combinations of atomic orbitals. For example the bond in H_2 can be understood as an orbital formed by combining the $1s$ of each hydrogen to form a molecular orbital which holds the two electrons of the bond. A useful starting point is thus to consider the origin of each of the two electrons in a bond and specify the bond orbital as a combination of each of the two orbitals, one on each atom, which originally held the electron. This approach leads to certain conclusions about the shape of the bond orbital, the strengths of bonds, and the geometry of molecules. If one imagines bonds as imaginary straight lines between atoms, then any two bonds from the same atom will intersect to form a bond angle.

Experiments readily demonstrate that this simple picture requires modification. For example, the two O—H bonds in water should be formed from a $2p_x$ and a $2p_y$ orbital each overlapping with a $1s$ of hydrogen. Since $2p_x$ and $2p_y$ are perpendicular, a bond angle for   H—O—H   is predicted to be 90°. It is found to be 104.5°. A great many such observations can be accounted for by postulating the existence of hybridization or the mixing of atomic orbitals. A hybrid orbital can be described as a blend of atomic orbitals. Thus if the $2s$ and a $2p$ orbital of an atom are mixed, two new hybrid orbitals are formed, each a blend of one part $2s$ and one part $2p$. Such orbitals are called $2sp$. If the $2s$ and the two $2p$ orbitals are mixed, three hybrid orbitals are formed, each a blend of one part $2s$ and two parts $2p$. Such orbitals are called $2sp^2$. Similarly the $2s$ and the three $2p$ orbitals form four $2sp^3$ hybrid orbitals. It is possible to calculate the most favored geometry associated with each of these types of hybrid orbitals. Two sp orbitals lead to a bond angle of 180° or a linear geometry; three sp^2 orbitals lead to bond angles of 120° or planar geometry and four sp^3 orbitals lead to bond angles of 109°28′ or tetrahedral geometry.

If the hybridization of an atom can be determined it becomes possible to predict its approximate geometry. Often this determination can be made by counting the number of different atoms and nonbonded electron pairs which surround the atom of interest. For example, the C of CH_4 is surrounded by four atoms; the N of $:NH_3$ is surrounded by three atoms and one nonbonded electron pair, a total of four; the O of $H_2\overset{..}{O}$ is surrounded by two atoms and two nonbonded electron pairs, a total of four. Since there are four $2sp^3$ hybrid orbitals formed from four atomic orbitals, the hybridization of C, N, and O in these compounds is $2sp^3$ and their bond angles are exactly 109°28′ for CH_4 and close to 109° for NH_3 and H_2O. When the total of atoms and nonbonded electron pairs surrounding an atom is four, that atom will generally be sp^3 hybridized.

Compounds such as BF_3 or $\begin{smallmatrix} H \\ \diagdown \\ \diagup \\ H \end{smallmatrix} C = C \begin{smallmatrix} \diagup H \\ \\ \diagdown H \end{smallmatrix}$ involve $2sp^2$ hybridization of B and C since the total of atoms and nonbonded electron pairs around these atoms is three. Their geometry involves bond angles of or close to 120°. They are planar. Compounds such as $H—C{\equiv}C—H$ and $H—C{\equiv}N$ involve a total of two atoms and no nonbonded electron pairs around C. The bond angles here are 180°, the molecules are linear.

In most cases the actual geometry of a molecule is only approximated by the value indicated by its hybridization. Superimposed are a number of other effects which are generally related to the minimization of interelectronic repulsions. Thus the H—N—H bond angles in NH_3 are somewhat less than 109° because a nonbonded electron pair occupies more space than a bonded pair. The nonbonded pair tends to push the bonded pairs and therefore the bond angles closer together. In water the H—O—H angle is still smaller because of two nonbonded pairs pushing the bonds together. A number of other factors of this type operate. In $\begin{smallmatrix} H \\ \diagdown \\ \diagup \\ H \end{smallmatrix} C = O$, the $\begin{smallmatrix} \diagup C \\ H \diagdown \\ \diagdown H \end{smallmatrix}$ angle is predicted to be 120° ($2sp^2$ on C). It is found to be 125°. The electronegative oxygen holds the electrons in the C=O relatively closely so that repulsion between these electrons and the ones in the C—H bonds are not too serious. This allows the C—H bond electrons to move further from each other spread-g the angle. A similar effect is seen in CH_3Cl where the C H Cl angle is slightly smaller than 109° and the C H H angles are slightly larger.

In atoms with expanded octets mixing of d orbitals with s and p orbitals often occurs leading to more complex hybridization. However here too it is possible to recognize certain idealized geometries associated with a total of five or six atoms and nonbonded pairs surrounding a single atom. In the five case the groups lie at the five corners of a trigonal bipyramid corresponding to dsp^3 hybridization. In the six case the groups lie at the corners of a regular octahedron corresponding to d^2sp^3 hybridization.

Hybridization also helps to account for the formation of double and triple bonds. When an atom forms three sp^2 hybrid orbitals, there remains one unhybridized p orbital. This orbital can overlap with the unhybridized p orbital of another sp^2 hybridized atom to form a second bonding orbital and accommodate the second pair of electrons for the double bond.

In single bonds, the two atomic orbitals involved point toward each other along the internuclear line, while in the second bond of the double bond the unhybridized p orbitals are perpendicular to the internuclear line and parallel to each other. These two types of bonds are called σ and π bonds respectively. They can be diagrammed as shown for $H_2C{=}CH_2$:

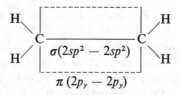

A triple bond can be thought of as consisting of a σ bond from overlap of sp orbitals and two π bonds from overlap of two pairs of unhybridized p orbitals.

PROBLEMS

18.22 What is the most likely hybridization of the starred atoms in each of the following:

(a) $*CH_2Cl_2$; (b) $*AlCl_3$ (c) $Br{-}C*{\equiv}N$

(d) $*CO_2$ (e) $HN*O_3$ (f) H_2O_2*

18.23 Fill in the blanks in the statements below for each of the following compounds:

(a) $H{-}C{\equiv}N$: (b) $H_2C{=}\dot{N}{-}H$ (c) $H_3C{-}NO_2$

(d) $H_3C{-}N{\equiv}C*$:

(1) The hybridization of the C is _____ .

(2) The hybridization of the N is _____ .

(3) The H—C—N angle is _____ .

18.24 Predict the most likely geometry for the following species:

(a) N_2O (b) AlF_3 (c) $SiCl_4$ (d) H_2S

18.25 What is the most likely geometry for the following ions:

(a) NO_3^- (b) CO_3^{--} (c) CH_3^- (d) NH_4^+

18.26 What is the most likely geometry for the following species involving expanded octets:

(a) PCl_5 (b) SF_6 (c) H_5IO_6 (d) SO_4^{--}

➡ **18.27** Considering only the positions of the atoms predict their positions in the following species which have expanded octets and which include nonbonded pairs of electrons:

(a) $BrCl_3$ (b) $BrCl_5$ (c) I_3^- (d) SCl_4

Solution: Work out the expected geometry in the usual way and then consider the most advantageous placement of the atoms and the nonbonded pairs in the available positions. In $BrCl_3$ the Br is surrounded by three atoms and two nonbonded pairs, a total of five corresponding to a trigonal bipyramid. The two nonbonded pairs are found to be in the equatorial plane, leaving the three Cl atoms at the corners of a distorted T.

18.28 Which pair of orbitals overlaps to form the π bond of $CH_2{=}O$?

18.29 Which pairs of orbitals overlap to form the two π bonds of CO_2?

18.30 If the hybridization in NO_3^- is imagined to be $2sp^2$ on each atom draw one overlap picture consistent with the description of NO_3^- by three contributing Lewis structures.

Nuclear reactions.

When a natural radioactive atom decays, its nucleus loses either an alpha particle or a beta particle. An alpha particle has a mass of 4 atomic mass units and carries 2 positive charges. A beta particle has the negligible mass of an electron and carries 1 negative charge. It follows, therefore, that when an alpha particle is emitted, the mass number of the atom decreases by 4 and the atomic number decreases by 2. The loss of a beta particle involves no change in the mass number of an atom but an increase of 1 in its atomic number.

The mass numbers and atomic numbers of elements are represented by the notation $^{238}_{92}U$ in which the number above and to the left of the symbol represents the mass number and the number below and to the left represents the atomic number. Since an alpha particle has the same mass as the helium atom, it is usually represented as ^4_2He. The beta particle, since it is an electron, is represented as $_{-1}^{0}e$; the -1 indicates that it has a single negative charge.

The reaction that takes place when an atom undergoes radioactive decay may be represented by an equation of the following type:

$$^{226}_{88}Ra = {}^{222}_{86}Rn + {}^4_2He$$

This reaction tells us that when radium decays, an alpha particle (^4_2He) is emitted and radon, with a mass number of 222 and an atomic number of 86, is formed.

The reaction

$$^{239}_{92}U = ^{239}_{93}Np + _{-1}^{0}e$$

tells us that the uranium isotope with a mass number of 239 and an atomic number of 92 emits a beta particle ($_{-1}^{0}e$) and is converted to neptunium, whose mass number is 239 and atomic number is 93. It should be noted that, in this reaction as in all decay and transmutation reactions, mass number and charge are both conserved. Thus,

$$239 = 239 + 0 \text{ and } 92 = 93 - 1$$

If the nucleus of an atom is bombarded by a certain particle such as an alpha particle ($_{2}^{4}He$), a beta particle ($_{-1}^{0}e$), a neutron ($_{0}^{1}n$), a proton ($_{1}^{1}H$), or a deuteron ($_{1}^{2}D$), the bombarding particle may be captured by the nucleus. As a result, the atom is converted into a new element or an isotope of the original element. The capture of the particle may or may not be accompanied by the expulsion from the nucleus of some other particle. For example, when nitrogen is bombarded by alpha particles, the nitrogen nucleus captures one alpha particle, but, in so doing, it also ejects a proton ($_{1}^{1}H$). The product is an atom of oxygen. The reaction involved in this transmutation may be represented as follows:

$$^{14}_{7}N + ^{4}_{2}He = ^{17}_{8}O + ^{1}_{1}H$$

(This is referred to as a (α, p) transmutation, since an alpha particle ($_{2}^{4}He$) is captured and a proton ($_{1}^{1}H$) is ejected.)

The rate of radioactive decay is generally expressed in terms of *half life*. The half life of a radioactive nuclide is the time required for one half of any given amount of the substance to decay.

PROBLEMS

19.1 A radioactive element A, whose mass number is 220 and atomic number 85, emits an alpha particle and changes to element B. Element B emits a beta particle and is converted to element C. What are the mass numbers and atomic numbers of elements B and C? Of what elements are B and C isotopes?

Solution: When A loses an alpha particle ($_{2}^{4}He$), its mass number decreases by 4 and its atomic number decreases by 2. Therefore, B has a mass number of 216 and an atomic number of 83. When B emits a beta particle ($_{-1}^{0}e$), its atomic number increases to 84 but the mass number remains the same, 216. Therefore C has a mass number of 216 and an atomic number of 84. Isotopes have the same atomic numbers but different atomic weights. Therefore, by consulting the table of atomic numbers, we find that B is an isotope of bismuth while C is an isotope of polonium.

19.2   Actinium, which has a mass number of 227 and an atomic number of 89, undergoes radioactive disintegration as follows: Actinium loses a beta particle and becomes radioactinium. Radioactinium loses an alpha particle and becomes actinium X. Actinium X loses an alpha particle and becomes actinon. For each of the 3 elements formed in the above series of changes, give the mass number, the atomic number, and the symbol of the element of which it is an isotope.

19.3   When boron (mass number 11, atomic number 5) is bombarded with alpha particles the following changes take place:

$$B + alpha\ particle = new\ element + neutron$$

(This is a (α, n) transmutation.) What is the mass number and atomic number of the new element and in what group of the periodic table will it fall?

Solution: The alpha particle has 2 positive charges and a mass of 4 amu. A neutron has no charge and a mass of 1 amu. Therefore, the mass number of the new element is 14 ($11 + 4 - 1$). Since the boron nucleus picks up 2 positive charges, the atomic number, which is the number of protons (positively charged units) in the nucleus, is increased by 2. Therefore, the atomic number of the new element is 7. This means that the new atom has 5 electrons in the valence shell. Therefore, it falls in group 5 of the periodic table.

19.4   Complete the following transmutation reactions, indicating, in each case, the symbol, mass number, and atomic number of the element formed:

(a) $^{9}_{4}Be + ^{4}_{2}He =$ _____ $+ ^{1}_{0}n$

(b) $^{28}_{14}Si + ^{2}_{1}D =$ _____ $+ ^{1}_{0}n$

(c) $^{27}_{13}Al + ^{1}_{0}n =$ _____ $+ ^{4}_{2}He$

19.5   Complete the following transmutation reactions, indicating, in each case, what particle, if any, was ejected:

(a) $^{14}_{7}N + ^{1}_{0}n = ^{11}_{5}B +$ _____

(b) $^{9}_{4}Be + ^{2}_{1}D = ^{10}_{5}B +$ _____

(c) $^{27}_{13}Al + ^{4}_{2}He = ^{30}_{15}P +$ _____

19.6   In the complete transformation of $^{238}_{92}U$ to $^{206}_{82}Pb$ how many alpha particles and how many beta particles are emitted for each atom of $^{206}_{82}Pb$ formed?

Solution: Since the mass number decreases by 32, 8 alpha particles must have been emitted; this would have decreased the atomic number by 16 units. Since the atomic number decreases by only 10 units, 6 beta particles must have been emitted.

19.7   In the complete radioactive decay of 1 mole of $^{235}_{92}U$ to form 1 mole of a stable solid element as the end product a total of 156.8 liters of helium gas (measured at STP) will be produced. In the course of this complete decay 4 beta particles will be emitted for each atom of stable element that is formed. Give the symbol, atomic number, and mass number of the end product.

19.8   An isotope of neptunium, $^{237}_{93}Np$, decays by alpha and beta emission. A 0.0100 mole sample of this isotope, if allowed to decay completely, would form 1.792 liters of helium gas measured at standard conditions. Six beta particles are given off in the radioactive decay for each atom of stable product formed. Write one balanced nuclear equation representing the overall decay process.

19.9   The nuclide $^{80}_{35}Br$ decays with a half life of 18 min. How much time will elapse before $\frac{7}{8}$ of a starting sample of the nuclide has decayed?

Solution: The half life of a radioactive nuclide is, by definition, the time required for one half of any given quantity of that nuclide to decay. Since the half life of $^{80}_{35}Br$ is 18 min, at the end of 18 min one half of the sample will have decayed and one half will remain. At the end of another 18-min period one half of the remaining half will have decayed and the other half of the half (or one fourth of the original sample) will remain. At the end of a third 18-min period one half of the remaining one fourth will have decayed and one half of that fourth, or one eighth of the original sample, will remain. That means that, at the end of 54 min, $\frac{7}{8}$ of a starting sample of $^{80}_{35}Br$ will have decayed.

19.10   When the radioactive element, A, decays to form the stable product, B, it emits an alpha particle. If the half life of A is 64 days and its atomic mass is 210, how many grams of A must you start out with in order to have collected 1.68 liters of helium gas, measured at 0°C and 2 atm pressure, at the end of 192 days?

19.11 $^{253}_{99}Es$ and $^{230}_{92}U$ both have 20-day half lives and both decay by alpha emission. In each case the daughters are beta emitters with very long half lives. If 1 g of pure $^{230}_{92}U$ and 1 g of pure $^{253}_{99}Es$ were placed in an evacuated 22.4 liter flask at 0°C, what would the pressure be at the end of 20 days?

➡ **19.12**   The metal M is in Group IIA. A particular isotope of M decays with a 20.0-day half life by alpha emission to form a product, X, whose half life is so large that, for the purposes of this problem, it can be considered to be nonradioactive.

Exactly 1.00 mole of MH_2, with all M initially present as the isotope described above, is placed in an evacuated 22.4-liter heavy-walled glass container maintained at 0°C. Calculate the total pressure in the vessel at the end of 40 days.

Solution: MH_2 is the hydride of a Group IIA element; therefore, it is a solid. When M decays by alpha emission to form new element X, this new element must fall in Group 0; it is a monatomic inert gas. Therefore, the reaction that takes place is:

$$MH_2 \text{ (solid)} = X \text{ (gas)} + H_2 \text{ (gas)} + He \text{ (gas)}$$

Since 2 half lives are involved, $\frac{3}{4}$ of the M will decay.

➡ **19.13** $^{38}_{17}Cl$ decays by beta emission with a half life of 40 min. Four moles of $H^{38}_{17}Cl$ are placed in an evacuated 62.4-liter heavy-walled container. After 80 min the pressure in the container is found to be 1650 mm of mercury. What is the temperature of the container in degrees Kelvin?

➡ **19.14** When ordinary BF_3 gas is bombarded with neutrons, the boron undergoes a (n, α) reaction but the fluorine is not affected. Two moles of BF_3 gas contained in a 62.4-liter flask at 27°C are bombarded with neutrons until half of the BF_3 has reacted. What is the pressure in the container, measured at 27°C, when half of the BF_3 has reacted?

➡ **19.15** When CaF_2 is bombarded with neutrons, the Ca atom captures 1 neutron and then emits 1 alpha particle, an atom of a new element being formed in the process. The F atoms are not affected by the bombardment.

A sample of CaF_2 contained in a 10-liter evacuated flask is bombarded with neutrons until one half of the Ca has undergone the reaction described above; the pressure in the flask, measured at 127°C, is then 1.105 atm. How many moles of CaF_2 were originally placed in the flask?

➡ **19.16** A sample of water was a mixture of ^1H_2O, ^2H_2O, and ^3H_2O; it contained ordinary oxygen, atomic weight 16.00, but contained some of each of the 3 hydrogen isotopes, ^1H (1.00 amu), ^2H (2.00 amu), and ^3H (3.00 amu). A total of exactly 1.00 faraday of electricity was required to electrolyze the sample; the mixture of ^1H, ^2H, and ^3H which was produced weighed exactly 1.75 g. On standing the ^3H all decayed according to the reaction, $^3_1H = ^3_2He + $ beta particle. By the time all of the ^3H had decayed a total of 1.5×10^{23} disintegrations had been counted. Calculate the mole percent of ^2H_2O in the sample. (The Avogadro number is 6.0×10^{23}.)

➡ **19.17** The isotope of hydrogen, tritium, represented by the symbol, ^3_1H or T, forms the weak acid, tritium fluoride, TF, in which the T^+ ion, formed in the ionization reaction, $TF \rightleftharpoons T^+ + F^-$, is equivalent in its properties to the H^+ ion. Tritium undergoes slow radioactive decay according to the equation

$$^3_1H = ^3_2He + \text{ one beta particle } (-^0_1e)$$

A freshly prepared aqueous solution of the weak acid, TF, has a pT (the equivalent of pH) of 1.50 and freezes at $-0.372°C$. If a 600-ml portion of this solution were allowed to stand for 24.8 yr a total of 4.50×10^{22} beta particles would be emitted.

The molecular weight of TF is 22.0. The freezing point constant for water is 1.86. The density of the above aqueous solution of TF is 1.0. The half life of tritium is 12.4 yr.

On the basis of the above data and facts calculate:

(a) The ionization constant for TF.

(b) The numerical value of the Avogadro number.

➡ **19.18** You are given a sample of pure "ammonia" gas in which all of the hydrogen that is present in the NH_3 molecule is the isotope of hydrogen, tritium (T), with mass 3. The formula of this ammonia can then be written NT_3.

It is a known fact that tritium undergoes radioactive decay according to the reaction:

$$_1^3T = {}_2^3He + {}_{-1}^0e$$

The ^3He, like ^4He, is a monatomic gas. Write the equation for the chemical reaction that occurs when the T in NT_3 decays.

In a series of precise experiments you observe that when a sample of pure NT_3 gas contained in a 1.00-liter sealed steel tank at 27°C and an unknown initial pressure is allowed to stand until exactly half of the ^3T has decayed to ^3He, the pressure in the tank rises to 2.46 atm. (The temperature is kept constant at 27°C.) Geiger counter readings show that, during the interval from the start of the experiment until the pressure reached a value of 2.46 atm, a total of 4.0×10^{22} beta particles were emitted. Calculate the value of the Avogadro number.

➡ **19.19** A student is given a sample of ammonia gas. He knows it has the formula NH_3, that the nitrogen has atomic weight 14.0, and that the hydrogen is partly the isotope of atomic mass 1.00 (^1H), and partly tritium (^3H or T, atomic mass 3.00). He knows the gas constant R. He is asked to redetermine Avogadro's number and to find the $^3H/^1H$ ratio in the sample.

He knows ^3H undergoes radioactive decay to ^3He.

He observes that a 1.00 liter sample at 1.50 atm pressure at 27°C increases in pressure to 3.75 atm on standing until all the ^3H decays. During this time 7.7×10^{22} disintegrations occur.

What does he calculate for the $^3H/^1H$ atom ratio and for Avogadro's number?

➡ **19.20** Calculate a value for the Avogadro number using only the following information:

(a) the atomic weight of chlorine is 35.5

(b) 1.00 coulomb of electricity will cause 3.61×10^{-4} g of chlorine to be liberated from melted NaCl

(c) a quantity of radioactive material emits 9.40×10^{10} alpha particles per sec; the current which accompanies this radioactivity is 2.80×10^{-8} coulombs/sec.

→ **19.21** The radioelement $^{24}_{11}$Na decays with a half life of 15 hours according to the reaction, $^{24}_{11}$Na $= ^{24}_{12}$Mg $+ ^{0}_{-1}$e.

(a) In how many hours will 25% of a sample of $^{24}_{11}$Na have decayed?

(b) What fraction of the sample will have decayed in 70 hours?

Solution: The rate of decay of any nuclide is represented by the equation

$$\log \frac{N_t}{N_0} = \frac{-kt}{2.303}$$

in which N_0 is the number of moles (or the fraction of the total moles) present at zero time (starting time), N_t is the number of moles (or the fraction of the initial total moles) present at time t, and k is the decay constant for the particular decay reaction. The decay constant, k, is numerically equal to 0.693 divided by the half life ($t_{1/2}$) of the particular nuclide.

$$k = \frac{0.693}{t_{1/2}}$$

Using the above formulas, the problem can be solved.

→ **19.22** Sixty percent of a radioactive nuclide was found to have decayed in 20 hours. Calculate its half life.

Appendix.

Table 1
Vapor pressure of water

Temperature (°C)	Pressure (mmHg)	Temperature (°C)	Pressure (mmHg)
0	4.6	21	18.5
1	4.9	22	19.8
2	5.3	23	20.9
3	5.6	24	22.2
4	6.1	25	23.6
5	6.5	26	25.1
6	7.0	27	26.5
7	7.5	28	28.1
8	8.0	29	29.8
9	8.6	30	31.5
10	9.2	31	33.4
11	9.8	32	35.4
12	10.5	33	37.4
13	11.2	34	39.6
14	11.9	35	41.9
15	12.7	36	44.2
16	13.5	37	46.7
17	14.4	38	49.4
18	15.4	39	52.1
19	16.3	40	55.0
20	17.4	100	760.0

Table 2

Ionization constants of acids and bases

Compound	Formula	K_i
Acetic	$HC_2H_3O_2$	1.8×10^{-5}
Arsenic	H_3AsO_4	$K_1 = 2.5 \times 10^{-4}$
		$K_2 = 5.6 \times 10^{-8}$
		$K_3 = 3.0 \times 10^{-13}$
Arsenious	H_3AsO_3	$K_1 = 6 \times 10^{-10}$
Boric	H_3BO_3	$K_1 = 6.0 \times 10^{-10}$
Carbonic	H_2CO_3	$K_1 = 4.2 \times 10^{-7}$
		$K_2 = 4.8 \times 10^{-11}$
Chromic	H_2CrO_4	$K_1 = 1.8 \times 10^{-1}$
		$K_2 = 3.2 \times 10^{-7}$
Formic	$HCHO_2$	2.1×10^{-4}
Hydrocyanic	HCN	4.0×10^{-10}
Hydrofluoric	HF	6.9×10^{-4}
Hydrogen sulfide	H_2S	$K_1 = 1.0 \times 10^{-7}$
		$K_2 = 1.3 \times 10^{-13}$
Hypochlorous	$HClO$	3.2×10^{-8}
Nitrous	HNO_2	4.5×10^{-4}
Oxalic	$H_2C_2O_4$	$K_1 = 3.8 \times 10^{-2}$
		$K_2 = 5.0 \times 10^{-5}$
Phosphoric	H_3PO_4	$K_1 = 7.5 \times 10^{-3}$
		$K_2 = 6.2 \times 10^{-8}$
		$K_3 = 1.0 \times 10^{-12}$
Sulfuric	H_2SO_4	$K_2 = 1.2 \times 10^{-2}$
Sulfurous	H_2SO_3	$K_1 = 1.3 \times 10^{-2}$
		$K_2 = 5.6 \times 10^{-8}$
Ammonium hydroxide	NH_4OH	1.8×10^{-5}
Water	H_2O	$K_i = 1.8 \times 10^{-16}$
		$K_{H_2O} = 1.0 \times 10^{-14}$

Table 3

Complex ion equilibria

Ligand	Equation	Instability Constant
Ammonia	$Cd(NH_3)_4^{++} \rightleftarrows Cd^{++} + 4\,NH_3$	$7.5 \times 10^{-8}\ M^4$
	$Cu(NH_3)_4^{++} \rightleftarrows Cu^{++} + 4\,NH_3$	$4.7 \times 10^{-15}\ M^4$
	$Co(NH_3)_6^{++} \rightleftarrows Co^{++} + 6\,NH_3$	$1.3 \times 10^{-5}\ M^6$
	$Co(NH_3)_6^{+++} \rightleftarrows Co^{+++} + 6\,NH_3$	$2.2 \times 10^{-34}\ M^6$
	$Ni(NH_3)_6^{++} \rightleftarrows Ni^{++} + 6\,NH_3$	$1.8 \times 10^{-9}\ M^6$
	$Ag(NH_3)_2^{+} \rightleftarrows Ag^{+} + 2\,NH_3$	$5.9 \times 10^{-8}\ M^2$
	$Zn(NH_3)_4^{++} \rightleftarrows Zn^{++} + 4\,NH_3$	$3.4 \times 10^{-10}\ M^4$
Cyanide	$Cd(CN)_4^{--} \rightleftarrows Cd^{++} + 4\,CN^-$	$1.4 \times 10^{-19}\ M^4$
	$Cu(CN)_2^{-} \rightleftarrows Cu^{+} + 2\,CN^-$	$5.0 \times 10^{-28}\ M^2$
	$Fe(CN)_6^{----} \rightleftarrows Fe^{++} + 6\,CN^-$	$1.0 \times 10^{-35}\ M^6$
	$Hg(CN)_4^{--} \rightleftarrows Hg^{++} + 4\,CN^-$	$4.0 \times 10^{-42}\ M^4$
	$Ni(CN)_4^{--} \rightleftarrows Ni^{++} + 4\,CN^-$	$1.0 \times 10^{-22}\ M^4$
	$Ag(CN)_2^{-} \rightleftarrows Ag^{+} + 2\,CN^-$	$1.8 \times 10^{-19}\ M^2$
	$Zn(CN)_4^{--} \rightleftarrows Zn^{++} + 4\,CN^-$	$1.3 \times 10^{-17}\ M^4$
Hydroxide	$Al(OH)_4^{-} \rightleftarrows Al^{+++} + 4\,OH^-$	$1.0 \times 10^{-34}\ M^4$
	$Zn(OH)_4^{--} \rightleftarrows Zn^{++} + 4\,OH^-$	$3.3 \times 10^{-16}\ M^4$
Chloride	$HgCl_4^{--} \rightleftarrows Hg^{++} + 4\,Cl^-$	$1.1 \times 10^{-16}\ M^4$
Bromide	$HgBr_4^{--} \rightleftarrows Hg^{++} + 4\,Br^-$	$2.3 \times 10^{-22}\ M^4$
Iodide	$HgI_4^{--} \rightleftarrows Hg^{++} + 4\,I^-$	$5.3 \times 10^{-31}\ M^4$
Thiosulfate	$Ag(S_2O_3)_2^{---} \rightleftarrows Ag^{+} + 2\,S_2O_3^{--}$	$3.5 \times 10^{-14}\ M^2$

Table 4
Solubility products at 20 °C

Compound	Product	K_{sp}
Aluminum hydroxide	$[Al^{+++}] \times [OH^-]^3$	5×10^{-33}
Barium carbonate	$[Ba^{++}] \times [CO_3^{--}]$	1.6×10^{-9}
Barium chromate	$[Ba^{++}] \times [CrO_4^{--}]$	8.5×10^{-11}
Barium sulfate	$[Ba^{++}] \times [SO_4^{--}]$	1.5×10^{-9}
Barium oxalate	$[Ba^{++}] \times [C_2O_4^{--}]$	1.5×10^{-8}
Bismuth sulfide	$[Bi^{+++}]^2 \times [S^{--}]^3$	1×10^{-70}
Cadmium hydroxide	$[Cd^{++}] \times [OH^-]^2$	2×10^{-14}
Cadmium sulfide	$[Cd^{++}] \times [S^{--}]$	6×10^{-27}
Calcium carbonate	$[Ca^{++}] \times [CO_3^{--}]$	6.9×10^{-9}
Calcium oxalate	$[Ca^{++}] \times [C_2O_4^{--}]$	1.3×10^{-9}
Calcium sulfate	$[Ca^{++}] \times [SO_4^{--}]$	2.4×10^{-5}
Chromium hydroxide	$[Cr^{+++}] \times [OH^-]^3$	7×10^{-31}
Cobalt sulfide	$[Co^{++}] \times [S^{--}]$	5×10^{-22}
Cupric hydroxide	$[Cu^{++}] \times [OH^-]^2$	1.6×10^{-19}
Cupric sulfide	$[Cu^{++}] \times [S^{--}]$	4×10^{-36}
Ferric hydroxide	$[Fe^{+++}] \times [OH^-]^3$	6×10^{-38}
Ferrous hydroxide	$[Fe^{++}] \times [OH^-]^2$	2×10^{-15}
Ferrous sulfide	$[Fe^{++}] \times [S^{--}]$	4×10^{-17}
Lead carbonate	$[Pb^{++}] \times [CO_3^{--}]$	1.5×10^{-13}
Lead chromate	$[Pb^{++}] \times [CrO_4^{--}]$	2×10^{-16}
Lead iodide	$[Pb^{++}] \times [I^-]^2$	8.3×10^{-9}
Lead sulfate	$[Pb^{++}] \times [SO_4^{--}]$	1.3×10^{-8}
Lead sulfide	$[Pb^{++}] \times [S^{--}]$	4×10^{-26}

Table 4
Solubility products at 20°C (cont.)

Compound	Product	K_{sp}
Magnesium carbonate	$[Mg^{++}] \times [CO_3^{--}]$	4×10^{-5}
Magnesium hydroxide	$[Mg^{++}] \times [OH^-]^2$	8.9×10^{-12}
Magnesium oxalate	$[Mg^{++}] \times [C_2O_4^{--}]$	8.6×10^{-5}
Manganese hydroxide	$[Mn^{++}] \times [OH^-]^2$	2×10^{-13}
Manganese sulfide	$[Mn^{++}] \times [S^{--}]$	8×10^{-14}
Mercurous chloride	$[Hg_2^{++}] \times [Cl^-]^2$	1.1×10^{-18}
Mercuric sulfide	$[Hg^{++}] \times [S^{--}]$	1×10^{-50}
Nickel hydroxide	$[Ni^{++}] \times [OH^-]^2$	1.6×10^{-16}
Nickel sulfide	$[Ni^{++}] \times [S^{--}]$	1×10^{-22}
Silver arsenate	$[Ag^+]^3 \times [AsO_4^{---}]$	1×10^{-23}
Silver bromide	$[Ag^+] \times [Br^-]$	5×10^{-13}
Silver carbonate	$[Ag^+]^2 \times [CO_3^{--}]$	8.2×10^{-12}
Silver chloride	$[Ag^+] \times [Cl^-]$	2.8×10^{-10}
Silver chromate	$[Ag^+]^2 \times [CrO_4^{--}]$	1.9×10^{-12}
Silver iodate	$[Ag^+] \times [IO_3^-]$	3×10^{-8}
Silver iodide	$[Ag^+] \times [I^-]$	8.5×10^{-17}
Silver phosphate	$[Ag^+]^3 \times [PO_4^{---}]$	1.8×10^{-18}
Silver sulfide	$[Ag^+]^2 \times [S^{--}]$	1×10^{-50}
Silver thiocyanate	$[Ag^+] \times [CNS^-]$	1×10^{-12}
Stannous sulfide	$[Sn^{++}] \times [S^{--}]$	1×10^{-24}
Zinc hydroxide	$[Zn^{++}] \times [OH^-]^2$	5×10^{-17}
Zinc sulfide	$[Zn^{++}] \times [S^{--}]$	1×10^{-20}

Table 5

Some standard oxidation potentials in acid solution*

	Half-reaction	$E°\ (V)$
1	$K\ (s) \rightleftarrows K^+ + e^-$	2.925
2	$Ca\ (s) \rightleftarrows Ca^{++} + 2\ e^-$	2.87
3	$Al\ (s) \rightleftarrows Al^{+++} + 3\ e^-$	1.66
4	$Mn\ (s) \rightleftarrows Mn^{++} + 2\ e^-$	1.18
5	$H_2O + H_2PO_3 \rightleftarrows H_3PO_4 + H^+ + e^-$	0.9
6	$Zn\ (s) \rightleftarrows Zn^{++} + 2\ e^-$	0.763
7	$P\ (s) + 2H_2O \rightleftarrows H_3PO_2 + H^+ + e^-$	0.51
8	$H_3PO_2 + H_2O \rightleftarrows H_3PO_3 + 2\ H^+ + 2\ e^-$	0.50
9	$Cr^{++} \rightleftarrows Cr^{+++} + e^-$	0.41
10	$H_3PO_3 + H_2O \rightleftarrows H_3PO_4 + 2\ H^+ + 2\ e^-$	0.276
11	$Ni\ (s) \rightleftarrows Ni^{++} + 2\ e^-$	0.250
12	$Sn\ (s) \rightleftarrows Sn^{++} + 2\ e^-$	0.136
13	$HS_2O_4^- + 2\ H_2O \rightleftarrows 2\ H_2SO_3 + H^+ + 2\ e^-$	0.08
14	$H_2\ (g) \rightleftarrows 2\ H^+ + 2\ e^-$	0.000
15	$PH_3 \rightleftarrows P\ (s) + 3\ H^+ + 3\ e^-$	−0.06
16	$H_2S \rightleftarrows 2\ H^+ + S + 2\ e^-$	−0.141
17	$Sn^{++} \rightleftarrows Sn^{++++} + 2\ e^-$	−0.15
18	$H_2SO_3 + H_2O \rightleftarrows SO_4^{--} + 4\ H^+ + 2\ e^-$	−0.17
19	$Cu\ (s) \rightleftarrows Cu^{++} + 2\ e^-$	−0.337
20	$S\ (s) + 3\ H_2O \rightleftarrows H_2SO_3 + 4\ H^+ + 4\ e^-$	−0.45
21	$2\ I^- \rightleftarrows I_2 + 2\ e^-$	−0.5355
22	$MnO_4^{--} \rightleftarrows MnO_4^- + e^-$	−0.564

Table 5

Some standard oxidation potentials in acid solution* (cont.)

	Half-reaction	$E°\ (V)$
23	$H_2O_2 \rightleftarrows O_2 + 2\ H^+ + 2\ e^-$	-0.682
24	$Fe^{++} \rightleftarrows Fe^{+++} + e^-$	-0.771
25	$Ag\ (s) \rightleftarrows Ag^+ + e^-$	-0.799
26	$NO_2 + H_2O \rightleftarrows NO_3^- + 2\ H^+ + e^-$	-0.80
27	$Hg\ (s) \rightleftarrows Hg^{++} + 2\ e^-$	-0.854
28	$NO + 2\ H_2O \rightleftarrows NO_3^- + 4\ H^+ + 3\ e^-$	-0.96
29	$NO + H_2O \rightleftarrows HNO_2 + H^+ + e^-$	-1.00
30	$2\ Br^- \rightleftarrows Br_2 + 2\ e^-$	-1.065
31	$Mn^{++} + 2\ H_2O \rightleftarrows MnO_2\ (s) + 4\ H^+ + 2\ e^-$	-1.23
32	$2\ Cr^{+++} + 7\ H_2O \rightleftarrows Cr_2O_7^{--} + 14\ H^+ + 6\ e^-$	-1.33
33	$2\ Cl^- \rightleftarrows Cl_2 + 2\ e^-$	-1.3595
34	$Cl^- + 3\ H_2O \rightleftarrows ClO_3^- + 6\ H^+ + 6\ e^-$	-1.45
35	$Mn^{++} + 4\ H_2O \rightleftarrows MnO_4^- + 8\ H^+ + 5\ e^-$	-1.51
36	$Mn^{++} \rightleftarrows Mn^{+++} + e^-$	-1.51
37	$Bi^{+++} + 3\ H_2O \rightleftarrows HBiO_3 + 5\ H^+ + 2\ e^-$	-1.70
38	$2\ H_2O \rightleftarrows H_2O_2 + 2\ H^+ + 2\ e^-$	-1.77
39	$2\ F^- \rightleftarrows F_2 + 2\ e^-$	-2.65
40	$2\ HF \rightleftarrows F_2 + 2\ H^+ + 2\ e^-$	-3.06

*For a complete list of oxidation potentials see *Oxidation Potentials,* Second Edition, by W. H. Latimer, Prentice-Hall, Inc., Englewood Cliffs, N.J., 1952.

Table 6

Some standard oxidation potentials in alkaline solution

	Half-reaction	$E°$ (V)
1	$Ca\ (s) + 2\ OH^- \rightleftarrows Ca(OH)_2\ (s) + 2\ e^-$	3.03
2	$H_2 + 2\ OH^- \rightleftarrows 2\ H_2O + 2\ e^-$	2.93
3	$K\ (s) \rightleftarrows K^+ + e^-$	2.925
4	$Al\ (s) + 4\ OH^- \rightleftarrows Al(OH)_4^- + 3\ e^-$	2.35
5	$P\ (s) + 2\ OH^- \rightleftarrows H_2PO_2^- + e^-$	2.05
6	$H_2PO_2^- + 3\ OH^- \rightleftarrows HPO_3^{--} + 2\ H_2O + 2\ e^-$	1.57
7	$Mn\ (s) + 2\ OH^- \rightleftarrows Mn(OH)_2\ (s) + 2\ e^-$	1.55
8	$Zn\ (s) + S^{--} \rightleftarrows ZnS\ (s) + 2\ e^-$	1.44
9	$Zn\ (s) + 4\ CN^- \rightleftarrows Zn(CN)_4^{--} + 2\ e^-$	1.26
10	$Zn\ (s) + 4\ OH^- \rightleftarrows Zn(OH)_4^{--} + 2\ e^-$	1.216
11	$HPO_3^{--} + 3\ OH^- \rightleftarrows PO_4^{---} + 2\ H_2O + 2\ e^-$	1.12
12	$S_2O_4^{--} + 4\ OH^- \rightleftarrows 2\ SO_3^{--} + 2\ H_2O + 2\ e^-$	1.12
13	$Zn\ (s) + 4\ NH_3 \rightleftarrows Zn(NH_3)_4^{++} + 2\ e^-$	1.03
14	$CN^- + 2\ OH^- \rightleftarrows CNO^- + H_2O + 2\ e^-$	0.97
15	$SO_3^{--} + 2\ OH^- \rightleftarrows SO_4^{--} + H_2O + 2\ e^-$	0.93
16	$Sn(OH)_4^{--} + 2\ OH^- \rightleftarrows Sn(OH)_6^{--} + 2\ e^-$	0.90
17	$PH_3 + 3\ OH^- \rightleftarrows P\ (s) + 3\ H_2O + 3\ e^-$	0.89
18	$Sn\ (s) + 4\ OH^- \rightleftarrows Sn(OH)_4^{--} + 2\ e^-$	0.76
19	$Ni\ (s) + 2\ OH^- \rightleftarrows Ni(OH)_2\ (s) + 2\ e^-$	0.72
20	$Fe(OH)_2\ (s) + OH^- \rightleftarrows Fe(OH)_3\ (s) + e^-$	0.56

Table 6

Some standard oxidation potentials in alkaline solution (cont.)

	Half-reaction	$E°$ (V)
21	$S^{--} \rightleftarrows S + 2\ e^-$	0.48
22	$Cr(OH)_4^- + 4\ OH^- \rightleftarrows CrO_4^{--} + 4\ H_2O + 3\ e^-$	0.13
23	$H_2O_2 + 2\ OH^- \rightleftarrows O_2 + 2\ H_2O + 2\ e^-$	0.076
24	$Mn(OH)_2\ (s) + 2\ OH^- \rightleftarrows MnO_2\ (s) + 2\ H_2O + 2\ e^-$	0.05
25	$Cu(NH_3)_2^+ + 2\ NH_3 \rightleftarrows Cu(NH_3)_4^{++} + e^-$	0.0
26	$Mn(OH)_2\ (s) + OH^- \rightleftarrows Mn(OH)_3\ (s) + e^-$	−0.1
27	$Co(NH_3)_6^{++} \rightleftarrows Co(NH_3)_6^{+++} + e^-$	−0.1
28	$Co(OH)_2\ (s) + OH^- \rightleftarrows Co(OH)_3\ (s) + e^-$	−0.17
29	$ClO_2^- + 2\ OH^- \rightleftarrows ClO_3^- + H_2O + 2\ e^-$	−0.33
30	$ClO_3^- + 2\ OH^- \rightleftarrows ClO_4^- + H_2O + 2\ e^-$	−0.36
31	$4\ OH^- \rightleftarrows O_2 + 2\ H_2O + 4\ e^-$	−0.401
32	$I^- + 2\ OH^- \rightleftarrows IO^- + H_2O + 2\ e^-$	−0.49
33	$Ni(OH)_2\ (s) + 2\ OH^- \rightleftarrows NiO_2\ (s) + 2\ H_2O + 2\ e^-$	−0.49
34	$MnO_4^{--} \rightleftarrows MnO_4^- + e^-$	−0.564
35	$MnO_2\ (s) + 4\ OH^- \rightleftarrows MnO_4^- + 2\ H_2O + 3\ e^-$	−0.588
36	$MnO_2\ (s) + 4\ OH^- \rightleftarrows MnO_4^{--} + 2\ H_2O + 2\ e^-$	−0.60
37	$ClO^- + 2\ OH^- \rightleftarrows ClO_2^- + H_2O + 2\ e^-$	−0.66
38	$Br^- + 2\ OH^- \rightleftarrows BrO^- + H_2O + 2\ e^-$	−0.76
39	$2\ OH^- \rightleftarrows H_2O_2 + 2\ e^-$	−0.88
40	$Cl^- + 2\ OH^- \rightleftarrows ClO^- + H_2O + 2\ e^-$	−0.89

Table 7
Standard heats of formation

Compound	ΔH_f° (kcal/mole)	Compound	ΔH_f° (kcal/mole)
$C_2H_{2(g)}$	54.19	$HBr_{(g)}$	−8.6
$C_2H_{4(g)}$	12.5	$HCl_{(g)}$	−22.06
$C_2H_{6(g)}$	−20.24	$H_2O_{(l)}$	−68.32
$C_3H_{8(g)}$	−24.82	$NH_{3(g)}$	−11.04
$CO_{2(g)}$	−94.05	$NH_4Cl_{(s)}$	−75.3
$CaCO_{3(s)}$	−288.4	$NO_{2(g)}$	8.1
$CaO_{(s)}$	−151.8	$N_2O_{4(g)}$	10.5
$Fe_2O_{3(s)}$	−196.5	$SO_{2(g)}$	−70.96

Table 8
Bond energies

Bond	Energy (kcal/mole)	Bond	Energy (kcal/mole)
C—C	82.6	H—H	104.2
C=C	145.8	H—Br	87.5
C≡C	199.6	H—Cl	103.2
C—Cl	80	H—I	71.5
C—H	98.7	H—N	93.5
Br—Br	46.1	H—O	110.6
Cl—Cl	58.1	N≡N	226
F—F	37.8	N=O	150
I—I	36.1	O=O	119

Table 9

Four-place logarithms

N	0	1	2	3	4	5	6	7	8	9	1	2	3	4	5	6	7	8	9
10	0000	0043	0086	0128	0170	0212	0253	0294	0334	0374	4	8	12	17	21	25	29	33	37
11	0414	0453	0492	0531	0569	0607	0645	0682	0719	0755	4	8	11	15	19	23	26	30	34
12	0792	0828	0864	0899	0934	0969	1004	1038	1072	1106	3	7	10	14	17	21	24	28	31
13	1139	1173	1206	1239	1271	1303	1335	1367	1399	1430	3	6	10	13	16	19	23	26	29
14	1461	1492	1523	1553	1584	1614	1644	1673	1703	1732	3	6	9	12	15	18	21	24	27
15	1761	1790	1818	1847	1875	1903	1931	1959	1987	2014	3	6	8	11	14	17	20	22	25
16	2041	2068	2095	2122	2148	2175	2201	2227	2253	2279	3	5	8	11	13	16	18	21	24
17	2304	2330	2355	2380	2405	2430	2455	2480	2504	2529	2	5	7	10	12	15	17	20	22
18	2553	2577	2601	2625	2648	2672	2695	2718	2742	2765	2	5	7	9	12	14	16	19	21
19	2788	2810	2833	2856	2878	2900	2923	2945	2967	2989	2	4	7	9	11	13	16	18	20
20	3010	3032	3054	3075	3096	3118	3139	3160	3181	3201	2	4	6	8	11	13	15	17	19
21	3222	3243	3263	3284	3304	3324	3345	3365	3385	3404	2	4	6	8	10	12	14	16	18
22	3424	3444	3464	3483	3502	3522	3541	3560	3579	3598	2	4	6	8	10	12	14	16	17
23	3617	3636	3655	3674	3692	3711	3729	3747	3766	3784	2	4	6	7	9	11	13	15	17
24	3802	3820	3838	3856	3874	3892	3909	3927	3945	3962	2	4	5	7	9	11	12	14	16
25	3979	3997	4014	4031	4048	4065	4082	4099	4116	4133	2	4	5	7	9	10	12	14	16
26	4150	4166	4183	4200	4216	4232	4249	4265	4281	4298	2	3	5	7	8	10	11	13	15
27	4314	4330	4346	4362	4378	4393	4409	4425	4440	4456	2	3	5	6	8	9	11	12	14
28	4472	4487	4502	4518	4533	4548	4564	4579	4594	4609	2	3	5	6	8	9	11	12	14
29	4624	4639	4654	4669	4683	4698	4713	4728	4742	4757	1	3	4	6	7	9	10	12	13
30	4771	4786	4800	4814	4829	4843	4857	4871	4886	4900	1	3	4	6	7	9	10	11	13
31	4914	4928	4942	4955	4969	4983	4997	5011	5024	5038	1	3	4	5	7	8	10	11	12
32	5051	5065	5079	5092	5105	5119	5132	5145	5159	5172	1	3	4	5	7	8	9	11	12
33	5185	5198	5211	5224	5237	5250	5263	5276	5289	5302	1	3	4	5	7	8	9	11	12
34	5315	5328	5340	5353	5366	5378	5391	5403	5416	5428	1	2	4	5	6	8	9	10	11
35	5441	5453	5465	5478	5490	5502	5514	5527	5539	5551	1	2	4	5	6	7	9	10	11
36	5563	5575	5587	5599	5611	5623	5635	5647	5658	5670	1	2	4	5	6	7	8	10	11
37	5682	5694	5705	5717	5729	5740	5752	5763	5775	5786	1	2	4	5	6	7	8	9	11
38	5798	5809	5821	5832	5843	5855	5866	5877	5888	5899	1	2	3	5	6	7	8	9	10
39	5911	5922	5933	5944	5955	5966	5977	5988	5999	6010	1	2	3	4	5	7	8	9	10
40	6021	6031	6042	6053	6064	6075	6085	6096	6107	6117	1	2	3	4	5	6	8	9	10
41	6128	6138	6149	6160	6170	6180	6191	6201	6212	6222	1	2	3	4	5	6	7	8	9
42	6232	6243	6253	6263	6274	6284	6294	6304	6314	6325	1	2	3	4	5	6	7	8	9
43	6335	6345	6355	6365	6375	6385	6395	6405	6415	6425	1	2	3	4	5	6	7	8	9
44	6435	6444	6454	6464	6474	6484	6493	6503	6513	6522	1	2	3	4	5	6	7	8	9
45	6532	6542	6551	6561	6571	6580	6590	6599	6609	6618	1	2	3	4	5	6	7	8	9
46	6628	6637	6646	6656	6665	6675	6684	6693	6702	6712	1	2	3	4	5	6	7	7	8
47	6721	6730	6739	6749	6758	6767	6776	6785	6794	6803	1	2	3	4	5	6	7	7	8
48	6812	6821	6830	6839	6848	6857	6866	6875	6884	6893	1	2	3	4	5	6	7	7	8
49	6902	6911	6920	6928	6937	6946	6955	6964	6972	6981	1	2	3	4	4	5	6	7	8
50	6990	6998	7007	7016	7024	7033	7042	7050	7059	7067	1	2	3	3	4	5	6	7	8
51	7076	7084	7093	7101	7110	7118	7126	7135	7143	7152	1	2	3	3	4	5	6	7	8
52	7160	7168	7177	7185	7193	7202	7210	7218	7226	7235	1	2	3	3	4	5	6	7	7
53	7243	7251	7259	7267	7275	7284	7292	7300	7308	7316	1	2	2	3	4	5	6	6	7
54	7324	7332	7340	7348	7356	7364	7372	7380	7388	7396	1	2	2	3	4	5	6	6	7
N	0	1	2	3	4	5	6	7	8	9	1	2	3	4	5	6	7	8	9

Table 9
Four-place logarithms   (cont.)

N	0	1	2	3	4	5	6	7	8	9	1	2	3	4	5	6	7	8	9
55	7404	7412	7419	7427	7435	7443	7451	7459	7466	7474	1	2	2	3	4	5	5	6	7
56	7482	7490	7497	7505	7513	7520	7528	7536	7543	7551	1	2	2	3	4	5	5	6	7
57	7559	7566	7574	7582	7589	7597	7604	7612	7619	7627	1	1	2	3	4	5	5	6	7
58	7634	7642	7649	7657	7664	7672	7679	7686	7694	7701	1	1	2	3	4	4	5	6	7
59	7709	7716	7723	7731	7738	7745	7752	7760	7767	7774	1	1	2	3	4	4	5	6	7
60	7782	7789	7796	7803	7810	7818	7825	7832	7839	7846	1	1	2	3	4	4	5	6	6
61	7853	7860	7868	7875	7882	7889	7896	7903	7910	7917	1	1	2	3	3	4	5	6	6
62	7924	7931	7938	7945	7952	7959	7966	7973	7980	7987	1	1	2	3	3	4	5	5	6
63	7993	8000	8007	8014	8021	8028	8035	8041	8048	8055	1	1	2	3	3	4	5	5	6
64	8062	8069	8075	8082	8089	8096	8102	8109	8116	8122	1	1	2	3	3	4	5	5	6
65	8129	8136	8142	8149	8156	8162	8169	8176	8182	8189	1	1	2	3	3	4	5	5	6
66	8195	8202	8209	8215	8222	8228	8235	8241	8248	8254	1	1	2	3	3	4	5	5	6
67	8261	8267	8274	8280	8287	8293	8299	8306	8312	8319	1	1	2	3	3	4	5	5	6
68	8325	8331	8338	8344	8351	8357	8363	8370	8376	8382	1	1	2	3	3	4	4	5	6
69	8388	8395	8401	8407	8414	8420	8426	8432	8439	8445	1	1	2	3	3	4	4	5	6
70	8451	8457	8463	8470	8476	8482	8488	8494	8500	8506	1	1	2	3	3	4	4	5	6
71	8513	8519	8525	8531	8537	8543	8549	8555	8561	8567	1	1	2	3	3	4	4	5	6
72	8573	8579	8585	8591	8597	8603	8609	8615	8621	8627	1	1	2	3	3	4	4	5	6
73	8633	8639	8645	8651	8657	8663	8669	8675	8681	8686	1	1	2	2	3	4	4	5	5
74	8692	8698	8704	8710	8716	8722	8727	8733	8739	8745	1	1	2	2	3	4	4	5	5
75	8751	8756	8762	8768	8774	8779	8785	8791	8797	8802	1	1	2	2	3	3	4	5	5
76	8808	8814	8820	8825	8831	8837	8842	8848	8854	8859	1	1	2	2	3	3	4	4	5
77	8865	8871	8876	8882	8887	8893	8899	8904	8910	8915	1	1	2	2	3	3	4	4	5
78	8921	8927	8932	8938	8943	8949	8954	8960	8965	8971	1	1	2	2	3	3	4	4	5
79	8976	8982	8987	8993	8998	9004	9009	9015	9020	9025	1	1	2	2	3	3	4	4	5
80	9031	9036	9042	9047	9053	9058	9063	9069	9074	9079	1	1	2	2	3	3	4	4	5
81	9085	9090	9096	9101	9106	9112	9117	9122	9128	9133	1	1	2	2	3	3	4	4	5
82	9138	9143	9149	9154	9159	9165	9170	9175	9180	9186	1	1	2	2	3	3	4	4	5
83	9191	9196	9201	9206	9212	9217	9222	9227	9232	9238	1	1	2	2	3	3	4	4	5
84	9243	9248	9253	9258	9263	9269	9274	9279	9284	9289	1	1	2	2	3	3	4	4	5
85	9294	9299	9304	9309	9315	9320	9325	9330	9335	9340	1	1	2	2	3	3	4	4	5
86	9345	9350	9355	9360	9365	9370	9375	9380	9385	9390	1	1	2	2	3	3	4	4	5
87	9395	9400	9405	9410	9415	9420	9425	9430	9435	9440	1	1	2	2	3	3	4	4	5
88	9445	9450	9455	9460	9465	9469	9474	9479	9484	9489	0	1	1	2	2	3	3	4	4
89	9494	9499	9504	9509	9513	9518	9523	9528	9533	9538	0	1	1	2	2	3	3	4	4
90	9542	9547	9552	9557	9562	9566	9571	9576	9581	9586	0	1	1	2	2	3	3	4	4
91	9590	9595	9600	9605	9609	9614	9619	9624	9628	9633	0	1	1	2	2	3	3	4	4
92	9638	9643	9647	9652	9657	9661	9666	9671	9675	9680	0	1	1	2	2	3	3	4	4
93	9685	9689	9694	9699	9703	9708	9713	9717	9722	9727	0	1	1	2	2	3	3	4	4
94	9731	9736	9741	9745	9750	9754	9759	9763	9768	9773	0	1	1	2	2	3	3	4	4
95	9777	9782	9786	9791	9795	9800	9805	9809	9814	9818	0	1	1	2	2	3	3	4	4
96	9823	9827	9832	9836	9841	9845	9850	9854	9859	9863	0	1	1	2	2	3	3	4	4
97	9868	9872	9877	9881	9886	9890	9894	9899	9903	9908	0	1	1	2	2	3	3	4	4
98	9912	9917	9921	9926	9930	9934	9939	9943	9948	9952	0	1	1	2	2	3	3	3	4
99	9956	9961	9965	9969	9974	9978	9983	9987	9991	9996	0	1	1	2	2	3	3	3	4
N	0	1	2	3	4	5	6	7	8	9	1	2	3	4	5	6	7	8	9

Table 10
Atomic weights of the common elements*

Element	Symbol	Atomic Weight
Aluminum	Al	27.0
Antimony	Sb	121.8
Arsenic	As	74.9
Barium	Ba	137.3
Bismuth	Bi	209.0
Boron	B	10.8
Bromine	Br	79.9
Cadmium	Cd	112.4
Calcium	Ca	40.1
Carbon	C	12.0
Chlorine	Cl	35.4
Chromium	Cr	52.0
Cobalt	Co	58.9
Copper	Cu	63.5
Fluorine	F	19.0
Gold	Au	197.0
Hydrogen	H	1.0
Iodine	I	126.9
Iron	Fe	55.8
Lead	Pb	207.2
Lithium	Li	6.9
Magnesium	Mg	24.3
Manganese	Mn	54.9
Mercury	Hg	200.6
Nickel	Ni	58.7
Nitrogen	N	14.0
Oxygen	O	16.0
Phosphorus	P	31.0
Potassium	K	39.1
Radium	Ra	226.0
Silicon	Si	28.1
Silver	Ag	107.9
Sodium	Na	23.0
Sulfur	S	32.1
Tin	Sn	118.7
Zinc	Zn	65.4

*These atomic weights have been rounded off to one significant figure to the right of the decimal point and should be used for solving all problems in this book. Table 11 is a complete table of exact international atomic weights.

Table 11
International atomic weights

	Symbol	Atomic Number	Atomic Weight		Symbol	Atomic Number	Atomic Weight
Actinium	Ac	89	227*	Mendelevium	Md	101	256*
Aluminum	Al	13	26.9815	Mercury	Hg	80	200.59
Americium	Am	95	243*	Molybdenum	Mo	42	95.94
Antimony	Sb	51	121.75	Neodymium	Nd	60	144.24
Argon	Ar	18	39.948	Neon	Ne	10	20.183
Arsenic	As	33	74.9216	Neptunium	Np	93	237*
Astatine	At	85	210*	Nickel	Ni	28	58.71
Barium	Ba	56	137.34	Niobium	Nb	41	92.906
Berkelium	Bk	97	247*	Nitrogen	N	7	14.0067
Beryllium	Be	4	9.0122	Nobelium	No	102	255*
Bismuth	Bi	83	208.980	Osmium	Os	76	190.2
Boron	B	5	10.811	Oxygen	O	8	15.9994
Bromine	Br	35	79.909	Palladium	Pd	46	105.4
Cadmium	Cd	48	112.40	Phosphorus	P	15	30.9738
Calcium	Ca	20	40.08	Platinum	Pt	78	195.09
Californium	Cf	98	251*	Plutonium	Pu	94	242*
Carbon	C	6	12.01115	Polonium	Po	84	210*
Cerium	Ce	58	140.12	Potassium	K	19	39.102
Cesium	Cs	55	132.905	Praseodymium	Pr	59	140.907
Chlorine	Cl	17	35.453	Promethium	Pm	61	145*
Chromium	Cr	24	51.996	Protactinium	Pa	91	231*
Cobalt	Co	27	58.9332	Radium	Ra	88	226*
Copper	Cu	29	63.54	Radon	Rn	86	222*
Curium	Cm	96	247*	Rhenium	Re	75	186.2
Dysprosium	Dy	66	162.50	Rhodium	Rh	45	102.905
Einsteinium	Es	99	254	Rubidium	Rb	37	85.47
Erbium	Er	68	167.26	Ruthenium	Ru	44	101.07
Europium	Eu	63	151.96	Samarium	Sm	62	150.35
Fermium	Fm	100	257*	Scandium	Sc	21	474.956
Fluorine	F	9	18.9984	Selenium	Se	34	8.96
Francium	Fr	87	223*	Silicon	Si	14	28.086
Gadolinium	Gd	64	157.25	Silver	Ag	47	107.870
Gallium	Ga	31	69.72	Sodium	Na	11	22.9898
Germanium	Ge	32	72.59	Strontium	Sr	38	87.62
Gold	Au	79	196.967	Sulfur	S	16	32.064
Hafnium	Hf	72	178.49	Tantalum	Ta	73	180.948
Helium	He	2	4.0026	Technetium	Tc	43	99*
Holmium	Ho	67	164.930	Tellurium	Te	52	127.60
Hydrogen	H	1	1.00797	Terbium	Tb	65	158.924
Indium	In	49	114.82	Thallium	Tl	81	204.37
Iodine	I	53	126.9044	Thorium	Th	90	232.038
Iridium	Ir	77	192.2	Thulium	Tm	69	168.934
Iron	Fe	26	55.847	Tin	Sn	50	118.69
Krypton	Kr	36	83.80	Titanium	Ti	22	47.90
Kurchatovium	Ku	104	260*	Tungsten	W	74	183.85
Lanthanum	La	57	138.91	Uranium	U	92	238.03
Lawrencium	Lr	103	256*	Vanadium	V	23	50.942
Lead	Pb	82	207.19	Xenon	Xe	54	131.30
Lithium	Li	3	6.939	Ytterbium	Yb	70	173.04
Lutetium	Lu	71	174.97	Yttrium	Y	39	88.905
Magnesium	Mg	12	24.312	Zinc	Zn	30	65.37
Manganese	Mn	25	54.9381	Zirconium	Zr	40	91.22

*Mass number of isotope of longest known half-life.

Answers to problems.

2.1(a). 22.2; **(b).** −28.9. **2.2(a).** 53.6; **(b).** −58.0 **2.3.** −40°. **2.4.** −34°. **2.5** 73°.

3.1(a). 2.1×10^{10}; **(b).** 7.6×10^2; **(c).** 2.7×10^{-3}; **(d).** 1.8×10^{-6}; **(e).** 1.0×10^{-1}. **3.2(a).** 2.69×10^{-1}; **(b).** 1.81×10^2. **3.3(a).** 2.2×10^{-6}; **(b).** 6.5×10^{-2}.

4.2. 30.42/40.08. **4.3.** 30.0. **4.5.** 89.7 g. **4.6.** 3.22 g. **4.7.** 45.1 g. **4.8.** 89.9 tons. **4.12.** 2.97×10^{24}. **4.13.** 74.2. **4.14.** 34.5. **4.15.** 7.53×10^{23}. **4.16.** 6.83×10^{-4}. **4.18.** 1.42×10^{28}. **4.20.** 99.98% of H, 0.02% of D.

5.7. 20/17. **5.9.** 1.69. **5.11.** 1.70. **5.12.** 34. **5.14.** 2.9. **5.16.** 2.24×10^{22}. **5.18.** 8.57×10^{-4}. **5.20.** 4.70×10^{22}. **5.22.** 12.0. **5.24.** 17. **5.26.** 162. **5.27.** 1.56. **5.29.** 3.72×10^{-4}. **5.30.** 545. **5.32.** 0.624. **5.33.** 3.34. **5.34.** 17.2. **5.36(a).** 48.0; **(b).** 50.0. **5.37(a).** 1.62; **(b).** 360; **(c).** 117; **(d).** 51.4; **(e).** 2.44; **(f).** 0.36; **(g).** 10.2; **(h).** 1.47×10^{22}; **(i).** 8.82×10^{22}; **(j).** 58.5; **(k).** $\frac{1}{6}$; **(l).** 0.0697; **(m).** 14.4. **5.38.** B. **5.39.** 47.3. **5.42.** $KClO_3$. **5.43.** yes. **5.45.** CuO. **5.47.** $Na_2S_2O_3$. **5.48.** $Mg_2P_2O_7$. **5.50.** $C_8H_{20}N_4$. **5.51.** $C_6H_{12}O_6$. **5.52.** C_3H_8. **5.53.** C_4H_8O. **5.54.** C_2H_3. **5.55.** C_5H_5N. **5.56.** C_3H_3ClN. **5.57.** CH_3NO_2. **5.58.** $C_6H_7NO_2S$. **5.59.** C_6H_9BrO. **5.60.** XY_3. **5.61.** XZ_2; X_2Z_5; 3.18 times as great.

6.2. 845. **6.3.** 12. **6.4.** 600. **6.5.** 280,000. **6.7.** 575 cc. **6.8.** 23.2 cu ft.
6.9. 606°C. **6.11.** 45.5 liters. **6.12.** 97°C. **6.13.** 11.8 atm. **6.14.** 645 mm.
6.15. 0.84 liter. **6.17.** 8.1 liters. **6.19.** 0.817. **6.21.** 24.5. **6.23.** 236.
6.25. 93.0. **6.26.** 34.9. **6.28.** 30. **6.29.** 31.6 liters. **6.30.** 9.27. **6.31.** 21.8.
6.32. 30. **6.38.** 356. **6.39.** 177. **6.40.** 14.6. **6.41.** 12. **6.42.** 46.
6.43. 1.205×10^{24}. **6.44.** 170. **6.46.** N_2. **6.47.** 2nd vol $= 7.79 \times$ 1st vol.
6.48. 2.2. **6.49.** 0.195. **6.51.** 1.25. **6.53.** 0.298 g/liter. **6.54.** O_2; 1.07.
6.55 0.13 g/liter. **6.56.** 0°C. **6.57.** 0.089 g/liter. **6.58.** 20 mm. **6.59.** 0.50.
6.60. 0.25. **6.61.** 148 mm. **6.62.** 178 mm. **6.63.** 1400 mm. **6.64.** 158.
6.68. 15 mm. **6.69.** 70. **6.70.** 0.320 g. **6.71(a).** 67.2 liters; **(b).** 17.5 liters;
(c). 6.25; **(d).** 168 g; **(e).** 1.96; **(f).** 81.8; **(g).** 200 mm; **(h).** 1.0. **6.73.** 109 cu ft.
6.74. 22 mm. **6.75.** 0.047 mole. **6.76.** 239. **6.78(a).** 0.664 atm; **(b).** 1.34 atm.
6.79. 112 mm. **6.81.** 0.020. **6.82.** 11.2 mm. **6.84.** 1.02 ft/min.
6.85. As 8 is to 9. **6.86.** 33.8. **6.87.** 36. **6.88.** 2.4. **6.89.** C_2H_6S.
6.90. $C_4H_{10}O$. **6.91.** 30.068. **6.92.** 16.042.

7.8. 4.20; 2.40; 3.60. **7.9.** 1.31; 0.524. **7.10.** 1.90. **7.11.** 8.48. **7.12.** 12.
7.14. 163. **7.15.** 3.70. **7.17.** 3000; 6000. **7.20.** 124. **7.21.** 54.8. **7.22.** 2.17.
7.23. 101. **7.24.** 80.0. **7.25.** 4.00%. **7.26.** 341.
7.27. $3 MnO_2 = Mn_3O_4 + O_2$. **7.28.** $Pb_3O_4 + 4 H_2 = 3 Pb + 4 H_2O$.
7.29. $2 NaNO_3 = 2 NaNO_2 + O_2$. **7.31.** 84.4. **7.32.** 61.7 g. **7.33.** $NaNO_3$.
7.34. 128. **7.36.** 108. **7.37.** 14.0; 107.9. **7.38.** $Zr + 4 HCl = ZrCl_4 + 2 H_2$.
7.40. 2.33 to 1. **7.41.** $2 H_2S = 2 H_2 + S_2$. **7.42.** C_3H_8. **7.43.** 42.0.
7.44. C_4H_{10}. **7.45.** $C_4H_8O_2$. **7.47.** 15.2.
7.48. $2 CrCl_3 + 3 H_2 = 2 Cr + 6 HCl$. **7.49.** 25%. **7.50.** 0.58. **7.51.** 21%.
7.52. 28.054.

8.2. 42.8%. **8.3.** 2.72. **8.6.** 13.6. **8.7.** 43.8. **8.8.** 44.5. **8.13.** 21.3.
8.14. 46.1. **8.15.** 58.0. **8.16.** 10.7. **8.17.** 29.7. **8.18.** 2.00. **8.19.** 33.9.
8.20. 21.7. **8.21.** 75. **8.22.** 13.6%. **8.23.** 16.7%. **8.24.** 128 mm.
8.25. 3.36. **8.26.** 9.35.

9.2. 80. **9.3.** 48,000 cal. **9.5.** 6000. **9.6.** 69. **9.7.** 731.6. **9.8.** 137.3 g.
9.9. 1.1×10^7 gal. **9.10.** 48.9 g. **9.14(a).** -42.2 kcal; **(b).** -5.7 kcal;
(c). -337 kcal; **(d).** -74.4 kcal. **9.15.** -530.6 kcal/mole.
9.16. -94.5 kcal/mole. **9.17(a).** 24.6 kcal; **(b).** -23.1 kcal.
9.18. 14.8 kcal/mole. **9.20.** -66.2 kcal. **9.21.** -70.2 kcal/mole.
9.22. -77.5 kcal. **9.25(a).** 24.7 kcal; **(b).** 104.2 kcal; **(c).** -26.4 kcal;
(d). 22.4 kcal; **(e).** -210 kcal. **9.26.** -38.7 kcal. **9.27.** 135.3 kcal/mole.
9.28. 168.5 kcal/mole. **9.30.** -5.1 kcal. **9.31.** 4250 cal. **9.34.** $\Delta E = \Delta H = 0$;
$q = w = 365$ cal. **9.35.** $\Delta E = \Delta H = 0$; $q = w = -1250$ cal.
9.36. $\Delta E = \Delta H = 0$; $q = w = 821$ cal. **9.37.** $\Delta E = \Delta H = 0$;
$q = w = -1230$ cal. **9.38.** $\Delta E = \Delta H = 0$; $q = w = 413$ cal.
9.41. $\Delta E = 745$ cal; $\Delta H = 1245$ cal; $q = 745$ cal; $w = 0$.
9.42. $\Delta H = q = 1242$ cal; $w = 497$ cal; $\Delta E = 745$ cal. **9.43(a).** 445 cal;

(b). 898 cal; **(c).** 701 cal; **(d).** 948 cal. **9.44.** $\Delta H = q = -1956$ cal;
$\Delta E = -1174$ cal; $w = -782$ cal. **9.45(a).** $w = 322$ cal, $q = 24$ cal,
$\Delta E = -298$ cal, $\Delta H = -496$ cal; **(b).** $w = 198$ cal, $q = -101$ cal,
$\Delta E = -299$ cal, $\Delta H = -497$ cal; **(c).** $w = 214$ cal, $q = -84$ cal,
$\Delta E = -298$ cal, $\Delta H = -497$ cal. **9.46.** $\Delta H = q = 540$ kcal;
$\Delta E = 499$ kcal; $w = 41.2$ kcal. **9.47.** $\Delta H = q = 10{,}140$ cal; $w = 760$ cal;
$\Delta E = 9380$ cal. **9.48.** $w = 0$; $\Delta H = \Delta E = q = -1340$ cal.
9.49. -13.7 kcal. **9.51.** $\Delta S = 6.09$ cal/deg; $\Delta G = -3135$ cal.
9.52. $\Delta S = 0.19$ cal/deg; $\Delta G = 311$ cal. **9.53.** $\Delta G = \Delta S_{\text{sys}} = 0$.
9.55. $2503°$K. **9.56.** 27.2 cal/deg. **9.57.** $\Delta S = 0.2$ cal/deg; $\Delta G = -51$ cal.
9.58. q: $+$, $+$, $-$, $+$, $+$; w: $+$, $+$, 0, 0, 0; ΔE: 0, $+$, $-$, $+$, $+$;
ΔH: 0, $+$, $-$, $+$, $+$; ΔS: $+$, $+$, $-$, $+$, $+$; ΔG: $-$, $+$, $-$, $+$, 0.

10.9. 11.7 **10.10.** 18. **10.11.** 12.5. **10.12.** 10. **10.15.** 0.021; 1.16 *m*.
10.16. 5.4 g. **10.17.** 0.0036. **10.18.** 1.72 *m*. **10.19.** 0.0176. **10.25.** 2.7.
10.26. 6.6. **10.27.** 0.27. **10.28.** 8.5. **10.29.** 4.57. **10.30.** 1.21 g/ml.
10.31. 1275. **10.33.** 0.0227; 13.3%; 1.29 *m*. **10.34.** 0.635 *M*. **10.37.(a).** 0.48;
(b). 0.24. **10.38.** 7.6. **10.40.** 4.40. **10.41.** 2.00. **10.43.** 1.83. **10.44.** 0.32.
10.45. 128; 4.59. **10.46.** 1500. **10.47.** 4.0. **10.48.** 0.250; 0.125. **10.49.** 0.69.

11.2. 129. **11.4.** 141. **11.5.** 0.81°C. **11.6.** 119. **11.7.** 46. **11.8.** 5.
11.9. 32.042. **11.10.** 100.42°C. **11.12.** 101.12°C. **11.13.** 10.4 g.
11.14. 3 moles. **11.15.** dimerized. **11.17.** 0.0498. **11.18.** -1.614°C.
11.21. 42.4 mm. **11.22.** 31.3 mm. **11.24.** $P_{\text{C}_2\text{H}_6\text{O}} = 21.9$ mm;
$P_{\text{H}_2\text{O}} = 22.6$ mm; $f_{\text{H}_2\text{O}} = 0.509$. **11.25.** 0.295; 0.145.
11.26. $P_{\text{C}_7\text{H}_8} = 37.1$ mm; $P_{\text{C}_6\text{H}_6} = 117.9$ mm. **11.28.** 251,000. **11.29.** 54.9 atm.
11.30. 1.52 g. **11.31.** 25.7 atm.

12.2. 0.2 PCl_3; 0.2 Cl_2; 0.8 PCl_5. **12.5.** 0.041. **12.6.** 267. **12.7.** 13.3.
12.8. 30. **12.9.** 16. **12.10.** 33.4. **12.13.** 3.3. **12.14.** 9.3.
12.17(a). $K_1 = 6.9$ liters²/moles²; **(b).** $K_2 = 0.38$ mole/liter;
$K_2 = 1/\sqrt{K_1}$. **12.18.** $[\text{N}_2\text{O}_5] = 0.94$; $[\text{N}_2\text{O}_3] = 1.62$; $[\text{N}_2\text{O}] = 1.44$.
12.19. 0.77. **12.20.** 0.23. **12.21.** 1.48×10^{-4}. **12.22.** 0.96. **12.23.** 0.1.
12.24. 0.68. **12.25.** 0.7. **12.30.** 0.12. **12.31.** 0.313. **12.32.** $K_p = 0.0534$ atm;
$K_c = 1.33$ liters/mole. **12.33.** 1 atm. **12.34.** $K_c = 1.33 \times 10^{-2}$ mole/liter;
$K_p = 0.665$ atm. **12.35.** 0.0343. **12.36.** 1.15×10^4 mm. **12.37(a).** 0.814 atm;
0.372 atm. **(b).** 1.78 atm. **12.40.** 6.18×10^5; 7.86×10^2. **12.41.** 1.41×10^{33}.
12.42. 1.41×10^{-17}. **12.43.** -5.08×10^4 cal. **12.44.** 1.33×10^4 cal.
12.45. 1.65×10^{-3} cal.

13.2(a). 1.7×10^{-5} *M*; **(b).** 1.8×10^{-5} *M*; **(c).** 4×10^{-10} *M*.
13.3. 1.18×10^{-2} *M*. **13.5.** 0.10 *F*; 1.36×10^{-3} *M*. **13.7.** 1.3×10^{-3} *M*.
13.10. 1.3×10^{-13} *M*. **13.11.** 3.2×10^{-7} *M*. **13.14(a).** 1.7×10^{-7};
(b). 5.6×10^{-10}. **13.15(a).** 5.0×10^{-10}; **(b).** 5.0×10^{-12}. **13.18(a).** 8;
(b). 2.7; **(c).** 2.5; **(d).** 10; **(e).** 11.3; **(f).** 3.4; **(g).** 11.6; **(h).** 12; **(i).** 6.96.

13.21(a). 3.2×10^{-2}; **(b).** 2.5×10^{-14}. **13.22(a).** 4.0×10^{-11}; **(b).** 1.6×10^{-8}.
13.23. (b). **13.24.** 10%. **13.26.** 1.8×10^{-5}. **13.27.** $6.0 \times 10^{-6} M$.
13.28. 8×10^{-5}. **13.30.** $1.3 \times 10^{-19} M$. **13.31.** 5.0×10^{-4}; 2.8×10^{-10};
$8.4 \times 10^{-22} M$. **13.33.** 8.3. **13.34.** $7.5 \times 10^{-6} M$. **13.35.** $0.11 M$;
$2.2 \times 10^{-5} M$. **13.36.** 6×10^{-8}. **13.44.** 0.04. **13.45.** 800. **13.46.** 1.6.
13.47. 29. **13.48.** 0.83. **13.49.** 83.0. **13.50.** 218 g. **13.51.** 37.5 ml.
13.52. 2.24. **13.53.** 0.5. **13.54.** 8.72. **13.55.** 5.22. **13.56.** 11.15.
13.57. 3.19. **13.58(a).** 7; **(b).** 5; **(c).** 11. **13.59.** 13.0, 12.5, 11.7, 10.7, 9.7, 7.0.
13.60. 11.2, 9.4, 8.4, 7.4, 6.4, 5.3. **13.62.** $7 \times 10^{-6} M$.
13.63. F of $NaC_2H_3O_2 = 18 \times F$ of $HC_2H_3O_2$. **13.64.** HF. **13.65.** 7.2.
13.66. 0.1 mole. **13.67.** 1.38 mole; -0.14. **13.69.** 12.5. **13.70.** 132 NaOH,
868 $HC_2H_3O_2$. **13.73.** 11.8. **13.74.** 5×10^{-2}. **13.75.** 5×10^{-4}.
13.76(a). 4.6; **(b).** 9.6; **(c).** 3.7. **13.77.** $4 \times 10^{-8} M$.
13.78. $[Na^+] = [NH_4^+] = 1.2 \times 10^{-3} M$; $[HCN] = 1.7 \times 10^{-3} M$.
13.79. $[Na^+] = [C_2H_3O_2^-] = 0.20 M$; $[NH_3] = 0.20 M$;
$[NH_4^+] = [OH^-] = 1.9 \times 10^{-3} M$; $[HC_2H_3O_2] = 5.9 \times 10^{-8} M$;
$[H^+] = 5.3 \times 10^{-12} M$.
13.80. $2.8 \times 10^{-10} M$. **13.81.** $5.00 F$. **13.82.** 0.36.
13.83. $[H^+] = 1.0 \times 10^{-4} M$; $[OH^-] = 1.0 \times 10^{-10} M$; $[H_2C_2O_4] = 1.9 \times 10^{-5} M$; $[HC_2O_4^-] = 1.2 \times 10^{-2} M$; $[C_2O_4^{--}] = 7.6 \times 10^{-3} M$. **13.84.** 42.5.
13.86. $0.70 F$. **13.87(a).** 11.2; **(b).** 7.1. **13.88.** 0.33. **13.89(a).** 4.0;
(b). 0.050 mole; **(c).** 8.0; **(d).** 0.26 mole. **13.90.** 0.134. **13.91.** 12;
$3.3 \times 10^{-5} M$; $2.14 M$; $2.25 M$; $4.2 \times 10^{-15} M$.

14.3. $1.8 \times 10^{-18} M^4$. **14.4.** $1.5 \times 10^{-32} M^5$. **14.5.** 1.78×10^{-18}.
14.6. 4.92×10^{-9}. **14.7(b).** 8.8×10^{-17}; **(c).** 3.3×10^{-13}; **(d).** 2.0×10^{-10};
(f). 8.38×10^{-9}. **14.9.** $3 \times 10^{-11} M$. **14.10(a).** 4×10^{-36}; **(b).** 2×10^{-16};
(c). 6×10^{-24}; **(d).** 2×10^{-23}. **14.11.** $0.008 M$. **14.12.** 1.7×10^{-7}.
14.13. $Fe(OH)_3$ ppts, $BaSO_4$ not. **14.14.** 9.3×10^{-8} mole. **14.15.** 4×10^{-6}.
14.17. 4.0×10^{-5}. **14.18.** 1.6×10^{-5}. **14.19(b).** 1.7×10^{-5}; **(d).** 1.4×10^{-2};
(e). 4×10^{-9}. **14.20.** 13.6 times as great. **14.22.** $1 \times 10^{-9} M$.
14.24. 1.9×10^{-12}. **14.25.** 1×10^{-5}. **14.26.** 2.8×10^{-8}. **14.27.** $2.0 \times 10^{-8} M$.
14.29. Cl^-; $0.0096 M$. **14.30(a).** Ag^+; $1.2 \times 10^{-5} M$; **(b).** $4.0 \times 10^{-4} M$;
(c). $2.5 \times 10^{-5} M$. **14.31.** $2.7 \times 10^{-6} M$. **14.32.** 1.00×10^{-8}.
14.33. 9.9×10^{-3}; 9.9×10^{-9}. **14.34(a).** $4 \times 10^{-6} M$; **(b).** $1.5 \times 10^{-3} M$;
(c). $1.0 \times 10^{-7} M$; **(d).** 1.15×10^{-2}; **(e).** 1.11×10^{-2};
(f). $(1.11 \times 10^{-2}) + (2.5 \times 10^{-8})$ mole; **(g).** 22.2. **14.35.** 2.2×10^{-6}.
14.36. 6.5×10^{-11}; $Pb(ClO_4)_2$. **14.38.** $2.0 \times 10^{-15} M$. **14.39.** $0.10 M$,
$0.050 M$, and 8.3×10^{-6}. **14.40.** 0%. **14.42.** yes; $Mg(OH)_2$. **14.43.** Reverse.
14.44(b). $1.8 \times 10^4 M$; **(c).** $0.18 M$; **(d).** $3.6 \times 10^{-2} M$; **(e).** $3.6 \times 10^{-3} M$;
(f). 5.7×10^{-6}; **(g).** $1.3 \times 10^{-7} M$. **14.46.** 0.66. **14.47.** 3×10^{-15}.
14.48. 4.0. **14.49.** no. **14.50.** $11 M$. **14.51.** 1.3×10^{-22}.
14.53. $5.9 \times 10^{-3} M$. **14.54.** 1.2 g. **14.55.** 14 g. **14.56.** 1.2 g.
14.57. $4.6 \times 10^5 M$; no. **14.58.** 2.2. **14.59.** $6 \times 10^{-11} M$.
14.60(a). 6.0×10^{-23}; **(b).** 1.2×10^{14} liters. **14.61.** 4×10^{-3}.
14.62. 1×10^{-4}; 1×10^{-8}. **14.63.** $1.0 \times 10^{-6} M$. **14.64.** 8.4×10^{-10}.
14.65. 3.2×10^{-6}. **14.66.** $5 \times 10^{-9} M$. **14.67.** $0.35 M$.

14.68(a). $2.5 \times 10^{-4}\ M$; **(b).** $(2.5 \times 10^{-4} + Y)\ F$. **14.69(a).** $a = 0.20$, $b = 0.25$, $c = 0.30$; **(b).** 51.0. **14.70.** $Ga(OH)_4^-$. **14.71.** $Co(NH_3)_6^{++}$. **14.72.** 0.55.
14.73. 8×10^{-8}. **14.74(a).** $1.34 \times 10^{-5}\ M$; **(b).** 3.9×10^{-8}; **(c).** 1.0×10^{-3};
(d). $0.35\ M$. **14.75.** 4.1×10^{-2}. **14.76.** 3×10^{-16}.
14.77. $[Cd^{++}] = 3.2 \times 10^{-11}\ M$, $[CN^-] = 2.7 \times 10^{-3}\ M$,
$[HCN] = 2.7 \times 10^{-6}\ M$; $0.051\ F$. **14.78(a).** 1.0×10^{-16}; **(b).** 1.5×10^{-19}.
14.79. $1 \times 10^{-2}\ M$. **14.80.** $[H^+] = [Ag(CN)_2^-] = 1.1 \times 10^{-4}\ M$;
$[OH^-] = 9.3 \times 10^{-11}\ M$; $[Ag^+] = 7.0 \times 10^{-13}\ M$; $[HCN] = 0.10\ M$;
$[CN^-] = 3.7 \times 10^{-7}\ M$. **14.81.** $[H^+] = 0.1\ M$; $[OH^-] = 1 \times 10^{-13}\ M$;
$[Cl^-] = 0.05\ M$; $[Cu(CN)_2^-] = 0.05\ M$; $[HCN] = 5 \times 10^{-4}\ M$;
$[Cu^+] = 6 \times 10^{-5}\ M$; $[CN^-] = 2 \times 10^{-12}\ M$. **14.82.** 4.5.
14.83. 1.0×10^{-16}. **14.84.** 3.4×10^{-6}. **14.85.** 7.85 to 12.3.

15.35. 1.6. **15.36.** 96. **15.37.** 0.91. **15.38(a).** 0.600; **(b).** 0.0120; **(c).** 0.0120;
(d). 42.5 g; **(e).** 0.510; **(f).** 50.0. **15.42.** 12.2. **15.43.** $556,000$. **15.44.** 73.3.
15.45. 2.4. **15.46.** 24. **15.47.** 100. **15.48.** 1.60×10^{-19}. **15.49.** 140.
15.51. 55.9. **15.52.** 386. **15.53(a).** 9650; **(b).** 2.6 liters; **(c).** 1.2×10^{23};
(d). 800 ml. **15.54(a).** $Y_2(SO_4)_3$; **(b).** 27; **(c).** 189 cc. **15.55.** MCl_3.
15.56. 114. **15.57.** 126.9. **15.58.** 65.4. **15.60.** 4. **15.61(a).** 0.513 V;
(b). 1.997 V; **(c).** 0.517 V; **(d).** 0.739 V. **15.62(a).** 0.96 V; **(b).** 1.71 V;
(c). 1.12 V. **15.63(a).** yes; **(b).** yes; **(c).** no; **(d).** yes; **(e).** yes. **15.64(a).** no;
(b). no; **(c).** yes; **(d).** no. **(e).** yes. **15.65.** 0.793 V. **15.66.** -0.806 V.
15.67(a). 1.09 V; **(b).** 1.00 V; **(c).** 0.885 V; **(d).** 1.315. **15.68(a).** 0.103;
(b). 0.354. **15.69(a).** 0.0827 V; **(b).** -0.200 V. **15.70.** $7.3\ M$.
15.71(a). 1×10^{10}; **(b).** 1×10^8; **(c).** 1×10^{120}; **(d).** 1×10^{37}.
15.72. -0.36 V. **15.73.** -1.47 V. **15.74.** 0.80 V. **15.76.** yes;
$3\ H_2SO_3 = S + 2\ SO_4^{--} + 4\ H^+ + H_2O$. **15.77(a).** -0.48 V; **(b).** yes;
$3\ ClO^- = ClO_3^- + 2\ Cl^-$. **15.78.** yes. **15.79.** 0.140 V. **15.80.** -1.
15.81. -0.56 V. **15.82(a).** 0.06 V; **(b).** no; **(c).** yes; **(d).** no; **(e1).** $+2$;
(e2). $+4$; **(e3).** $+3$; **(f).** no. **15.83.** 1.41×10^9.

16.3. 0.010 moles/liter-min, 0.0096 moles/liter-min, 0.0090 moles/liter-min.
16.4. 0.12 mm/sec, 0.093 mm/sec, 0.070 mm/sec. **16.5.** 0.06 mm/sec,
0.047 mm/sec, 0.035 mm/sec. **16.6.** $NO = 14.2$ mm/min; $NH_3 = 14.2$ mm/min;
$O_2 = 17.8$ mm/min. **16.7.** 1.34 mm/min. **16.8.** 3rd order; 1st order in O_2,
2nd order in NO. **16.9.** 3/2 order, 1st order in H_2, 1/2 order in Br_2.
16.11. 0.0103 min^{-1}. **16.12.** 2.08×10^{-6} mm^{-1} sec^{-1}. **16.13.** 2nd order;
3.2×10^{-4} liter/mole-sec. **16.15.** 2.28×10^{-3} hr^{-1}.
16.16. 3.05×10^{-2} mm^{-1} min^{-1}. **16.18.** $73,200$ sec. **16.19.** 48.2 min.
16.20. 337 mm, 663 mm. **16.22.** 3.24×10^{-3} liters/mole-min.
16.23. 2.10×10^{-6} mm^{-1} sec^{-1}. **16.24.** 2500 sec. **16.25.** 27.3 mm.
16.28. 0.27 hr^{-1}, 2.6 hr, 0.83 hr. **16.29(a).** 0%; **(b).** 25%; **(c).** 33.3%.
16.30. 224 sec^{-1}. **16.31.** 10.6 kcal. **16.32.** $K = k_1/k_{-1}$

17.1. 5.8×10^{14} sec^{-1}. **17.2.** 0.05 cm. **17.3.** 3.3×10^4 cm/sec.
17.4. 2.7×10^{-7} erg. **17.5.** 6.6×10^{-12} erg. **17.6.** 1.5×10^{-11}.

17.7. 1.2×10^{20}. **17.8.** 3.3×10^{-12} erg; 46.2 kcal/mole. **17.9.** 9.9×10^{-12} erg.
17.10. 6.67×10^4 sec^{-1}. **17.11.** 3.7×10^7 cm/sec. **17.12.** 4.04×10^{-4} cm,
1.28×10^{-4} cm, 4.33×10^{-5} cm, 9.47×10^{-6} cm. **17.13.** 1.2×10^{-5} cm.
17.14. 3.03×10^{-12} erg/atom. **17.15.** 2.18×10^{-11} erg/atom.
17.16. 1.21×10^7 cm/sec. **17.17.** 1.26×10^{-9} cm. **17.18.** 1.64×10^{-32} cm.
17.19. $2d$, $1p$, $3f$. **17.20(a).** [Kr] $5s^2$; **(b).** [Ar] $3d^7 4s^2$; **(c).** [Ar] $3d^{10} 4s^2 4p^4$;
(d). [Xe] $6s^2$. **17.21(a).** [Ar] $3d^{10}4s^24p^6$; **(b).** [Kr] $4d^{10}5s^25p^6$;
(c). [Xe] $4f^{14}5d^{10}6s^26p^6$. **17.22(a).** N; **(b).** Ar; **(c).** Ni; **(d).** Si, S; **(e).** Cu.
17.23. 17, 35, 53. **17.24.** 34. **17.25(a).** Be; **(b).** Ti; **(c).** Cr; **(d).** Fe.
17.26(a). 5; **(b).** 3; **(c).** 4; **(d).** 0. **17.27(a).** $1s^22s^22p_x \uparrow 2p_y \downarrow$; **(b).** $1s^22s^22p^53s^1$;
(c). [Ar] $4p^1$. **17.28.** They are filling the $4f$ shell. The outer valence shell is $6s^2$.
17.29(a). $1s^22s^22p^43s^1$; **(b).** $1s^22s^22p^43s^23p^44s^1$; **(c).** He, 2; O, 8; Si, 14;
(d). K, Sc. **17.30.** F. **17.31(a).** $1s^32s^32p^1$; **(b).** $1s^32s^32p^93s^33p^1$;
(c). 15, 27, 54.

18.1. (e)(d)(a)(b)(c). **18.2.** (b)(g)(a)(f)(e)(c)(d). **18.3(a).** 9; **(b).** 19; **(c).** 35;
(d). 22, 32; **(e).** 34; **(f).** 15; **(g).** 15, 16; **(h).** 17.
18.4. Bonds = valence electrons when electrons < 4 and bonds = 8 − valence
electrons when electrons ≥ 4. **18.5.** (b), (d). **18.6.** (c), (e).

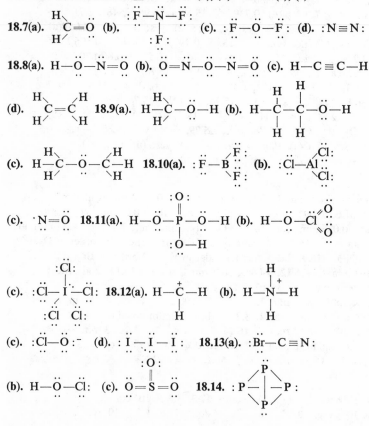

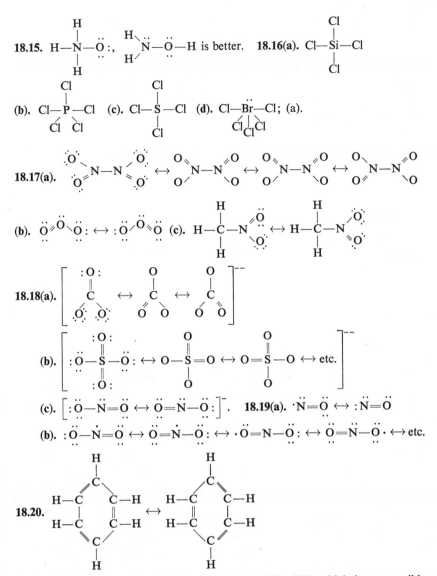

18.15. H—N—O:, N—O—H is better. **18.16(a).** Cl—Si—Cl

(b). Cl—P—Cl **(c).** Cl—S—Cl **(d).** Cl—Br—Cl; (a).

18.17(a).

(b). Ö⁄Ö\Ö: ↔ :Ö⁄Ö\Ö **(c).** H—C—N ↔ H—C—N

18.18(a).

(b).

(c). [:Ö—N=Ö ↔ Ö=N—Ö:]⁻. **18.19(a).** ·N=Ö ↔ :N=Ö

(b). :Ö—N=Ö ↔ Ö=N—Ö: ↔ ·Ö=N—Ö: ↔ Ö=N—Ö· ↔ etc.

18.20.

18.21. Resonance in the allyl cation ↔ ⁺CH₂—CH=CH₂ which is not possible without the double bond. **18.22(a).** $2sp^3$; **(b).** $3sp^2$; **(c).** $2sp$; **(d).** $2sp$; **(e).** $2sp^2$; **(f).** $2sp^3$. **18.23(a).** $2sp$, $2sp$, 180°; **(b).** $2sp^2$, $2sp^2$, 120°; **(c).** $2sp^3$, $2sp^2$, 109°; **(d).** $2sp$, $2sp$, 109°. **18.24(a).** linear; **(b).** planar; **(c).** tetrahedral; **(d).** planar. **18.25(a).** planar; **(b).** planar; **(c).** tetrahedral; **(d).** tetrahedral. **18.26(a).** trigonal bipyramidal; **(b).** octahedral; **(c).** octahedral; **(d).** tetrahedral. **18.27(b).** at 5 of the 6 corners of a distorted octahedron; **(c).** linear (the two apices of a trigonal bipyramid); **(d).** at the 4 corners of a trigonal bipyramid. **18.28.** $2p_x$ of C — $2p_x$ of O. **18.29.** $2p_x$ of C — $2p_x$ of O, $2p_y$ of C — $2p_y$ of other O. **18.30.** The unhybridized p orbital of the N overlaps with each of the unhybridized *p* orbitals of the three O atoms to form a bonding orbital spread over all four atoms which holds two electrons.

19.2. $^{227}_{90}$Th; $^{223}_{88}$Ra; $^{219}_{86}$Rn.   **19.4(a).** $^{12}_{6}$C; **(b).** $^{29}_{15}$P; **(c).** $^{24}_{11}$Na.
19.5(a). $^{4}_{2}$He; **(b).** $^{1}_{0}$n; **(c).** $^{1}_{0}$n.   **19.7.** $^{207}_{82}$Pb.   **19.8.** $^{237}_{93}$Np $= ^{205}_{83}$Bi $+ 8^{4}_{2}$He $+ 6_{-1}^{0}$e.   **19.10.** 35.5 g.   **19.11.** 4.15×10^{-3} atm.   **19.12.** 2.25 atm.   **19.13.** 300°K.
19.14. 900 mm.   **19.15.** 0.22.   **19.16.** 25.   **19.17(a).** 7.3×10^{-3}; **(b).** 6.0×10^{23}.
19.18. 6.0×10^{23}.   **19.19.** 1.5; 7.0×10^{23}.   **19.20.** 6.5×10^{23}.   **19.21(a).** 6.3;
(b). 0.96.   **19.22.** 15 hrs.

Index.

A

Absolute zero, 40
Activation energy, 238-239
Activity, 128, 145
Adiabatic process, 95
Alpha particle, 266
Ampere-hour, 218
Ampere-second, 218
Amu, 12
Atomic mass unit, 12
Atomic weights, 11
  International table, 287
  list, 286
Aufbau method, 250-252
Avogadro number, 13

B

Balancing redox equations, 203-215
Beta particle, 266

Boiling-point elevation, 115
Bond angle, 262-264
Bond energy, 88, 93
Boyle's law, 37
Brønsted-Lowry, 142
Buffers, 165-166

C

Calorie, 88
Celsius temperature, 4-5
Central ion, 196
Charles' law, 40-41
Chemical formula, 19-23
Colligative properties, 114
Common ion effect, 151
Complex ions, 196
Concentration:
  change in, 227
  effective, 128, 145